Pollination Biology

Pollination Biology

EDITED BY

Leslie Real

Department of Zoology
North Carolina State University
Raleigh, North Carolina

1983

ACADEMIC PRESS, INC.
Harcourt Brace Jovanovich, Publishers
Orlando San Diego New York
Austin Boston London Sydney
Tokyo Toronto

ACADEMIC PRESS, INC.
Orlando, Florida 32887

United Kingdom Edition published by
ACADEMIC PRESS, INC. (LONDON) LTD.
24/28 Oval Road, London NW1 7DX

Library of Congress Cataloging in Publication Data

Main entry under title:

Pollination biology.

Includes index.
1. Pollination. I. Real, Leslie.
QK926.P59 1983 582'.016 83-11873
ISBN 0-12-583980-4 (hardcover)
ISBN 0-12-583982-0 (paperback)

PRINTED IN THE UNITED STATES OF AMERICA

86 9 8 7 6 5 4 3 2

Dedicated to the Memory of Guy Palazzola

A Flower Does Not Talk

Silently a flower blooms,
In silence it falls away;
Yet here now, at this moment, at this place,
  the whole of the flower, the whole of
  the world is blooming.
This is the talk of the flower, the truth
  of the blossom;
The glory of eternal life is fully shining here.

Zenkei Shibayama

The pine tree lives a thousand years
The morning glory but a single day
Yet both fulfill their destiny.

Japanese poem

Contents

Contributors

Numbers in parentheses indicate the pages on which the authors' contributions begin.

HERBERT G. BAKER (7), Botany Department, University of California, Berkeley, California 94720

ROBERT I. BERTIN* (109), Department of Zoology, Miami University, Oxford, Ohio 45056

WILLIAM L. CREPET (29), Biological Sciences U-42, The University of Connecticut, Storrs, Connecticut 06268

STEVEN N. HANDEL (163), Department of Biology, Yale University, New Haven, Connecticut 06511

DAVID MULCAHY (151), Botany Department, University of Massachusetts, Amherst, Massachusetts 01003

BEVERLY RATHCKE (305), Division of Biological Sciences, University of Michigan, Ann Arbor, Michigan 48109

LESLIE REAL (1, 287), Department of Zoology, North Carolina State University, Raleigh, North Carolina 27650

ANDREW G. STEPHENSON (109), Department of Biology, Pennsylvania State University, University Park, Pennsylvania 16802

KEITH D. WADDINGTON (213), Department of Biology, University of Miami, Coral Gables, Florida 33124

*Present address: Biology Department, Bucknell University, Lewisburg, Pennsylvania 17837.

NICKOLAS M. WASER (241), Department of Biology, University of California at Riverside, Riverside, California 92521, and Rocky Mountain Biological Laboratory, Crested Butte, Colorado 81224

DONALD R. WHITEHEAD (97), Department of Biology, Indiana University, Bloomington, Indiana 47405

ROBERT WYATT (51), Department of Botany, University of Georgia, Athens, Georgia 30602

Foreword

The abundance of thought-provoking papers in this well-constructed volume testifies to the healthy state of the field of pollination biology in the late twentieth century. A field that has tended to become entirely descriptive and thus moribund has been revitalized through the introduction of experimental methods, as well as the construction of a theoretical basis against which to measure observational and experimental data. Like all aspects of population biology, however, studies of pollination systems are beset by problems that arise from their very complexity. Even the generalized statements that have been made about the specificity of pollination relationships in general have proved to be misleading as more information has accumulated; figs, yuccas, and specialized orchids have proved to be the exception rather than the rule!

Taken as a whole, the present volume indicates a healthy contemporary coupling between studies of the morphological and chemical properties of flowering plants and the behavioral and sensory features of their visitors, on the one hand, with the ecological setting in which the interactions involved take place, on the other hand. Just as for population biology as a whole, the burgeoning science of ecology and the generalities that are being derived within it are infusing pollination biology with new life and providing a context in which new generalities can be derived.

In order to demonstrate evolution of any kind, however, it is necessary to concentrate upon genetically determined variation in features that can clearly be shown to be adaptive, as well as the ways in which selective pressures influence the differential reproduction of the organisms that possess these features. Whether the features concerned are the color of the flowers of the angiosperms, their relative proportions and sizes, the chemical constituency of their nectar, or any other feature, on the one hand, or the size, configuration, hairiness, behavioral attributes, or any other feature of their visitors, on the other, it will ultimately be necessary to demonstrate for these variable traits that they do, in populations, actually contribute to the reproductive success of individual organisms in a way that can be quantified and shown to have a genetic basis.

Without such studies, pollination biology will continue to emphasize traits as if they were properties of species rather than of individuals, and our advance in understanding the evolutionary implications of the systems involved will be correspondingly slow. We are accustomed to thinking of the evolution of ecotypes of flowering plants as resulting, like all evolution, from the differential reproduction of individuals with different genotypes—this being the very definition of natural selection—but we are not accustomed to carrying out studies of the coevolving participants in pollination systems in such a way as to emphasize these features.

Nevertheless, there are some approaches to studies of this kind in the present volume. The excellent paper by Waser, by stressing the importance of the experimental testing of individual hypotheses in evaluating pollination systems, makes a strong statement that is directly relevant to such an approach. Experimental studies of this kind, carried out in a well-understood ecological context and on a community basis of the sort explored effectively in the essay by Real, will, I believe, provide the key to the coming advances in studies of pollination biology in particular and population biology in general.

The syndromes of features that characterize the plants and animals involved in coevolved pollination systems can be understood only in the perspective of their occurrence in communities. The energetics of the systems has provided a valuable clue as to the ways in which they function and the meaning of such characters, formerly thought to be neutral, as flower number per branch and per individual, spacing of individual plants, and the like. The way in which these plants are organized in communities has been explored brilliantly by Bernd Heinrich, and his investigations have proved invaluable in arriving at a better understanding of the ways in which outcrossing flowering plants and their animal visitors occur together in natural communities. In the present book, Rathcke provides an effective contribution to this field, considering the competitive interactions that occur between flowering plants in relation to their pollinators. Given the sorts of generalized pollination systems, involving a whole spectra of visitors, that seem to occur in most natural situations, it can be concluded with certainty that an understanding of community interactions of this kind will eventually prove to be of even more fundamental importance in exploring the role of many adaptive features associated with the functioning of pollination systems in nature.

Approaching the matter from a different standpoint, the well-known influence of genetics on evolutionary studies from the 1930s onward led to a veritable revolution in our understanding of the way in which the evolution of populations of plants and animals proceeds. By linking the genetic characteristics of individuals in populations with an increased understanding of the context in which these systems were operating, researchers were able to make significant advances that now are taken for granted in understanding the dynamics of plant and animal populations.

In many fields, however, and plant physiological ecology is, unfortunately, a classic example, little is done to assess individual variation. It is difficult enough in a complex field like physiological ecology to assess the parameters of interest for a given population with any degree of certainty, much less to be able to ascribe these features differentially to a series of individuals and then test the reproductive success of these individuals in a meaningful fashion. If it be difficult thus to assess the characteristics of individuals for these features, even when intensive coevolutionary interactions are not involved, how much more difficult it will be to do so in systems of the type that are the main concern of pollination biologists.

Notwithstanding these points, it is my belief that such studies must be made if we truly are to understand the evolution of the populations of angiosperms and their animal visitors that are involved in the fascinating kinds of systems explored in such detail in this present outstanding volume. In population biology as a whole, we are now apparently on the verge of a new wave of understanding in which the principles of ecology become fused with those of systematics and genetics, in much the same way as the principles of genetics infused the field with new life 50 years ago. The combination is a difficult one, but many of the perceptive articles in this book begin to point the way. Above all, the authors have, by the way they have approached their subject, made it clear that pollination biology will be an important area of investigation for many years to come.

Peter H. Raven

Preface

Every evolutionary biologist realizes that the field of pollination biology is undergoing a tremendous expansion and rapid evolution. Almost every recent journal in ecology or evolution contains at least one or two articles pertaining to pollination. Symposia and conferences devoted to pollinator–plant interactions are growing in number and attendance. And yet there are very few books available today that adequately survey the many new interests in this area. This volume will, I hope, provide such a survey.

The book is designed to provide useful material for advanced undergraduate and graduate students wishing to familiarize themselves with modern pollination biology and also to provide new insights into specific problems for those already engaged in pollination research. The book is intended to be used for both teaching and research. The chapters generally contain thorough reviews of broad topics on the evolution and ecology of pollination systems, and, in addition, each contributor has presented new results and/or models that advance our understanding of each of these topics. By combining these two features we hope to indicate the direction of future research in pollination biology based upon our current understanding.

A project such as this critically depends upon the willingness of authors to "step back" and observe their own work and thoughts. Without their generosity, keen interest, and unfailing determination, this book would not have been possible. I thank all of them for their efforts.

I wish to thank Janis Antonovics, James Hardin, Nickolas Waser, Beverly Rathcke, John Staddon, Tom Caraco, and the students of my pollination biology class for reviewing manuscripts and stimulating discussion and criticism. The editorial staff of Academic Press has shown remarkable patience. My wife, Damienne, has been a constant source of encouragement throughout this long and, at times, frustrating experience. To all of these people I owe my gratitude.

Leslie Real

CHAPTER 1

Introduction

LESLIE REAL
Department of Zoology
North Carolina State University
Raleigh, North Carolina

Interest in pollination biology is as ancient as civilized culture. Farmers have always shown a keen interest in the reproductive biology of plants and mechanisms of fruit production. The first record of interest in pollination comes from a bas relief found at Layard at Nimrod in ancient Assyria, dating to about 1500 B.C., depicting the artificial cross-pollination of the date palm, *Phoenix dactylifera,* by winged devine creatures with human forms and eagles' heads. So much of our agricultural production depends upon pollination that it is no wonder that civilized peoples have long been interested in this biological phenomenon.

Interest in the pollination of flowers has not been restricted to agricultural applications. Pollination systems have long been recognized as a model for understanding the interplay between natural selection and evolution. Charles Darwin wrote several books and papers on pollination systems and plant reproduction, and many of his conclusions are still valid today.

In numerous instances, the selective forces operating on plant populations and their pollinators can be fairly well described. Plants, with their often elaborate and specialized floral designs, and animals, with their complex systems of reproduction and behavior, are ideal models for exploring the process of adaptation. Much of the early research in pollination biology was devoted to describing these adaptations and the selective forces presumed to bring them about.

The past 15 years have seen a remarkable change in the style of pollination research. Prior to this time, most research was rather descriptive. Long lists of visitors to flowers were compiled, the individual breeding systems were evaluated, and mechanisms of pollen transfer guaranteeing specific cross-pollination

POLLINATION BIOLOGY

ISBN 0-12-583980-4 (hardcover)
ISBN 0-12-583982-0 (paperback)

were considered. Often, pheromonal species-specific mutualisms were uncovered—such as those between the different fig species and their chalcid wasp pollinators or *Yuccas* and their moth pollinator, *Tegaticula*. Unfortunately, little attention was paid to discovering generalizable ecological and evolutionary principles that could be tested experimentally, but this situation has changed significantly. Field biologists have now recognized in pollination systems a unique opportunity to test many of the current hypotheses presumed to hold generally for ecological and evolutionary processes. The peculiarities of the pollination system, which so amazed early naturalists, are no longer the central focus. Emphasis is now placed on uncovering general principles that underlie all ecological and evolutionary changes. Whether we shall find such principles remains to be seen. But the search is certainly on!

In Chapter 2, Herbert Baker traces the historical trends in pollination research and the development of the two styles of pollination biology. The older, more descriptive style constitutes the Old Testament of anthecology, and the more recent experimental work, directed toward uncovering generalizations, constitutes the New Testament. Using this distinction, Baker examines the areas of ecology and evolution that have profited from New Testament pollination biology, including the evolution of breeding systems, animal energetics, phylogenetic relationships and constraints, temporate and tropical comparisons of ecosystem structure and function, and population structure and dynamics. The subsequent chapters expand on these areas and provide extensive reviews and a discussion of current research topics.

The controversy over the mode and tempo of macroevolutionary change has not been restricted to paleontological investigations. Whether evolutionary change is gradualistic or punctuated is being debated by paleobotanists as well. In Chapter 3, William Crepet reviews our understanding of the evolution of the angiosperms and suggests that the punctuated equilibrium model may better account for observed patterns in the fossil record. Flowering plant evolution may be associated with two periods of rapid evolution—one associated with the origin of the flower per se, and the other with adaptive radiations coupled with the diversification of insect forms and potential pollination mechanisms.

The evolution of sexuality has received considerable attention over the past few years. Plants often show elaborate adaptation to promote cross-pollination. In Chapter 4, Robert Wyatt summarizes current studies on the evolution of plant-breeding systems, emphasizing their relation to general theories of sex allocation. Plants are also famous for their ability to self-fertilize, and Wyatt compares the selective forces that may promote autogomy as opposed to dichogamy. Relying upon his work on *Arenaria* species that occur on granite outcrops in the southeastern United States, he points out the interaction between ecological and evolutionary considerations in the evolution of autogamy.

Most pollination studies have focused on animal-mediated pollen transfer.

Many species, however, are wind pollinated, and often this breeding habit is a derived condition from insect-pollinated ancesters. What conditions cause wind pollination to prevail over insect pollination, and what are the morphological adaptations of flowers that increase the effectiveness of wind pollination? Much of the early research on these questions was initiated in a seminal paper by Donald Whitehead in 1969. In Chapter 5, he updates his earlier analysis and pays particular attention to the geographical correlations between breeding habit, climate, and mode of pollen transfer. The evolution of dioecy, monoecy, and wind pollination is linked to environmental factors controlling the spatial distribution of plant species, insect diversity and specialization, and flower and leaf phenologies. In addition, he discusses the adaptations for capturing pollen found in individual wind-pollinated flowers based on aerodynamic considerations.

When Darwin postulated his theory of evolution, he distinguished among several kinds of selection. Perhaps the most controversial of these, then and now, was what he called "sexual selection." In this form of selection, morphological and behavioral adaptations among individuals in a population contribute directly to their ability to mate with members of the opposite sex. These adaptations are maintained by the intrinsic benefits associated with increased matings and are not necessarily coupled with any adaptations which provide the individual with an increased ability to survive or be buffered against harsh environmental conditions. Indeed, many of the traits presumed to result from sexual selection seem to handicap individuals with regard to survivorship; for example, bright coloration in male birds may make them more susceptible to predation. Most research on sexual selection has been directed toward animal populations. The traits associated with mating success are often obvious in animals; bright coloration, elaborate courtship displays, and male combat are common. Yet, plants may also compete for mates, and in Chapter 6 Andrew Stephenson and Robert Bertin review the growing literature on sexual selection in plants. Such selection may occur as competition between male gametes for fertilization of ovules or as selective fertilization and fruit maturation by female recipients. These concepts will undoubtedly contribute to our understanding of the evolution of breeding systems and sexuality in both plants and animals.

In Chapter 7, David Mulcahy presents a computer model simulating the process of sexual selection through gametic competition in *Geranium maculatum*. The model is used to evaluate the evolutionary consequences of altering different floral traits (e.g., time between pollinator visits, grains per pollinator visit, style length, and ovules per flower) for the outcome of gametic competition.

The movement of pollinators in and among plant individuals governs the pattern of pollen/gene movement within the population. Consequently, our understanding of pollinator-mediated pollen/gene flow is critical to uncovering population structure and differentiation. The ability of populations to become locally differentiated, to spread favorable mutations, and to colonize new habitats to some degree depends on patterns of pollen movement and dispersal. After reviewing the variety of techniques now available for assaying pollen movement, Steven Handel in Chapter 8 analyzes the consequences of different gene movement patterns for plant population structure. Handel examines the importance of population shape, floral density and spatial distribution, and pollinator foraging based on the characterization of the plant resource being exploited. Through carefully controlled plantings of crop species with known spatial distributions and densities, Handel demonstrates how a more experimental analysis of gene flow should be carried out. The generalizations we reach through such experimental manipulations will guide our analysis of field populations.

The pattern of gene flow depends greatly upon the foraging patterns of pollinators, and these, in turn, depend upon underlying sensory and behavioral mechanisms. In Chapter 9, Keith Waddington summarizes the important features of insect physiology and behavior that pertain to sensory information and integration, learning, and memory. Cost-benefit models predicting how a forager should integrate and weigh the information extracted from the environment are compared. Throughout his exposition, Waddington attempts to wed the mechanistic analysis of behavior to the global properties and implications of these behaviors in pollination systems. Waddington's work on the transfer and integration of information within honeybee colonies is a model of such analysis.

It is often assumed that the great variety of forms, colors, and patterns we observe in flowering plants constitutes true adaptation, the products of years of natural selection among individuals. But what evidence do we have that these are truly adaptations? In Chapter 10, Nickolas Waser undertakes the heroic task of criticizing our normal approaches to the analysis of floral adaptations and indicates the need for carefully controlled experiments to evaluate the adaptiveness of different floral traits. Waser examines the evidence for adaptedness in floral color and shape, inflorescence size, and nectar production patterns as they affect the frequency of pollination visits and concomitant reproductive success in the plants visited. The experimental manipulation of plant species to determine the effects of altering floral traits on the plant's reproductive success is central to Waser's research program.

Pollination biology has contributed significantly to research in animal behavior on one level and community structure on another. However, these two levels should be integrated to provide a unified theory of communities. From a mostly theoretical perspective, Leslie Real in Chapter 11 indicates how such an integration of levels may occur. Using a basic pollination search model, Real shows how differential attractiveness among habitats can lead to the convergence or divergence of flowering times.

In Chapter 12, Beverly Rathcke summarizes the evidence for competitive interactions among flowering plants for pollinators. She points out that many plant communities show facilitative as well as competitive interactions. Determining cause-and-effect relationships in community phenologies may be difficult, but Rathcke provides a general model on the evolution of flowering time that may help us distinguish epiphenomena from the underlying behavioral and ecological mechanisms from which they emerge.

The chapters could be said to range from mostly evolutionary to mostly ecological in focus (i.e., breeding systems appear to be a more evolutionary topic then does community structure). Yet, even a cursory reading of the chapters should reveal that this distinction, so prevalent during the early phases of the post-Darwinian synthesis, is rapidly deteriorating. Evolutionary processes do not occur outside of an environmental context, and ecological interactions are not devoid of history. Together they form the nexus of a new evolutionary-ecological synthesis, and pollination biology will undoubtedly continue to be a model for articulating this new paradigm.

Reference

Whitehead, D. R. (1969). Wind pollination in the angiosperms: Evolutionary and environmental considerations. *Evolution* **23,** 28–35.

CHAPTER 2

An Outline of the History of Anthecology, or Pollination Biology*

HERBERT G. BAKER

Botany Department
University of California at Berkeley
Berkeley, California

I. Origins of Pollination Biology

Before there was agriculture, there must have been human awareness of the reproduction of flowering plants by seed. Jack Harlan (1975) has provided evidence that the gatherer-hunters of the preagricultural communities of the Near

*Based on, but expanded and updated from, a paper published in the N.Z. J. Bot. (Baker, 1979), by permission of the editor.

POLLINATION BIOLOGY

ISBN 0-12-583980-4 (hardcover)
ISBN 0-12-583982-0 (paperback)

East and Middle East did not have to devote almost all their time to food winning but had the opportunity to observe many natural phenomena relatively dispassionately. The naturally observant gatherer of wild seeds must soon have become aware that some of them sprouted to produce seedlings. And an appreciation of vegetative propagation as an alternative to the reproduction of plants by sowing seeds was necessary for the gradual development of ''root crop'' agriculture in the humid tropics, where it may have preceded seed-based agriculture (Harris, 1967).

The practice of reproducing desirable plants by seed need not necessarily mean that there was an appreciation of the sexual processes that are usually involved, although there is a famous Mesopotamian bas relief from the ninth century B.C., reproduced, for example, in Meeuse (1961), in which men are shown shaking flowering branches from a staminate date palm over the flowers of a pistillate palm. The writings of Theophrastus (fourth century B.C.) refer to this artificial pollination of date palms in terms of ''male'' and ''female'' contributions but, as Proctor and Yeo (1973) point out, even Theophrastus had little idea of the real sexuality of plants.

The ''caprification'' of figs in the Mediterranean region also has a very long history (see references in Condit and Enderud, 1956), so it may be assumed, even in the absence of substantiating literature, that with observations such as these and with the perceived occurrence of pollen showers from catkin-bearing and cone-bearing trees, there should have been at least a suspicion that seed production by seed plants involves more than just a swelling of a potential fruit.

From ancient times to the present day, there has been an accumulation of information (not always without misinterpretation) on pollination biology (a subject which was appropriately named ''anthecology'' by Charles Robertson in 1904). ''Pollination biology'' and ''anthecology'' will be used interchangeably in this chapter. To me, it seems convenient to divide the history of pollination biology (or anthecology) into two stages—the ''Old Testament,'' in which anecdotal and observational investigations greatly exceeded experiments and quantitative treatments, and the quantitative and experimental studies of the ''New Testament'' (Baker 1979).

II. The Old Testament of Anthecology

The Old Testament of anthecology received contributions from ancient times to the early decades of the twentieth century. Then, pollination biology attracted little attention from scientists (or even naturalists) until the realization of the potentialities of the ''synthetic'' study of evolution brought biologists from many disciplines together in the 1940s and it became essential to investigate the modes of reproduction of plants to understand their population dynamics and evolution.

But the hiatus between the two phases of anthecology was quite distinct, and each phase needs to be looked at in some detail. Proctor and Yeo (1973) and Lorch (1978) have provided useful summaries of various aspects of the history of the first phase of pollination biology and an excellent bibliography has been compiled by Schmid (1975).

In Europe, the earliest statement that sexual processes occurred in the generation of seeds (involving stamens and pistil) is usually attributed to Nehemiah Grew (1682), but there is evidence that Thomas Millington, of Oxford University, made the first suggestion of this in discussions with Grew. Rudolf Camerarius, a professor at the University of Tübingen, experimented on the removal of stamens or styles from plants of *Ricinus communis* and *Zea mays* and made isolation experiments with pistillate plants of several dioecious species. His results which showed the necessity of both "male" and "female" organs for seed setting, were made public in 1694.

In 1700, Willem Bosman, a Dutch traveler, visited the coast of the Gulf of Guinea, West Africa, and in an epistle to a correspondent in Holland wrote, "There grow multitudes of papay-trees [*Carica papaya*] all along the Coast; and these are of two sorts, viz. the male and female, or at least they are so called, on account that those named males bear no fruit but are continually full of blossoms, consisting of long white flowers; the female also bears the same blossoms, though not so long nor so numerous.

"Some have observed that the females produce their fruits in greatest abundance when the males grow near them; you may, Sir, believe what you please: but if you do not, I shall not charge you with heresy" (Bosman, translation in Pinkerton, 1814).

Lest it be thought that the elucidation of the mysteries of pollination, fertilization, and seed setting took place continuously and positively, it should be reported that there was active opposition, exemplified by scornful denunciation of the sexual idea in 1700 by Joseph Tournefort (Proctor and Yeo, 1973). Lorch (1978) quotes a series of authors with various impressions of pollination (and the function of nectar secretion in the process). However, the view of Grew and Camerarius gradually prevailed, with notable observations and experimental contributions from Richard Bradley (1717), Philip Miller (Blair, 1721), Paul Dudley (1724), James Logan (1736), and Arthur Dobbs (1750). Tulips and maize appear to have been the plants most in favor for study.

In 1735, Carolus Linnaeus put forward his "sexual system" of seed plant classification. With this classification, the characteristic androecial and gynoecial structures of flowers were recognized as being constant within taxa and, therefore, utilizable for classification. But it did not imply a true understanding of the functions of these groups of organs.

In the years of 1761–1766, Joseph Kölreuter published studies of pollinatory activities of insects and made experimental cross-pollinations which established the necessity of the transfer of pollen from stamen to stigma in some species by insects before seed setting would occur. However, he thought of this process in terms of insect-mediated self-pollination (Baker, 1979). Kölreuter also clarified the distinction between nectar (which bees collected) and honey (which they made from it), a distinction that took a long time to become prevalent in the literature of pollination—which, even in the nineteenth century, contains references to the presence of "honey" in flowers. Kölreuter observed pollen grain ornamentation and was able to see some correspondence between their number and the number of ovules in the ovary.

Most flowering plants have "hermaphrodite" flowers, and it was hard for the eighteenth- and very early nineteenth-century pollination biologists to see how this arrangement could promote anything but self-pollination (the exception, of course, being the minority of dioecious species).

When Christian Konrad Sprengel published his pioneering work, *Das entdeckte Geheimniss der Natur im Bau und in der Befruchtung der Blumen,* in 1793, he presented plenty of evidence that flowers attract insects and reward them with nectar. He also showed that wind pollination occurs in some species, demanding in those cases a copious supply of pollen. Whatever the mode of pollination, he wrote, "It seems that nature is unwilling that any flower should be fertilized by its own pollen" (translation in Lovell, 1918, p. 12). But it seems that he did not appreciate the biological consequences of this insistence on cross-pollination. Incidentally, Sprengel spent so much time on his botanical studies that he was removed from his position as rector at Spandau.

Sprengel's work lay almost unnoticed or unappreciated for more than 50 years. Hermann Müller, himself a great anthecologist, pointed out a reason for this neglect (Müller, 1883, p. 3): "It is remarkable in how many cases Sprengel recognized that the pollen is carried of necessity to the stigmas of *other* flowers by the insect-visitors, without suspecting that therein lies the value of insect visits to the plant." Müller blamed Sprengel's failure to show an advantage to the plants in cross-pollination for the neglect of his work in subsequent years. Unfortunately, Sprengel seems to have been unaware of the work of Thomas Knight, president of the Horticultural Society of London. In 1799, Knight reported on the results of many crosses that he had made between varieties of garden peas and showed that cross-fertilization had unexpected results: "Among the progeny of my hybridizations I found a numerous variety of new kinds produced, many of which were in size and in every other respect, much superior to the original white kind, and grow with excessive luxuriance" (Knight, 1799). He then made his famous statement "No plant self-fertilizes itself for a perpetuity of generations."

Similar experiments were made later by Carl Friedrich von Gärtner (1844, 1849), William Herbert (1837), Friederich Hildebrand (1867) and, of course, Charles Darwin (especially Darwin, 1876). This experimentation, in turn, led to a vast amount of data collecting by naturalists intent on demonstrating the operation of outcrossing mechanisms. Fritz Müller, Hermann Müller, Anton Kerner, T. F. Cheeseman, George Malcolm Thomson, Frederico Delpino, and others contributed notably to the assemblage.

How did these "natural history" observations and experiments tally with laboratory investigations of the structure and development of stamens, pollen grains, pollen tubes, carpels, embryo sacs, and embryos? In 1830, Giovani Amici demonstrated that pollen grains germinating on a stigma sent pollen tubes down the style to the ovules. But it was commonly believed that the "seed" was in the pollen, whence it traveled down the pollen tube, was implanted in the ovule, and developed into an embryo-containing true seed. This reasoning was negated completely by an observation by John Smith (1841) in a paper presented to the Linnean Society in London in 1839. Smith was first employed at the Edinburgh Botanic Garden, but moved to the Royal Botanic Gardens at Kew in 1822.

At Kew, Smith received a shipment of five plants of what we should now call *Alchornea ilicifolia* Muell. Arg., in the Euphorbiaceae. These plants came from Moreton Bay in Queensland, Australia. He noted that all five of them were pistillate plants, with no sign of an androecium. Nevertheless, he observed them to set full crops of germinable seed at Kew, and he raised plants from this seed that he used as a demonstration to the Linnean Society. At the end of a beautiful three-page paper he wrote, "I shall conclude by merely observing that the absence of pollen in this instance is irreconcilable with the theory that every grain of pollen furnishes a germ, and that the ovulum is merely a matrix to receive and nourish it till it becomes a perfect seed."

In a few lines, he not only demonstrated apomixis but also dealt a death blow to the once widely held belief about the parentage of embryos, both in plants and, by extrapolation, in animals and human beings, too.

A. Heresies

The Old Testament of pollination biology could now be assembled, with plenty of field workers to supply the raw material. In general, there was agreement on the objectives and interpretations of the data that were collected, but it must be pointed out that some biologists were unconvinced of the truth of the alleged advantages of cross-pollination (Baker, 1965). Thus, in England, the Reverend George Henslow, after originally agreeing with Charles Darwin on the benefits of cross-pollination, objected strongly (Henslow, 1876 *et seq.*) that some of the most successful species are weeds that have penetrated the farthest reaches of the world. For example, *Poa annua, Oxalis corniculata, Stellaria media, Capsella bursa-pastoris,* and *Anagallis arvensis* are usually self-pollinated. Henslow believed that flower shapes and colors were caused by the "irritation" of the insects that visited the flowers. Twelve years after his first protest, he wrote: "our previous ideas were all wrong and *we must reverse them.* The most conspicuous flowers . . . fertilized by insects are not best off; *but they cannot help themselves.* Whatever an insect does to them, they must yield to it and grow in adaptation to it; but while they are thus stimulated to become what *we* may choose to call finer flowers—all this is secured at a sacrifice of fertility. They neither set seed in anything like the proportion that "weedy" plants produce, nor can they hold their own so well when they find themselves transported to distant countries. . . . Those plants which cannot have plenty of well-nourished seeds, and are dependent on insects—perhaps one or two kinds only—will ultimately die out and disappear altogether" (Henslow, 1891, pp. 147–148).

Although Henslow's ideas of insect irritation of flowers will hardly stand up to examination, he made a number of valuable observations in his many publications, and his championing of the selfers is backed up by the recognition that a capacity for self-pollination is advantageous to the establishment of a seed-reproducing colony after long-distance dispersal of a single propagule (Baker, 1955, 1974). The twentieth-century view is that the breeding mechanisms of many plants are flexible and adaptable to environmental circumstances, and that cross-pollination is not always advantageous.

In North America, Thomas Meehan (1877) took the same line as Henslow, and a rather similar view was put forward by a Dutchman, W. Burck (1909).

George Henslow was the son of John Stevens Henslow, who inspired Charles Darwin when he was a Cambridge undergraduate and who had a strong influence in getting Darwin to join the voyage of the H.M.S. *Beagle.* Also, Joseph Dalton Hooker, Darwin's botanical right-hand man, had married George Henslow's sister. Consequently, in the British social system of the Victorians, these two facts alone would have been enough to enable George Henslow to escape public criticism from Darwin's camp. Thus, he was just ignored. Thomas Meehan was not so fortunate: At Darwin's instigation, he received a smart dressing down from Asa Gray (Baker, 1965).

Less well known is the heresy of the great French botanist, Gaston Bonnier. Bonnier, who published a big study of floral nectaries and nectar in 1879, along with some other papers on pollination, decided that the purpose of nectar, often produced by glands in the vicinity of the ovary, is to be reabsorbed as nourishment for the developing seeds, provided that it is not stolen by insects. He denied that its presence in flowers resulted from selective pressure for cross-pollination. What Bonnier never explained, however, was why nectar could also be found in purely staminate flowers. Linnaeus (1751) had long before pointed out that staminate flowers of *Salix* produce nectar.

The idea that the exudation of nectar from a nectary is just a means of getting sugar and water out of the way temporarily can be traced back to Giulio Pontedera in 1720, and was subsequently espoused by several botanists (see a list in Lorch, 1978). Charles Darwin's grandfather, Erasmus Darwin, published a remarkable book, *Phytologia,* in 1800. Erasmus Darwin believed that plant parts could be compared with animal organs. He thought that the leaves of a plant and the petals of the flowers functioned as lungs for the oxygenation of the vegetable blood (its sap). The veins of the leaves corresponded to the arterial and venous systems of an animal, and he even suggested that the soft pith in the center of a stem might be brain tissue. He wrote (E. Darwin, 1800, p. 109): "honey—is secreted by an appropriate gland from the blood after its oxygenation in the corol—and is absorbed for nutrient by the sexual parts of the flower." By sexual parts he meant the ovary *and* stamens, for he was well aware of the existence of nectar-producing staminate as well as pistillate flowers.

Later, in 1907, Burck stated just the opposite. Aware that nectar exudation often coincided with anthesis and particularly with anther dehiscence, he suggested that it is a kind of leakage in the flowers during the process of withdrawing water from the anthers, causing them to dehisce. He thought that the nectar somehow protects the ovary from desiccation (Burck, 1909). Other heretical views on the function of nectar are reviewed by Lorch (1978).

B. Progress in the Mainstream

In the second half of the nineteenth century, the Darwinian theory of evolution by natural selection caught on, and there was much activity by naturalists to elucidate the *selective* advantage, in promoting cross-pollination, of every floral character (Baker, 1979). Careful studies of the morphology and behavior of flower parts were made. Long lists of insect visitors to flowers were drawn up, often without much concern for whether all of these visitors were effective and frequent carriers of pollen between plants.

The greatest contributor of all was Hermann Müller, who in 1873 published *Die Befruchtung der Blumen durch Insekten.* This book was translated into English and published in London in 1883. It contains an enormous amount of descriptive information but is also remarkable for some ideas that were ahead of their time, such as habitat correlations with flower color, chemical contributions to insect diet by pollen and nectar (see Baker, 1979), and the relative proportions of different kinds of insects associated with various communities of flowering plants. He also indicated whether the various flower visitors were taking nectar, collecting pollen, or engaging in other activities.

But although he contributed so much material to it, Hermann Müller did not write what may be considered the actual Old Testament of pollination biology (or anthecology). There was a notable handbook by Anton Kerner (1878) and a two-volume compilation by Ernst Loew in 1895, which were succeeded by a grand compilation of information in three volumes by Paul Knuth (1898–1905)—his *Handbuch der Blütenbiologie.* Knuth dedicated this work to Sprengel and Hermann Müller, truthfully indicating whence came the inspiration for his own observations and compilation.

Two of Knuth's three volumes were translated into English by J. Ainsworth Davis and published as three volumes in Oxford in 1906–1909. The English edition never included Knuth's original third volume, which dealt largely with extra-European items, and for many botanists who have seen only the English edition, the third volume might never have been written (Schmid and Schmid, 1970).

This was the end of an era in pollination biology.

To some extent, the publication of these apparently comprehensive surveys may have suggested that the job was done (Baker, 1979), but there must have been other reasons for the decline in the amount of active research that occurred in the first four decades of the twentieth century. The rediscovery of Mendelism tended to put a damper on overenthusiastic Darwinism, with its ascription of easily appreciated selective advantages to each morphological character. Also, the rise of physics, chemistry, and other indoor sciences tended to promote laboratory studies rather than field natural history as the means of winning acclaim. The taxonomists, who were active, made little use of floral *function,* even though they were giving close attention to floral *structures* and were making use of them in classification.

Even outdoors, the new discipline of ecology was in a phase in which the description of vegetation took most of the time of its exponents.

Although in decline, anthecology was not dead. Books were published, e.g., by Oskar von Kirchner (1911) and H. Cammerloher (1931). *The Flower and the Bee* (which dealt with North American flowers and various insect visitors) was written by John H. Lovell (1918) and beautifully illustrated by photographs. In some ways Lovell's book, along with Charles Robertson's (1928) list of midwestern North American flowers and their pollinators, helped to redress the European imbalance of Knuth's handbook. It also looked to the future with its entomological as well as botanical concern, and with its stress on floral biology on a community basis, as well as its concern with individual species.

Also, during this period, there were a few substantial experimental studies by F. Knoll (1921, etc.), Frederick E. Clements and F. L. Long (1923), E. Daumann (1932), and the great studies with honey bees by Karl von Frisch (1950, etc.). In the tropics and elsewhere, Otto Porsch (1909, etc.) was traveling and observing pollination mechanisms, especially those involving birds (1924, *et seq.*). Porsch appears to have been the first to give serious attention to bat pollination (and pollination by some other mammals) as a significant feature of tropical anthecology (Porsch, 1931).

III. The New Testament of Anthecology

The rise of the synthetic study of evolution (also known as ''neo-Darwinism'') in the late 1930s and during World War II gave anthecology renewed importance. If gene flow, or the lack of it, is important evolutionarily, then knowledge of breeding systems and anthecology is needed. And to be able to provide something that is needed is a great human stimulus to work.

When the new anthecology began, there was a reaction against the traditional long lists of insect visitors. Syndromes of characters were drawn up whereby any flower could be identified as a butterfly flower, a hawkmoth flower, a beetle flower, and so on. Syndromes are described particularly in the works of van der Pijl (1960–1961), Faegri and van der Pijl (1966, 1971, 1978), Baker and Hurd (1968), and Kevan and Baker (1982). However, there is danger in assuming that the needs of only one kind of pollinator are catered to by the flower. An illustration would be a species of *Aloë* from South Africa. Red, scentless, broadly tubular flowers with exserted stamens and styles, a plentiful supply of only slightly viscous nectar at the base of the corolla tube, and a standing place on the inflorescence for a nectar taker that will contact anthers and stigmas all add up to a sunbird pollination syndrome in Africa. But honey bees also visit these flowers (to collect pollen), and they must play some role in the pollination process, too.

Even the classic case of strict mutualism on a 1:1 basis of the Yuccas and the Yucca moths (*Tegeticula*) is being modified to some extent. Galil (1973) has found nectar in the flowers of cultivated *Yucca aloifolia* which are visited by honey bees when these are available.

A. Tropical Studies

The old anthecology was largely made up of temperate zone studies, but now there is a growing emphasis on tropical investigations.

The foraging patterns of diurnally active bees, birds, and butterflies, and of nocturnal moths and bats are often more easily studied in tropical forests and savannas, and such phenomena as trap lining by some of these animals have come to light there (cf. Janzen, 1971; Baker, 1973). Some bats forage singly and others in groups (Heithaus *et al.*, 1974; Sazima and Sazima, 1977, 1978), and more is being learned about these patterns of behavior by the use of sequences of photographs (Baker and Harris, 1957; Baker, 1970; Baker *et al.*, 1971), night-viewing devices (Ayensu, 1974), and radio telemetry (Fleming *et al.*, 1977; Heithaus and Fleming, 1978).

Start and Marshall (1976) and Gould (1978) in Malaysia, and Heithaus and Fleming (1978) and Howell (1978) in Central America, have shown that bats may forage in trees many kilometers from their roosts. Sussmann and Raven

(1978) have theorized that before flower-visiting bats inhabited the Paleotropics, nonvolant mammals performed pollinatory functions, but have now largely been replaced by the bats. Lemurs may still be pollinators in Madagascar. In Australia, Turner (1982) has provided evidence that marsupial pollination may have preceded the arrival and radiation of pollinatory birds, particularly the honeyeaters (Meliphagidae). Wiens and Rourke (1978) have studied the pollination by rodents of Proteaceae that produce their flowers at ground level in South Africa, and ground-foraging birds (of the Meliphagidae) have been identified as the pollinators of some Australian Leguminosae (Ford *et al.*, 1979; Armstrong, 1979).

Pollination in the tropics by hummingbirds and by insects is also covered in recent compendia.

B. Family Studies

A very important development is the increasing attention that is being given to the *comparative* anthecology of some of the larger plant families, following the example set by Grant and Grant (1965) with the Polemoniaceae. Thus, the Bignoniaceae (Gentry, 1974), the Lecythidaceae (Prance, 1976, etc.) and the Onagraceae (Raven, 1979) have been treated with advantage.

IV. Special Topics

A. Ecosystem Aspects

Where there is an abundance of flowers with pollen and nectar "rewards" available to flower visitors, it is likely that an appropriately large cadre of visitors will be built up, many of which will be potential cross-pollinators. But in climatically stressful situations there may be an inadequate supply of pollinators, and competition for their services may occur. Hocking (1953, 1968) and Kevan (1972) have surveyed this situation in the Arctic, and Mosquin (1971) initiated the study of pollinator–flower relationships on a seasonal basis in Canada. Evidence of complementarity in flowering time of plants growing in the same habitat, with a consequent sharing of pollinators, has been investigated by a number of authors following these pioneer studies (see Boucher *et al.*, 1982).

The cooperation of flowering plant species in providing a sequence of food sources that will maintain a population of long-lived pollinators (e.g., bats, birds, and some bees) in a community throughout the year was claimed first for tropical regions (van der Pijl, 1960–1961; Baker, 1961; Baker *et al.*, 1971). Whether or not there has been coevolution or merely the selection of fortuitous preadaptations in many of these cases is debatable (see Janzen, 1980).

Research of this sort may already be expanding into studies of community (ecosystem) anthecology, although this represents a quantum leap in the complexity of the study required. A start has been made by studies such as those by Moldenke (1975, 1976) for Californian communities and by Moldenke and Lincoln (1979) for montane Colorado. Yorks (1980) has initiated the study of floral rewards in relation to pollinator needs on a seasonal basis for a chaparral community in California. At a more general level, a review by Regal (1982) considers the circumstances of occurrence and proportions in floristic lists of anemophilous and entomophilous species worldwide. Regal shows that anemophily becomes more common at high latitudes and high altitudes and less common in desert communities.

In Utah, Ostler and Harper (1978) have used data from earlier ecological surveys to reveal patterns of predominant flower color variation between ecosystems, an approach which was pioneered in the European Alps by Hermann Müller (1873, 1883, pp. 596–597) and also hinted at by Lovell (1918). Even at this level of ecosystem analysis, much work is required; for total investigation of flower–pollinator interactions in only one ecosystem, complete dedication by the investigator will be needed.

Consequently, we may expect the move toward ecosystem studies to be very gradual. It may be that the important basic information will be obtained independently, only subsequently taking its place in a synthesis. Thus, comparative phenological studies such as those in Costa Rican forest regions by Frankie *et al.* (1974) and Opler *et al.* (1980) may lay the basis of the ecosystem study of pollination, which could continue by adding studies such as that of Heithaus (1973, 1974) comparing insect–plant relations in a range of Costa Rican ecosystems, noting whether the flower visitors are generalists or specialists, whether they are long-lived or short-lived, and, if they have a free-living larval stage, what its requirements are.

A very important aspect of community anthecology is the energetics of the flower visitors, which affect their foraging patterns. The seminal paper in this area was published by Heinrich and Raven (1972), and good reviews of the subject, with special attention to bumble bees, were written by Heinrich (1975, 1979). Optimal foraging theory has been applied to flower visitation and has helped in understanding the patterns of flower visitor behavior (Pyke, 1978, 1979; Kevan and Baker, 1982).

Important in connection with vegetation analysis is the necessity of recognizing that there are microclimates to be taken into consideration. Thus, the observation by Beattie (1971) on the value to the flowers of *Viola sempervirens,* on the floor of redwood forests in California, of the passage of sunflecks is interesting. They have the property of raising ambient temperatures several degrees, thus attracting pollinating bees to flowers of the violets.

Solar tracking by flowers has long been known (sunflowers, for example, do it), but this has been shown to be important in boosting flower temperatures in colder regions such as the arctic (Hocking and Sharplin, 1965; Kevan, 1975) and the Andean paramó (Smith, 1975). The maintenance of humidity inside a tubular flower may be very important in preventing the nectar from evaporating to a consistency inappropriate to the pollinator, as would occur more frequently in a bowl-shaped flower. Park (1929) and Corbet (1978) have treated this matter, and Corbet (1978) has pointed out that a monomolecular layer of lipid on the surface of a pool of nectar in the flower may also diminish evaporation of the water in it.

It has become clear that a widespread flowering plant species, growing in a variety of communities, is likely to show "pollination ecotypes" that are serviced by different pollinators in the different habitats. The first attempt to demonstrate this was the study of various populations of *Gilia splendens* in California (Grant and Grant, 1965, pp. 69–73), where hummingbirds and flies of the genera *Bombylius* and *Eulonchus* are variously responsible for pollination, and where there is also a self-pollinating race. A tropical example of pollination ecotypes may be provided by the genus *Luehea* in Costa Rica (Haber and Frankie, 1982).

B. Attractants and Rewards

Ever since the nineteenth century, it has been recognized that size, shape, color (including color patterns), and scent of flowers are all part of the syndrome of attractants that bring visitors to the flowers.

The enormous subject of flower color, which is such an important part of the attraction, is ably reviewed by Kevan (1978, 1979, 1983), and flower shape as an evolutionary phenomenon has been treated several times by Leppik (1977). The bibliography of flower morphology is enormous and cannot be reviewed here. Almost as voluminous is the literature on the rewards that flower visitors receive—pollen, nectar, and occasionally special food bodies. Even the removal of nectar without pollination has been given its own terminology by Inouye (1980): Flower visitors that manage to slip past the stamens and stigmas without causing pollination are called "nectar thieves," whereas those that bite holes in the floral envelope and get at the nectar that way are called "nectar robbers."

1. Pollen Studies

Kerner (1878) reckoned that a very important feature of floral morphology was the protection given to the pollen (exposed as it is by the dehiscence of the anthers) by the perianth. The shape and attitude in space of the flower, the presence of hairs in the corolla tube, and other factors were accorded this function. However, Lidforss (1896) postulated that the pollen itself might be less easily caused to swell and burst in some cases (see also Jones, 1967). Zeroni and

Galil (1965) and Eisikowitch and Woodell (1974) provided evidence that substances in the anther walls can protect pollen from ruination by wetting.

The internal chemistry of pollen grains has much to tell us about the pollination biology of the flowers that produce it (reviewed in Baker and Baker, 1979, 1982, 1983a). Pollen grains of flowers that are predominantly self-pollinated have oil-rich contents if they are small and starch reserves if they are larger than about 25 nm in diameter. Among cross-pollinated species the pollen grains of bee- and fly-pollinated species tend to be oil-rich, whereas those of bird- and lepidopteran-pollinated species tend to be starch-storing, in these cases related to the food use of pollen by bees and flies and its nonutilization by most birds and lepidoptera. Pollen of some gymnosperms contains callose in enough quantity to be an energy reserve (Baker and Baker, 1983a).

2. *Nectar Studies*

Nineteenth-century botanists knew of the presence of sugars in nectars (cf. Bonnier, 1879) and their value to pollinators of the flowers. Recent publications by Baker and Baker (1977, 1982, 1983b) have reviewed the history of the discovery that nectar is more than just sugar water. The presence of amino acids in concentrations sufficient to be nutritionally significant or at least taste-affecting was first noted by Ziegler (1956). Since then the presence of lipids, proteins, organic acids (including ascorbic acid, or vitamin C), phenolics, saponins, glycosides, and alkaloids has been demonstrated in many nectars, and we are just beginning to understand their significance in relation to the nectar foraging (and, consequently, pollinatory activities) of the wide range of anthophilous insects, birds, and mammals. There is also significance in the ratio of the amount of the disaccharide sucrose to that of the hexose sugars fructose and glucose (Baker and Baker, 1982, 1983b), which is high in the cases of pollination by large bees, lepidoptera and hummingbirds, variable for small bees, and low for passerine birds and (at least neotropical) bats.

Lipids have been found to be quite common nectar constituents (mostly in emulsion form in the aqueous nectar) (Baker and Baker, 1975; Baker, 1978), and they may be part of the energy provision by the flowering plant that is used by the visitors. Floral gland exudates that are almost completely lipid have been found in tropical and subtropical Malpighiaceae, Krameriaceae, Scrophulariaceae, and Orchidaceae (Vogel, 1974; Simpson *et al.*, 1979), as well as in the Melastomataceae (Buchmann and Buchmann, 1981). These exudates are collected by bees, especially those of the genus *Centris*, which appear to use them to feed their larvae. In north temperate *Lysimachia* (Primulaceae), a similar exudation of lipids occurs (Vogel, 1976).

The occurrences of potentially toxic constituents of nectar are reviewed by

Baker and Baker (1975, 1982) and by Baker (1977, 1978). They include alkaloids, nonprotein amino acids, phenolics, and glycosides. They seem to be more common in tropical regions than in temperate (especially alpine) zones.

Several authors have shown that the rewards provided by staminate and pistillate flowers in monoecious and dioecious species may not be strictly equal. Pistillate plants usually do not provide a pollen reward, although exceptions occur when nonfunctional pollen grains are produced by these flowers in *Actinidia* (Schmid, 1978) and *Solanum* (Anderson, 1979). But even the nectar rewards of pistillate and staminate flowers may not be equivalent, being greater in the former in several species (Bawa and Opler, 1975) and more profuse in the latter in others (Baker, 1976). In a genus such as *Carica* (Caricaceae), the female flowers produce no nectar and visits by pollinatory hawkmoths represent "mistake pollination" (Baker, 1976), a situation that may be more widespread in the tropics than has been recognized in the past. Another form of mistake pollination has been described for nectarless, hermaphrodite flowers of some vines in the Bignoniaceae which produce small numbers of flowers at well-spaced intervals during the year. Gentry (1974) suggests that these are visited by "naive" bees that have to learn that they are unrewarding. This is only one of the strategies for pollination in the Bignoniaceae described by Gentry.

C. Breeding Systems

Obviously, the pollination biology of plants must be interpreted in the light of what can be found out about their breeding systems. Self-incompatibility or dioecism requires cross-pollination between compatible plants to occur before seed can be set. And herkogamy, dichogamy, and gynodioecism encourage outcrossing.

Self-incompatibility of several sorts (Lewis, 1979; Ganders, 1979; Pandey, 1960, 1979) occurs commonly in the angiosperms, more frequently than dioecism. It was appreciated by Darwin (summed up by Darwin, 1877) as a force providing the benefits (he thought mostly of heterosis) of cross-pollination. Many studies have been made of the circumstances in which self-incompatibility is replaced by self-compatibility (reviewed in Jain, 1976)—one of which is a shortage of pollinators (Baker, 1966). Similarly, the structure of the flower and the timing of maturation of the androecium and gynoecium may be altered and achieve self-pollination.

Darwin (1877) was aware of the logical extreme that this process could take and described several cases of cleistogamy, in which flowers regularly set seed without opening, in contrast to chasmogamy (normal opening). He also discovered that the cleistogamous flowers show morphological as well as physiological differences, even from chasmogamous flowers on the same plant (e.g.,

Viola odorata). The subject of cleistogamy has been reviewed recently by Lord (1981).

Autogamy (regular self-fertilization) is one way of ensuring seed set and the perpetuation of a genotype very much like that of the parent plant. Apomixis, the substitution of asexual for sexual reproduction, provides total reproduction of the parental genotype, including such heterozygosity as may be present. Apparently, the first to become aware of apomixis in 1839 was John Smith (see page 11).

In recent years, it has become clear that the existence of a hermaphrodite flower does not necessarily imply that it will contribute genes to the next generation equally through pollen and ovules. The leader in recognizing that the sex of a flower (hermaphrodite in this case) may differ from its effective gender (male or female) has been David Lloyd (Lloyd, 1979), who has investigated the implications of this fact in relation to environmentally sensitive strategies in flowering plants. Much of modern reproductive biology is based on the recognition that such differential contributions occur and that the gender of the flowers, even in a single inflorescence, may change from female or hermaphrodite to male as successive flowers, all ostensibly hermaphrodite, enter anthesis (e.g., *Catalpa speciosa;* Stephenson, 1979; also see Willson and Price, 1977). This is a rapidly expanding area of investigation, and any attempt to enlarge upon it here would be soon outdated. Journals such as *Evolution,* the *American Naturalist, Heredity,* and *Acta Biotheoretica* are likely sources of new material and ideas.

1. Sex Lability

Sex changes in flowering plants were noted in the early years of this century as a rare result of infection by fungi (e.g., *Ustilago violacea,* which causes anthers to develop in pistillate flowers of *Silene alba* and *S. dioica* and then fills the anthers with spores that are dispersed by the normal pollinating insects; Baker, 1947). This process provided evidence of some sex lability in flowering plants.

More recently, however, sex changes without the intervention of parasites, although often with an environmental cause or trigger, have been reported and substantiated in a wide variety of dioecious and monoecious flowering plant species (see reviews in Freeman *et al.,* 1976, 1980).

2. Dioecism

In addition to its function as an outbreeding mechanism (see Bawa, 1980, for review), dioecism is now being recognized as having significance in extending the environmental range of a species (usually with staminate plants able to function in more extreme environments, especially droughty ones). Dioecism, which appears to be especially common in woody plants, particularly tropical forest trees (Bawa, 1980), is sometimes linked to wind pollination (Whitehead,

1969) or to pollination by small bees (Bawa and Opler, 1975; Bawa, 1980). Undoubtedly, the study of dioecism still has much to offer the investigator, and there is a great deal of research on the subject at present.

D. Phylogenetic Constraint

In all features of floral morphology and the chemistry of pollen and nectar, allowance for the likelihood of some "phylogenetic constraint" must be made. This may be defined as "the limitation on present adaptation that is imposed by genetic features inherited from ancestors" (Baker and Baker, 1979, 1982, 1983a,b). Phylogenetic constraint may result in less than 100% adaptation between plant and pollinator, but the fit is close enough to be successful. For example, although wind-pollinated flowering plants usually have starchy pollen, in the Compositae, where pollen grains are apparently always oil-rich, there are no starch grains in mature pollen grains of wind-pollinated *Artemisia* or *Ambrosia* (Baker and Baker, 1979). Similar examples could be cited for other floral features in other taxa. Nevertheless, in general there is such a close fit between the evolved flowers and the biotic and abiotic factors influencing them that the evolution of pollination phenomena will undoubtedly be successfully studied even more actively in the future than it has been in the past.

Acknowledgments

Obviously, in the treatment of a subject as large as the history of anthecology, selection rather than total coverage of topics and publications is necessary. I tender my apologies to those authors whose deserving work was not quoted.

I offer my thanks to the editor of the *New Zealand Journal of Botany* for giving permission to use verbatim parts of a paper of mine that was published in that journal. Also, I thank most sincerely my teammate in pollination biology, Irene Baker.

References

Amici, G. B. (1830) see von Sachs, J. (1890). "History of Botany (1530–1860)." Oxford Univ. Press (Clarendon), Oxford (transl. by H. E. F. Garney).

Anderson, G. J. (1979). Dioecious *Solanum* species of hermaphrodite origin is an example of a broad convergence. *Nature (London)* **282,** 836–838.

Armstrong, J. A. (1979). Biotic pollination mechanisms in the Australian flora: A review. *N. Z. J. Bot.* **17,** 467–508.

Ayensu, E. S. (1974). Plant and bat interactions in West Africa. *Ann. M. Bot. Gard.* **61,** 702–727.

Baker, H. G. (1947). Infection of species of *Melandrium* by *Ustilago violacea* (Pers.) Fuckel and the transmission of the resultant disease. *Ann. Bot. N.S.* **11,** 333–348.

Baker, H. G. (1955). Self-compatibility and establishment after "long-distance" dispersal. *Evolution (Lawrence, Kans.)* **9,** 347–348.

Baker, H. G. (1961). The adaptation of flowering plants to nocturnal and crepuscular pollinators. *Q. R. Biol.* **36,** 64–73.

Baker, H. G. (1965). Charles Darwin and the perennial flax—a controversy and its implications. *Huntia* **2,** 141–161.

Baker, H. G. (1966). The evolution, functioning and breakdown of heteromorphic incompatibility systems. I. Plumbaginaceae. *Evolution (Lawrence, Kans.)* **20,** 349–368.

Baker, H. G. (1970). Two cases of bat pollination in Central America. *Rev. Biol. Trop.* **17,** 187–197.

Baker, H. G. (1973). Evolutionary relationships between flowering plants and animals in American and African tropical forests. *In* "Tropical Forest Ecosystems in Africa and South America: A Comparative Review" (B. J. Meggers, E. S. Ayensu, and W. D. Duckworth, eds.), pp. 145–159. Random House (Smithsonian Inst. Press), New York.

Baker, H. G. (1974). The evolution of weeds. *Annu. Rev. Ecol. Syst.* **5,** 1–24.

Baker, H. G. (1976). "Mistake" pollination as a reproductive system with special reference to the Caricaceae. *In* "Tropical Trees: Variation, Breeding and Conservation" (J. Burley and B. T. Styles, eds.), pp. 161–169. Academic Press, New York.

Baker, H. G. (1977). Non-sugar constituents of nectar. *Apidologie* **8,** 349–356.

Baker, H. G. (1978). Chemical aspects of the pollination of woody plants in the tropics. *In* "Tropical Trees as Living Systems" (P. B. Tomlinson and M. H. Zimmermann, eds.), pp. 57–82. Cambridge Univ. Press, London and New York.

Baker, H. G. (1979). Anthecology: Old testament, new testament, apocrypha. *N. Z. J. Bot.* **17,** 431–440.

Baker, H. G., and Baker, I. (1975). Studies of nectar-constitution and pollinator-plant coevolution. *In* "Animal and Plant Coevolution" (L. E. Gilbert and P. H. Raven, eds.), pp. 100–140. Univ. of Texas Press, Austin.

Baker, H. G., and Baker, I. (1977). Intraspecific constancy of floral nectar amino acid complements. *Bot. Gaz. (Chicago)* **138,** 183–191.

Baker, H. G., and Baker, I. (1979). Starch in angiosperm pollen grains and its evolutionary significance. *Amer. J. Bot.* **66,** 591–600.

Baker, H. G., and Baker, I. (1982). Chemical constituents of nectar in relation to pollination mechanisms and phylogeny. *In* "Biochemical Aspects of Evolutionary Biology" (M. H. Nitecki, ed.), pp. 131–172. Univ. of Chicago Press, Chicago, Illinois.

Baker, H. G., and Baker, I. (1983a). Some evolutionary and taxonomic implications of variation in the chemical reserves of pollen. *In* "Pollen: Biology and Implications for Plant Breeding" (D. L. Mulcahy and E. Ottaviano, eds.), pp. 43–52. Elsevier Biomedical, New York.

Baker, H. G., and Baker, I. (1983b). Floral nectar sugar constituents in relation to pollinator type. *In* "Handbook of Experimental Pollination Biology" (C. E. Jones and R. J. Little, eds.), pp. 117–141. Van Nostrand-Reinhold, Princeton, New Jersey (in press).

Baker, H. G., and Harris, B. J. (1957). The pollination of *Parkia* by bats and its attendant evolutionary problems. *Evolution (Lawrence, Kans.)* **11,** 449–460.

Baker, H. G., and Hurd, P. D., Jr. (1968). Intrafloral ecology. *Annu. Rev. Entomol.* **13,** 385–414.

Baker, H. G., Cruden, R. W., and Baker, I. (1971). Minor parasitism in pollination biology and its community function. The case of *Ceiba acuminata*. *BioScience* **21,** 1127–1129.

Baker, I, and Baker, H. G. (1982). Some chemical constituents of floral nectars of *Erythrina* in relation to pollinators and systematics. *Allertonia* **3,** 25–38.

Bawa, K. S. (1980). Evolution of dioecy. *Annu. Rev. Ecol. Syst.* **11,** 15–40.

Bawa, K. S., and Opler, P. A. (1975). Dioecism in tropical forest trees. Evolution *(Lawrence, Kans.)* **29,** 167–179.

Beattie, A. J. (1971). Itinerant pollination in a forest. *Madroño* **21,** 120–124.

Blair, P. (1720). "Botanick Essays." London.

Bonnier, G. (1879). Les nectaires. *Ann. Sci. Nat., Bot. Biol. Veg.* **8,** 1–213.

Boucher, D. H., James, S., and Keeler, K. H. (1982). The ecology of mutualism. *Annu. Rev. Ecol. Syst.* **13,** 315–348.

Bradley, R. (1717). "New Improvements of Planting and Gardening." London.

Buchmann, S. L., and Buchmann, M. D. (1981). Anthecology of *Mouriri myrtilloides* (Melastomataceae: Memecycleae), an oil flower in Panama. *Biotropica* Suppl., **13,** 7–24.

Burck, W. (1907). On the influence of the nectaries and other sugar containing tissues in the flower on the opening of the anthers. *Rec. Trav. Bot. Neerl.* **3,** 163–172.

Burck, W. (1909). On the biological significance of the secretion of nectar in the flower. *Proc. Sect. Sci. K. Acad. Wet. Amsterdam* **2,** 445–449.

Cammerloher, H. (1931). "Blütenbiologie," Vol. I. Borntrager, Berlin.

Clements, F. E., and Long, F. L. (1923). "Experimental Pollination." Carnegie Institution, Washington.

Condit, I. J., and Enderud, J. (1956). A bibliography of the fig. *Hilgardia* **15,** 1–663.

Corbet, S. A. (1978). Bees and the nectar of *Echium vulgare. In* "Pollination of Flowers by Insects" (A. J. Richards, ed.), pp. 21–30. Academic Press, New York.

Darwin, C. R. (1876). "The Effects of Cross and Self Fertilisation in the Vegetable Kingdom." Murray, London.

Darwin, C. R. (1877). "The Different Forms of Flowers on Plants of the Same Species." Murray, London.

Darwin, E. (1800). "Phytologia." Johnson, London.

Daumann, E. (1932). Über postflorale Nektarabscheidung. *Beih. Bot. Zentralbl.* **49,** 720–734.

Dobbs, A. (1750). Concerning bees and their method of gathering wax and honey. *Phil. Trans. R. Soc. London* **46,** 536–549.

Dudley, P. (1724). Observations on some of the plants in New England. *Phil. Trans. R. Soc. London* **33,** 194.

Eisikowitch, D., and Woodell, S. R. J. (1974). Effect of water on pollen germination. *Evolution (Lawrence, Kans.)* **28,** 692–694.

Faegri, K., and van der Pijl, L. (1966, 1971, 1978). "The Principles of Pollination Ecology." Editions 1, 2, and 3. Pergamon, Oxford.

Fleming, T. H., Heithaus, E. R., and Sawyer, W. B. (1977). An experimental analysis of the food location behavior of frugivorous bats. *Ecology* **58,** 619–627.

Ford, H. A., Paton, D. C., and Forde, N. (1979). Birds as pollinators of Australian plants. *N. Z. J. Bot.* **17,** 509–519.

Frankie, G. W., Baker, H. G. and Opler, P. A. (1974). Comparative phenological studies of trees in tropical wet and dry forests in the lowlands of Costa Rica. *J. Ecology* **62,** 881–919.

Freeman, D. C., Klikoff, L. G., and Harper, K. T. (1976). Differential resource utilization by the sexes of dioecious plants. *Science* **193,** 597–599.

Freeman, D. C., Harper, K. T., and Charnov, E. L. (1980). Sex change in plants: Old and new observations and new hypotheses. *Oecologia* **47,** 222–232.

Galil, J. (1973). Topocentric and ethodynamic pollination. *In* "Pollination and Dispersal" (N. B. M. Brantjes and H. F. Linskens, eds.), pp. 85–100. Dept. of Botany, Univ. of Nijmegen, Nijmegen.

Ganders, F. R. (1979). The biology of heterostyly. *N. Z. J. Bot.* **17,** 607–636.

Gentry, A. H. (1974). Coevolutionary patterns in Central American Bignoniaceae. *Ann. M. Bot. Gard.* **61,** 728–759.

Gould, E. (1978). Foraging behavior of Malaysian nectar-feeding bats. *Biotropica* **10,** 184–192.

Grant, V., and Grant, K. A. (1965). "Flower Pollination in the Phlox Family." Columbia Univ. Press, New York.

Grew, N. (1682). "Anatomy of Plants." Rawlins, London.

Haber, W. A., and Frankie, G. W. (1982). Pollination of *Luehea* (Tiliaceae) in Costa Rican deciduous forest. *Ecology* **63,** 1740–1750.

Harlan, J. (1975). "Crops and Man." American Society of Agronomy: Crop Science Society of America, Madison.

Harris, D. R. (1967). New light on plant domestication and the origins of agriculture. A review. *Geogr. Rev.* **57,** 90–107.

Heinrich, B. (1975). Energetics of pollination. *Annu. Rev. Ecol. Syst.* **6,** 139–170.

Heinrich, B. (1979). "Bumblebee Economics." Harvard Univ. Press, Cambridge, Massachusetts.

Heinrich, B., and Raven, P. H. (1972). Energetics and pollination ecology. *Science* **176,** 597–602.

Heithaus, E. R. (1973). Species diversity and resource partitioning in four neotropical plant-pollinator communities. Ph.D. Thesis, Biology, Stanford Univ., Stanford, California.

Heithaus, E. R. (1974). On the role of plant-pollinator interactions in determining community structure. *Ann. M. Bot. Gard.* **61,** 675–691.

Heithaus, E. R., and Fleming, T. H. (1978). Foraging movements of a frugivorous bat *Carollia perspicillata* (Phyllostomatidae). *Ecol. Monogr.* **48,** 127–143.

Heithaus, E. R., Opler, P. A., and Baker, H. G. (1974). Bat activity and pollination of *Bauhinia pauletia;* plant/pollinator coevolution. *Ecology* **55,** 412–419.

Henslow, G. (1876). Self-fertilisation of plants. *Nature (London)* **14,** 543–544.

Henslow, G. (1891). "The Making of Flowers." Society for Propagating Christian Knowledge, London.

Herbert, W. (1837). "Amaryllidaceae, with a Treatise on Cross-Bred Vegetables." London.

Hildebrand, F. H. G. (1867). "Die Geschlechtsverteilung bei den Pflanzen." W. Engelmann, Leipzig.

Hocking, B. (1953). The intrinsic range and speed of flight of insects. *Trans. R. Entomol. Soc. London* **104,** 223–345.

Hocking, B. (1968). Insect-flower association in the high Arctic with special reference to nectar. *Oikos* **19,** 359–388.

Hocking, B., and Sharplin, C. D. (1965). Flower basking by arctic insects. *Nature (London)* **206,** 215.

Howell, D. J. (1978). Time sharing and body partitioning in bat-pollination systems. *Nature (London)* **270,** 509–510.

Inouye, D. W. (1980). The terminology of floral larceny. *Ecology* **61,** 1261–1263.

Jain, S. K. (1976). The evolution of inbreeding in plants. *Annu. Rev. Ecol. Syst.* **7,** 469–495.

Janzen, D. H. (1971). Euglossine bees as long-distance pollinators. *Science* **171,** 203–205.

Janzen, D. H. (1980). When is it coevolution? *Evolution (Lawrence, Kans.)* **34,** 611–613.

Jones, C. E. (1967). Some evolutionary aspects of water stress in flowering in the tropics. *Turrialba* **17,** 188–190.

Kerner von Marilaun, A. (1878). "Flowers and their Unbidden Guests." Kegan Paul, London (transl. and edited by W. Ogle).

Kevan, P. G. (1972). Insect pollination of high arctic flowers. *J. Ecol.* **60,** 831–867.

Kevan, P. G. (1975). Sun tracking solar furnaces in high arctic flowers: Significance for pollination and insects. *Science* **189,** 723–726.

Kevan, P. G. (1983). Flower colors through the insect eye: What they are and what they mean. *In* "Handbook of Experimental Pollination Biology" (C. E. Jones and J. Little, eds.), pp. 5–30. Van Nostrand-Reinhold, Princeton, New Jersey.

Kevan, P. G. (1979). Vegetation and floral colors using ultraviolet light: Interpretational difficulties for functional significance. *Am. J. Bot.* **66,** 744–751.

Kevan, P. G. (1983). Flower colors through the insect eye: What they are and what they mean. *In* "Handbook of Experimental Pollination Biology" (C. E. Jones and J. Little, eds.), Van Nostrand-Reinold, Princeton, New Jersey (in press).

Kevan, P. G., and Baker, H. G. (1982). Insects as flower visitors and pollinators. *Annu. Rev. Entomol.* **28,** 407–453.

Knight, T. (1799). Experiments on the fecundation of vegetables. *Phil. Trans. R. Soc. London* **89,** 195–204.
Knoll, F. (1921). *Bombylius fuliginosus* und die Farbe der Blumen. *Abh. Zool. Bot. Gesellsch. Wien* **12,** 117–119.
Knuth, P. (1898–1905). "Handbuch der Blütenbiologie," 3 Vols. W. Engelmann, Leipzig.
Knuth, P. (1906–1909). "Handbook of Flower Pollination," 3 Vols. (I, 1906; II, 1908; III, 1909). Oxford Univ. Press (Clarendon), Oxford (Transl. by J. A. Davis).
Kölreuter, J. G. (1761–1766). "Vorläufige Nachricht von einigen das Geschlecht der Pflanzen betreffenden Versuchen und Baobachtung." Leipzig.
Leppik, E. E. (1977). "Floral Evolution in Relation to Pollination Ecology." Today and Tomorrow's Print and Publ., New Delhi.
Lewis, D. (1979). Genetic versatility of incompatibility in plants. *N. Z. J. Bot.* **17,** 637–644.
Lidforss, B. (1896). Zur Biologie des Pollens. *Jahrb. Wiss. Bot.* **24,** 1–38.
Linnaeus, C. (1735). "Systema Naturae." Leyden.
Linnaeus, C. (1751). "Philosophia Botanica." Stockholm.
Lloyd, D. G. (1979). Parental strategies of angiosperms. *N. Z. J. Bot.* **17,** 595–606.
Loew, E. (1895). "Einführung in der Blütenbiologie auf historischer Grundlaga." F. Dümmler, Berlin.
Logan, J. (1736). Experiments concerning the impregnation of the seed of plants. *Phil. Trans. R. Soc. London* **39,** 192.
Lorch, J. (1978). On the discovery of nectaries and its relation to views on flowers and insects. *Isis* **69,** 514–533.
Lord, E. M. (1981). Cleistogamy: A tool for the study of floral morphogenesis, function, and evolution. *Bot. Rev.* **47,** 421–450.
Lovell, J. H. (1918). "The Flower and the Bee: Plant Life and Pollination." Scribner's, New York.
Meehan, T. (1877). *Gentiana Andrewsii. Bull. Torrey Bot. Club* **6,** 189.
Meeuse, B. J. D. (1961). "The Story of Pollination." Ronald Press, New York.
Moldenke, A. R. (1975). Niche specialization and species diversity along a California transect. *Oecologia* **21,** 214–242.
Moldenke, A. R. (1976). California pollination ecology and vegetation types. *Phytologia* **34,** 305–361.
Moldenke, A. R., and Lincoln, P. G. (1979). Pollination ecology in montane Colorado: A community analysis. *Phytologia* **42,** 349–379.
Mosquin, T. (1971). Competition for pollinators as a stimulus for evolution of flowering time. *Oikos* **22,** 398–402.
Müller, H. (1873). "Die Befruchtung der Blumen durch Insekten und die gegenseitigen Anpassungen beider." W. Engelmann, Leipzig.
Müller, H. (1883). "The Fertilisation of Flowers." Macmillan, London transl. by D'A.W. Thompson).
Opler, P. A., Frankie, G. W., and Baker, H. G. (1980). Comparative phenological studies of treelet and shrub species in tropical wet and dry forests in the lowlands of Costa Rica. *J. Ecology* **68,** 167–188.
Ostler, W. K., and Harper, K. T. (1978). Floral ecology in relation to plant species diversity in the Wasatch Mountains. *Ecology* **59,** 848–861.
Pandey, K. K. (1960). Evolution of gametophytic and sporophytic systems of self-incompatibility in angiosperms. *Evolution (Lawrence, Kans.)* **14,** 98–115.
Pandey, K. K. (1979). Overcoming incompatibility and promoting genetic recombination in flowering plants. *N. Z. J. Bot.* **17,** 645–663.
Park, O. W. (1929). The influence of humidity upon sugar concentration in the nectar of various plants. *J. Econ. Entomol.* **22,** 534–544.

Pinkerton, J., ed. (1814). "Bosman's Guinea," A general collection of the Best and Most Interesting Voyages and Travels in All Parts of the World, Vol. 16, pp. 337–547. Longman, London.
Pontedera, G. (1720). "Anthologia sire de floris natura, libri tres." Padua.
Porsch, O. (1909). Neuere Untersuchungen über die Insektenlochangsmittel der Orchideenblüte. *Mitt. Naturwisse. Ver. Stiermark* **45,** 346–370.
Porsch, O. (1924). Vogelblumenstudien, I. *Jahr. Wiss. Bot.* **63,** 553–706.
Porsch, O. (1931). *Crescentia,* eine Fledermäusblume. *Österr. Bot. Zeitung* **80,** 31–44.
Prance, G. T. (1976). The pollination and anthophore structure of some Amazonian Lecythidaceae. *Biotropica* **8,** 235–241.
Proctor, M., and Yeo, P. (1973). "The Pollination of Flowers," The New Naturalist Series, No. 54. Collins, London.
Pyke, G. H. (1978). Optimal foraging in bumblebees and coevolution with their plants. *Oecologia* **36,** 281–293.
Pyke, G. H. (1979). Optimal foraging in bumblebees: Role of movement between flowers within inflorescences. *Anim. Behav.* **27,** 1167–1181.
Raven, P. H. (1979). A survey of reproductive biology in the Onagraceae. *N. Z. J. Bot.* **17,** 575–594.
Regal, P. J. (1982). Pollination by wind and animals: Ecology of geographic patterns. *Annu. Rev. Ecol. Syst.* **13,** 497–524.
Robertson, C. (1904). The structure of flowers and the mode of pollination of the primitive angiosperms. *Bot. Gaz. (Chicago)* **37,** 294–298.
Robertson, C. (1928). "Flowers and Insects: Lists of Visitors to Four Hundred and Fifty-Three Flowers. Publ. by the author, Carlinville, Illinois.
Sazima, I., and Sazima, M. (1977). Solitary and group-foraging; two flower-visiting patterns of the Lesser Spear-nosed bat *Phyllostomus discolor. Biotropica* **9,** 213–215.
Sazima, M., and Sazima, I. (1978). Bat pollination of the passion flower *Passiflora mucronata,* in southeastern Brazil. *Biotropica* **10,** 100–108.
Schmid, R. (1975). Two hundred years of pollination biology: An overview. *Biologist* **57,** 26–35.
Schmid, R. (1978). Reproductive anatomy of *Actinidia chinensis. Bot. Jahrb. Syst.* **100,** 149–195.
Schmid, R., and Schmid, M. (1970). Knuth's often overlooked "Handbuch der Blütenbiologie, III Band." *Ecology* **51,** 357–358.
Simpson, B., Seigler, D. S., and Neff, J. L. (1979). Lipids from the floral glands of *Krameria. Biochem. Syst. Ecol.* **7,** 193–194.
Smith, A. P. (1975). Insect pollination and heliotropism in *Oritrophium limnophilum* (Compositae) of the Andean paramó. *Biotropica* **7,** 284–286.
Smith, J. (1841). Notice of a plant which produces perfect seeds without any apparent action of pollen. *Trans. Linn. Soc. London* **18,** 509–512.
Sprengel, C. K. (1793). "Das Entdeckte Geheimniss der Natur im Bau und in der Befruchtung der Blumen." Friedrich Vieweg, Berlin.
Start, A. N., and Marshall, A. G. (1976). Nectarivorous bats as pollinators of trees in West Malaysia. *In* "Tropical Trees: Variation, Breeding, and Conservation" (J. Burley and B. T. Styles, eds.), pp. 145–150. Academic Press, New York.
Stephenson, A. G. (1979). An evolutionary examination of the floral display of *Catalpa speciosa. Evolution (Lawrence Kans.)* **33,** 1200–1209.
Sussmann, R. W., and Raven, P. H. (1978). Pollination by lemurs and marsupials: An archaic coevolutionary system. *Science* **200,** 731–736.
Turner, V. (1982). Marsupials as pollinators in Australia. *In* "Pollination and Evolution" (J. A. Armstrong, J. M. Powell, and A. J. Richards, eds.), Roy. Bot. Gard., Sydney, Sydney.
Vogel, S. (1974). Ölblumen und ölsammelnde Bienen. *Akad. Wiss. Lit. Mainz, Trop. Subtrop. Pflanz.* **7,** 1–547.

Vogel, S. (1976). *Lysimachia:* Oelblumen der Holarktis. *Naturwissenschaften* **63,** 44–45.

van der Pijl, L. (1960–1961). Ecological aspects of flower evolution. I and II. *Evolution* (*Lawrence, Kans.*) **14,** 403–416; **15,** 44–59.

von Frisch, K. (1950). ''Bees, their Vision, Chemical Senses and Language.'' Cornell Univ. Press, Ithaca, New York.

von Gärtner, C. F. (1844). ''Beitrage zur Kentniss der Befruchtung.'' Stuttgart.

von Gärtner, C. F. (1849). ''Versuche über die Bastarderzeugung im Pflanzenreich.'' Stuttgart.

von Kirchner, O. (1911). ''Blumen und Insekten.'' Teubner, Leipzig.

Whitehead, D. R. (1969). Wind pollination in the angiosperms. Evolutionary and environmental considerations. *Evolution* (*Lawrence, Kans.*) **23,** 28–35.

Whitehead, D. R. (1969). Wind pollination in the angiosperms. Evolutionary and environmental considerations. *Evolution* (*Lawrence, Kans.*) **23,** 28–35.

Wiens, D., and Rourke, J. P. (1978). Rodent pollination in southern Africa: *Protea* spp. *Nature* (*London*), **276,** 71–73.

Willson, M. F., and Price, P. W. (1977). The evolution of inflorescence size in *Asclepias* (Asclepiadaceae). *Evolution* (*Lawrence, Kans.*) **31,** 495–511.

Yorks, P. F. (1980). Chemical constituents of nectar and pollen in relation to pollinator behavior in a California ecosystem. Ph.D. Thesis (Botany), Univ. of California, Berkeley, California.

Zeroni, M., and Galil, J. (1965). Pollen germination inhibitory substances in the anthers of *Androcymbium palestinum* (Boiss.) Bak. *Proc. Bot. Soc. Isr.* **14,** 205.

Ziegler, H. (1956). Untersuchungen über die Leitung und Sekretion der Assimilate. *Planta* **47,** 447–500.

CHAPTER 3

The Role of Insect Pollination in the Evolution of the Angiosperms

WILLIAM L. CREPET
Biological Sciences
The University of Connecticut
Storrs, Connecticut

Relations between flowers and insect pollinators are archetypes of the results of coevolutionary interactions. They are superficially fascinating, yet their mystique intensifies on further analysis. This is especially evident in the ubiquity of insect pollination mechanisms in the angiosperms, their frequent bilateral complexity, and their uncertain evolutionary history.

POLLINATION BIOLOGY

ISBN 0-12-583980-4 (hardcover)
ISBN 0-12-583982-0 (paperback)

The importance of insect pollination mechanisms in the establishment of present angiosperm diversity may be inferred from correlative evidence. The angiosperms are predominantly, and virtually uniquely, insect pollinated (Baker and Hurd, 1968). Stebbins (1981), for example, has recently noted that several of the largest families of angiosperms are associated with a diversity of insect pollinators.[1]

In addition to correlative evidence suggesting that insect pollination has been important in angiosperm success, there are a variety of possible tangible advantages to insect pollination both at the species and at higher taxonomic levels. These include (1) energy conservation through reliable directional pollination (Pohl, 1937; Cruden, 1977); (2) the potential of more outcrossing than wind pollination, especially in populations of relatively widely dispersed individuals (Regal, 1977); and (3) successful pollination under conditions that are unfavorable for wind pollen dispersal—in effect, opening new niches for colonization and potentially allowing more species/unit area.

Pollination by faithful pollinators (referred to below as constant pollination for convenience) has additional advantages. This too is suggested by correlative evidence. For example, the family in which pollinator constancy has reached a peak, the Orchidaceae, is the largest family of vascular plants. Certain specific advantages may also be suggested that follow from constant pollination mechanisms. Constant pollination might allow a maximum number of plant species/unit area, since with constant pollination each plant species might have an efficient pollinator in spite of the relatively high species density. Constant pollination mechanisms are maximally efficient, since minimizing visits to other species reduces pollen waste. Finally, as initially suggested by Grant (1949) and elaborated in regard to punctuated equilibrium theory (Crepet, 1982, 1983), faithful pollinators might provide a mechanism for restricting gene flow from parental populations to isolated demes or might actually initiate the isolation of demes, thus augmenting the speciation process in a way consistent with the rapid evolution of new species.

Although the actual impact of certain of these suggested advantages of insect pollination may be better assessed in light of further empirical data, evidence presently available does suggest that insect pollination has been extremely important in the success of the angiosperms. Most botanists view insect pollination as the plesiomorphic condition within the flowering plants (Baker and Hurd, 1968). It is interesting to consider whether insect pollination has always been an exclusively angiospermous phenomenon and, if it has not, whether

[1]Notable exceptions include the wind-pollinated Poaceae, but the grasses are based on sets of characters that were partially accumulated by insect-pollinated ancestors and cannot be completely divorced from insect pollination in an evolutionary perspective. Grasses have many unique reproductive features, some of which optimize the probability of speciation, including somatic mutations associated with clonal growth (Silander, 1983).

insect pollination represents an apomorphy at the level Angiospermae or is a continuation of a plesiomorphic condition in a pre-angiospermous lineage.

To consider these and other questions regarding the history of insect pollination and to appreciate the relationship between insect pollination and angiosperm success, it is necessary to regard insect pollination in the context of the fossil record. This has been traditionally difficult because appropriate data (e.g., the history of floral structure in angiosperms) have not always been available to paleontologists. Recent studies of floral morphology from various geological horizons have provided important data on the history and significance of various pollination mechanisms in the angiosperms (Crepet, 1979a, 1982; Dilcher, 1979; Dilcher and Crane, 1982; Tiffney, 1977; Friis, 1982). Other paleobotanical investigations have added to our recent appreciation of angiosperm history (Doyle and Hickey, 1976; Hickey and Doyle, 1977; Walker and Walker, 1982; Brenner, 1976). Furthermore, careful analyses of pre-angiospermous plants from reproductive and paleoecological perspectives have made it possible to speculate on the origin and nature of early insect pollination (Crepet, 1974; Taylor and Millay, 1979, 1981a; Taylor and Scott, 1983; Scott and Taylor, 1983; Phillips and DiMichele, 1981).

Fundamental to speculations on early insect pollination mechanisms has been recent progress in understanding insect evolutionary patterns (Carpenter, 1976; Michener, 1979; MacKay, 1977) based on the fossil record and on historical interpretations of contemporary biogeographical data.

Considering the possible origin of insect pollination and the nature of early insect pollination mechanisms involves framing questions in the context of neontological observations and generating hypotheses by attempting to answer them in the context of fossil evidence. Such an approach results in the creation of plausible hypotheses that may not represent unique explanations of available data. Although this limit confronts the paleontologist who ventures into certain areas of inquiry, such hypotheses are important in directing the analysis of the fossil record. Furthermore, these hypotheses are testable inasmuch as they may be rejected on the basis of new fossil evidence. With this caveat, I will consider the possible origin, history, and significance of insect pollination in angiosperms based on available fossil evidence.

I. The Origin of Insect Pollination

To appreciate the circumstances and taxa likely to have been involved in the events culminating in the origin of insect pollination, it is necessary to place this phenomenon in a temporal perspective.

The first definitive evidence of insect pollination occurs in the Upper Carboniferous. The size of the (pre)pollen grains of certain species of the pteridospermous family Medullosaceae (up to 600 μm) suggests that effective dispersal by wind was not likely (Taylor, 1981a). Circumstantial evidence supports this conclusion. Medullosans grew either on floodplains, often in association with dominant tree ferns (*Psaronius*), or in coal swamps where they were understory plants (Phillips and DiMichele, 1981). In both instances, their distribution was patchy relative to the distributions of other taxa (Phillips and DiMichele, 1981), suggesting that wind dispersal of pollen would have been much less effective since it is often associated with large stands of particular species (Whitehead, 1969). Further, in the latter environment, the remarkably reduced velocity of wind in the understory relative to the canopy today (Scheihing, 1980) suggests that the large pollen grains, probably not wind dispersed under the most favorable aeolic conditions, would have traveled insignificant distances in available air currents. Finally, pollen-bearing organs (e.g., *Dolerotheca*) were adapted in ways that cannot be considered to improve pollen dispersal by wind (to the contrary), but that can be interpreted as enhancing their attractive value to archaic pollinators (i.e., more obvious with a higher concentration of reward).

Thus, we might assume that insect pollination was a *fait accompli* in the pteridosperm line by the Upper Carboniferous. We may now consider contemporary and earlier plant and arthropod taxa and the conditions likely to have existed during the transition from wind to insect pollination in certain lineages.

A. Plants

Knowledge of Carboniferous forests is extensive, and is based on floodplain deposits where plant remains are preserved as compressions and on coal swamp deposits that represent wetter environments. Coal swamp plants are usually preserved by calcium carbonate petrifications or coal balls. Although many taxa have been found in common in these two deltaic environments, floodplain communities tended to be more heterogeneous and to differ somewhat in the nature of their associations (Phillips and DiMichele, 1981). The most significant components of the Carboniferous forests were the lycopod trees (Lepidodendrales), the articulates (Sphenophyta), the Cordaitales (conifer sister group and probable ancestors), the first true conifers (Voltziales), the ferns (Filicales), and the seed ferns (Pteridospermales). Typical coal-age forests were divided into canopy and understory taxa, with the specific species composition depending on the time within the Carboniferous and the local ecological setting. Frequently, the coal swamp canopy was composed of lepidodendroid trees or cordaitaleans (Phillips and DiMichele, 1981). In other instances, tree ferns shared the canopy with pteridosperms.

B. Reproduction in Carboniferous Plants

Spore-bearing plants were significant and dominant during the Carboniferous. The lycopods, true ferns, and articulates were all spore-bearing, although the seed habit was approached in both the Sphenophyta (*Calamocarpon*) and the Lepidodendrales (*Lepidocarpon*) (Taylor, 1981b). Dispersal of spores must have occurred predominantly by wind and water, but there is some evidence that certain arthropods, including insects, were spore feeders (Taylor and Scott, 1983). In some instances, this undoubtedly involved omnivorous feeding on fallen cones, but it is possible that certain arthropods had an active role in spore dispersal. A variety of Carboniferous coprolites are known to have been rich in spores, including those that were almost exclusively lepidodendroid spores (*Lycospora*). A spore-dispersing role for some arthropods represented by coprolites may be hypothesized in light of Chaloner's experiments showing significant spore viability after passing through the gut of a saltatorian (grasshopper) (Chaloner, 1976). Such animal-mediated spore dispersal may represent the earliest mutualistic plant–animal interactions (Scott and Taylor, 1983).

Of course, in considering pollination per se, the seed habit is involved by definition. The Cordaitales, conifers, and seed ferns are the major seed-bearing plants of the Carboniferous. The seed habit first appeared in the Upper Devonian and proliferated during the Lower Carboniferous, when it characterized the Cordaitales and seed ferns. In the cordaitalean–conifer line the seed habit was clearly associated with wind pollination. Pollen was well adapted to wind dispersal by the formation of air sacs resulting from the separation of the two pollen wall layers (nexine and sexine).

In the seed ferns, it is obvious that various strategies evolved for the dispersal and reception of pollen. Certain taxa, the Callisophytaceae with saccate pollen, for example, retained the apparently plesiomorphic condition of wind dispersal (Taylor, 1981b), whereas in at least one other taxon, the Medullosaceae, it is obvious that animal interactions most easily accounted for pollen dispersal. In still other Carboniferous pteridosperms, pollination mechanisms cannot be easily inferred from the reproductive morphology. It is likely that the medullosans were the culmination of adaptation for insect pollination in Carboniferous pteridosperms and that their ancestors and contemporaries were variously adapted. Consistent with the possibility that at least several pteridospermous taxa were insect pollinated is the patchiness implied by the observation that although the pteridosperms were never numerically dominant during the Carboniferous, they were the most abundant plants of the Carboniferous in terms of numbers of taxa.

C. Arthropods

A variety of terrestrial arthropods existed during the Carboniferous, including mites, millipedes, the apterygote orders Collembola and Thysanura, and the true insects. The 10 orders of Carboniferous true insects included 5 palaeopterous orders (non-wing-folding insects) and 10 neopterous orders (wing-folding insects). The Paleoptera included the Megasecoptera, Palaeodictyoptera, and Diaphanoptera, Protodonata, and Emphemeroptera (extant mayflies). The Neoptera included the Blattodea (cockroaches—extant), Saltatoria (extant), Protorthoptera, Miomoptera, and Caloneurodea. The Megasecoptera, Palaeodictyoptera, and Diaphanoptera had sucking-piercing mouthparts and are considered to have been sap feeders (Carpenter, 1976). The Protodonata are the largest known insects, with wing spans of up to 75 cm, and are considered to have been aggressive predators (Carpenter, 1976). The Ephemeroptera, the only extant order of Carboniferous Paleoptera, had well-developed mouthparts and may well have been predatory even though modern Ephemeroptera do not feed as adults. The Carboniferous Neoptera include omnivorous (Blattodea), phytophagous (Saltatoria), and predacious species. At least some of the Protorthoptera seem to have been well adapted for predation, with chewing mouthparts and raptorial forelegs (Carpenter, 1976). The Miomoptera and Caloneurodea also had mandibulate mouthparts and are considered to have been likely predators (C. S. Henry, personal communication).

D. Arthropod–Plant Interactions and Likely Early Pollinators

There are several kinds of evidence for arthropods interacting directly with the reproductive structures of plants. Coprolites of various-size classes have been found containing vegetative material and pollen, in addition to spores, as discussed earlier. *Monoletes* (the prepollen of the Medullosaceae), lycopod and sphenophyte spores, lycopod megaspore fragments, and various types of cordaitalean pollen have all been found in coprolites. The sources of each type of coprolite are unknown, but there are size classes consistent with those of the coprolites produced by the millipedes, Collembola, Blattodea, and Saltatoria. Nothing is known, of course, about the coprolites of extinct taxa. Seeds and megaspores have been discovered, although rarely, with holes bored in the seed coats or megaspore walls (Scott and Taylor, 1983). Other evidence suggestive of animal interactions with the reproductive organs of plants involves pollen or spores found on or in the guts of fossil arthropods. Recently, an orthopteran was discovered with small, undescribed spores between its folded wings (Richardson, 1980) and a millipede is known with several monolete (medullosan prepollen) spores on the posterior edge of a leg (Richardson, 1980).

There is no question that arthropods capable of consuming pollen existed during the Carboniferous. The question involves which of these taxa were most important in the origin of insect pollination. It is more than possible that certain of these taxa were responsible for the first insect pollination and that they were succeeded by more efficient pollinators. Alternatively, the origin and early process of insect pollination may have been associated with only one group of arthropods.

The behavior of extant taxa with Carboniferous relatives provides a correlative, although imperfect means of assessing the behavior of the extinct forms.

Structural features such as mandibulate mouthparts provide important clues to behavior that might be related to pollination. It seems likely, by analogy with contemporary Apterygota, that the Collembola and Thysanura were generalist litter feeders and not likely to have been important pollinators. The absence of wings would certainly make them highly ineffective in gaining access to and transporting pollen. Indeed, these insects might have consumed pollen or spores encountered on the forest floor, which might account for the presence of such spores in Collembola-like coprolites.

Palaeopterous insects cannot be ignored as possible pollinators, but some taxa are more likely than others. The Palaeodictyoptera, Megasecoptera, and Diaphanoptera had piercing-sucking mouthparts and are considered to have been sap feeders (Carpenter, 1976). Nonetheless, it is possible that pollen was injested whole or, if large enough, after penetration. Seed feeding by penetration is common today in the Hemiptera, although these insects are not significant pollen consumers or pollinators. The Protodonata were apparently aggressive predators, (Carpenter, 1976), and the extremely large size of their wings suggests that they would have been ill-equipped for moving in the foliage in search of seeds and pollen. The Ephemeroptera seem likely to have been more generalized predators.

The Neoptera of the Carboniferous include omnivores (Blattodea), which are usually ground feeders, and phytophagus insects (Saltatoria), and neither seems likely to have been effective as a pollinator. The remaining three orders—Protorthoptera, Miomoptera, and Caloneurodea—are considered to have been generalist predators (Carpenter, 1976; C. S. Henry, personal communication). Clearly, no one group stands out as likely pollinators on the basis of the feeding syndrome we are able to deduce from the fossils.

One logical way to speculate on early pollinators is to imagine which of the available insect groups were best preadapted for an effective role in pollination. This depends on certain behavioral characteristics, including feeding. Before hypothesizing the best set of preadaptations for insect pollination, however, it is necessary to speculate on the nature of early attractants and rewards. In its most basic form, the plant–pollinator relationship is one in which the pollinator receives nutrition and the plant receives pollination. The most obvious initial attractants are nutritionally rich pollen and ovules (Janzen, 1971; Stanley and Linskins, 1974). Seed plants were either monoecious or dioecious at that time and if either pollen or ovules separately were the only attractant, successful pollination would depend on fortuitous contact of the pollinator with the organ that it did not visit for nourishment. Although such contact might have occurred during the search for the appropriate reproductive organ/nourishment source, it would certainly have been less than optimal.

The nature of early attractants/rewards suggests something about the feeding adaptations of early pollinators. Insects with mandibulate mouthparts would have been well adapted to consume both pollen and ovules. Arthropod taxa with nonmandibulate mouthparts could have fed on seeds (e.g., modern Hemiptera), but seem less likely to have been pollen feeders or effective pollinators (e.g., modern Hemiptera). Of course, other insects with nonmandibulate mouthparts are pollinators today, but these are primarily adapted to consume nectar, a clearly derived reward (e.g., Diptera, Lepidoptera). Taylor and Millay (1979) have suggested that the glands on the cupules of certain Carboniferous pteridosperms might have produced nectar as a reward to early pollinators. However, these glands occur on all epidermal surfaces of the taxon involved (*Lyginopteris*), except the inside of the cupule nearest the ovule, and can also be plausibly explained as defense mechanisms. Taylor and Scott (1983), however, note certain ovule modifications in apparently pteridospermous taxa that are consistent with the role of attractant/reward.

On the basis of the likely nature of early rewards and the disposition of the reproductive structures of Carboniferous pteridosperms (i.e., separate ovules and pollen-bearing organs borne considerable distances from the ground), it is possible to speculate further on optimal preadaptations for insect pollination. Flight seems important for gaining access to pollen and ovules and for transporting pollen significant distances. Wing-folding ability (as in neopterous insects) might have been important if gaining access to ovules and pollen depended on moving through dense foliage. Finally, a behavior pattern like that of predacious insects would have been an important preadaptation for effective foraging on ovules and pollen. In fact, the similarities between the foraging patterns of modern pollinators and predatory insects have recently been pointed out by Levin (1979). One ideally preadapted incipient pollinator can then be considered to have been a general predator, and several Carboniferous insect groups fit that category.

E. A Scenario for the Origin of Insect Pollination

Certain predacious insects preadapted by their foraging behavior, flight, chewing mouthparts, and nutritional requirements (presumably high in protein) began to feed on ovules and pollen after repeatedly encountering them while searching pteridospermous foliage for prey items. The cryptic wing venation patterns of Carboniferous insects that mimic the venation of certain pteridosperm pinnules (e.g., *Neuropteris, Odontopteris*) can be interpreted as camouflage against predation and regarded as evidence that predacious insects searched pteridosperm foliage for prey (Taylor and Scott, 1983). As feeding on ovules and pollen became a regular part of the behavior of such predators, and feeding on noncryptic, nonmotile food sources that were rich in nutritive value would have been obviously advantageous, the frequency of incidental pollination events would presumably have increased according to the distance between ovules and pollen organs, the sequence of organs visited, and the elapsed time between visits. As the frequency of successful pollinations increased in response to the archetypal pollinators, including more and more ovules and pollen in their diets, the plants may have responded by producing better attractants/rewards—such as complex pollen-bearing synangia (e.g., *Dolerotheca,* Medullosaceae)—and thus insect pollination would have become established.

Although this scenario seems plausible on the basis of the available data, the possibility that early selective pressure for the establishment of insect pollination was due partly to less well preadapted arthropods (i.e., those that seem less well suited by today's standards) cannot be ruled out. In fact, several generalist taxa might have been important cocontributors to early animal pollination mechanisms. In that eventually, it is possible to speculate that later in the Carboniferous these early pollinators would have been replaced by more reliable and better-adapted taxa.

II. Selective Pressure Imposed by Insect Pollination

Following the origin of insect pollination, it seems likely that selective pressure for increased efficiency would have become significant. If both reproductive structures, pollen and ovules, served as early attractants, then the percentage of successful pollination events would have been related to the average time between a pollinator's visit to an ovule and its last visit to a pollen-bearing organ of the appropriate species. The longer the time and the greater the frequency of intervening visits to other ovules or to the reproductive organs of other taxa, the less likely it is that the ovule in question would have been successfully pollinated. If either pollen or ovules alone were the earliest attractant, then successful pollination would still have been affected by the time between the pollinator's visit to an ovule and an appropriate pollen source. To reiterate, successful pollination in this instance would be far less likely because contact with the appropriate opposite organ would have been fortuitous and not directed. In each of these cases, the distance between the pollen sources and the ovules represents a significant factor in the effectiveness of the pollination mechanism. The closer together the pollen and ovules, the more likely it is that a particular visit to an ovule will have been immediately or recently preceded by a visit to an appropriate pollen-bearing organ. Therefore, the ultimate in efficiency would be the bisporangiate or hermaphroditic condition. With hermaphroditic fructifications the pollinator might have deposited pollen from an exogenous source while picking up new pollen. At the same time, the chances of carrying pollen from another individual of the same species would have been maximized since any previous visit to the fructifications of other individuals of the same species would involve pollen pickup.

Selective pressure for increasing the efficiency of insect pollination, then, would have favored the juxtaposition of ovules and pollen-bearing organs and would have followed from the origin of insect pollination in presumably monoecious or dioecious taxa. It seems most likely that the hermaphroditic condition evolved from monoecious ancestors because it is evolutionarily parsimonious.

A. The Permo-Triassic and the Appearance of the Hermaphroditic Condition

The interval between the development of insect pollination and the appearance of the hermaphroditic condition in the Jurassic (i.e., the Permo-Triassic) is an interesting and important one in the evolution of seed plants, insects, and pollination mechanisms.

There were drastic changes in the flora and fauna during the Permo-Triassic due, in part, to an overall climatic drying. With the end of the Paleozoic era came the end of the dominance of the lycopods and sphenophytes, and many arborescent forms disappeared entirely. Although vascular cryptogams survived in a subordinate role, seed plants flourished. Conifers, already present by the end of the Carboniferous (the Voltziales), became modern and diversified during the Permo-Triassic (Taylor, 1981b).

One of the most significant events in vascular plant evolution is represented by the Permian radiation of the division Pteridospermophyta that resulted in four new orders of pteridosperms—the Glossopteridales (Permian–Triassic), the Corystospermales (Triassic), the Peltaspermales (Triassic), and the Caytoniales (Triassic–Cretaceous)—and two new divisions of seed plants—the Cycadophyta (Triassic–Recent) and Cycadeoidophyta (Triassic–Cretaceous) (Taylor, 1981b). The Permian and post-Permian Pteridospermophyta all had partially enclosed ovules and, as far as can be determined, simpler fronds and synangia than their ancestors. The Pteridospermophyta was the only Carboniferous taxon that later diversified into dramatically different new orders or divisions. Because of the unique association of the pteridosperms with insect pollination, it is tempting to speculate that other taxa were trapped on adaptive peaks associated with canalization and stabilizing selection. Insect pollination might be regarded as an escape route that was especially effective during times of significant environmental change.[2]

The Permian was also an extremely important time of transition in the evolution of the arthropods. Insects were diverse in a unique way during the Permian because many paleopterous orders, although in decline, coexisted with the wax-

[2]Suggesting that the Pteridospermophyta had some intrinsic quality that allowed them to diversify more significantly than other taxa that survived the Permian involves the assumption that they represent a monophyletic group.

ing neopterous orders that would replace them (Carpenter, 1976). One of the four most important orders of contemporary anthophilous insects, the Coleoptera (beetles), appeared during the Permian (Carpenter, 1976). Modern beetles usually have chewing mouthparts and include taxa that feed on pollen, ovules, and other floral parts (Borror and DeLong, 1981).

Insect pollination appears to have been well established at the beginning of the Permian in monoecious or dioecious taxa. The Permian emergence of the Coleoptera (important anthophilous insects today and presumably more efficient than previous pollinators) suggests a possible increase in the selective pressure favoring the hermaphroditic condition. The appearance of protected ovules in many taxa during the Permian, whether or not related to possible insect pollination, is an index of increasing ovule or seed predation (Crepet, 1979b). Finally, conditions that would apparently favor the origin of the hermaphroditic condition existed against a background of environmental conditions that would have presumably facilitated rapid evolutionary change.

B. The Hermaphroditic Condition: The Cycadeoidophyta

Consistent with apparently increasing selective pressure for the bisporangiate condition is the first fossil evidence of unequivocal hermaphroditic reproductive structures in Jurassic cycadeoids. It is my opinion, however, that cycadeoids were primitively hermaphroditic and that the bisporangiate condition may have originated in the Triassic. Known unisexual Triassic cycadeoid fructifications probably represent dispersed abscised microsporangiate complexes or ovulate receptacles preserved after the abscision of their microsporangiate organs.

Cones of the cycadeoids are complex structures consisting of conical or dome-shaped ovulate receptacles bearing hundreds of ovules among protective interseminal scales. These are subtended by whorls of pinnate to variously modified microsporophylls that may have had anywhere from four to hundreds of synangia depending on the taxon involved (Crepet, 1974). The entire cone complex was surrounded by helically borne sterile bracts and was either exposed on slender branches or buried among persistent leaf bases, again depending on the specific taxon. Early cycadeoid cones were exposed and open, whereas late Mesozoic cycadeoid cones were protected and closed (Crepet, 1974). Observed evolutionary trends in the position and structure of cycadeoid cones suggest increasing predation by insects (Crepet, 1974). Well-preserved cones of the Jurassic–Cretaceous genus *Cycadeoidea* support the role of insects in these evolutionary trends and have provided strong evidence that the cycadeoids were beetle pollinated (Crepet, 1972). The probability of insect pollination is based on the nature and frequency of observed damage to cycadeoid cones and on the details of the cycadeoid reproductive cycle inferred from cones exquisitely preserved at various ontogenetic stages. The probability of beetle pollination is based on the nature of the damage and on the variety of insects, including the Coleoptera that were available at that time (Crepet, 1974).

The tendency for ovules to become enclosed in sterile tissue, noted in the Permian, continues in the Mesozoic in taxa other than the cycadeoids and provides further evidence of growing insect involvement in seed plant reproductive systems. Naturally, not all of these relationships involved insect pollinators/pollination.

The cycadeoids are also important in illustrating the type of reproductive morphology likely to have been associated with early coleopteran pollinators (i.e., radial symmetry, dish-bowl shape, numerous parts). Finally, cones of the cycadeoidaleans allow possible consequences of the attainment of the hermaphroditic condition to be explored.

C. Possible Selective Consequences of the Bisporangiate Condition

One of the possible consequences of the hermaphroditic condition might have been selective pressure to change the reward structure. If both ovules and pollen had to be attractants for early insect pollination mechanisms to be effective, then with the advent of the hermaphroditic condition the situation might have changed. It seems likely that since an attractant/reward existed in all hermaphroditic fructifications (ovules and pollen), the one more expensive to produce (ovules) might have been conserved and the attractive role completely shifted to pollen. The bisporangiate fructifications of the cycadeoidaleans are consistent with this possibility. The tiny seeds are extremely well protected by sterile interseminal scales which expand into tough capitate heads (Crepet, 1974). The result is an ovulate receptacle surface that is armored by expanded interseminal scale heads, with only the micropyles exposed.

III. The Angiospermae

The details of angiosperm origin are still clouded by uncertainty and missing paleontological data, but the timing of angiosperm origin and early radiation is now becoming clear. Recognizable angiosperm fossils first appear in the lower Cretaceous, and careful studies of fossil pollen and leaves suggest that the first angiosperm radiation closely followed their initial appearance (Doyle, 1969, 1978; Doyle and Hickey, 1976; Hickey and Doyle, 1977).

Angiosperms are generally considered to have evolved from pteridosperms, and of the known seed ferns, the Caytoniales seem to be the most closely related to the flowering plants (Doyle, 1978). The Caytoniales appear at the proper time to have been involved in angiosperm origin, and they share a unique set of derived characters with the angiosperms (enclosed ovules, net-veined leaves, four-chambered microsporangiate synangia) (Thomas, 1925). Nonetheless, recent palynological data suggest that a direct caytonialean origin of the angiosperms is unlikely (Crepet and Zavada, 1983). These data reveal an apparently pteridospermous clade that has not yet been recognized on the basis of megafossils and that might itself be more closely related to the angiosperms. Evolutionary trends within this clade suggest that the transition to angiospermy began during the Triassic–Jurassic and culminated in the Lower Cretaceous (Zavada, 1982).

Whether or not this clade proves important in angiosperm history, it is clear that angiospermous characters occur in a variety of gymnospermous taxa during the Triassic–Jurassic (e.g., hermaphroditic condition, net-veined leaves, tectate columellate-like wall structure). Given the possibility of mosaic evolution, the characters now comprising the angiosperms could have originated in the angiosperm lineage at different times during the Triassic–Cretaceous and become consolidated during the Lower Cretaceous. Alternatively, the complex of characters that we associate with the angiosperms today might have originated and shortly thereafter become consolidated during the Upper Jurassic–Lower Cretaceous, consistent with punctuated equilibria theory (Gould and Eldredge, 1977). In either of these instances, what is known about the fossil record of insects and what has been inferred about the evolution of insect pollination mechanisms in seed plants other than the angiosperms suggest that the origin and assembly of the basic angiosperm floral characters took place during a time when insect pollen and ovule predation were significant and when insect pollinators were strong selective agents. Consequently, the assembly of basic angiosperm floral characters and their individual attributes may be related to coevolution with insect pollinators. Succinctly, the angiospermous condition, as defined by reproductive structures, may have been precipitated by the coevolution of insect pollinators with an as yet unidentified pteridospermous taxon.

The primitive angiosperm flower proposed on the basis of neontological studies is radially symmetrical, many-parted, hermaphroditic, and dish-bowl shaped (e.g., the Magnoliaceae; Takhtajan, 1969). As it is presently understood, the fossil record is consistent with this interpretation. Pollen of the Magnoliidae is diverse relatively early in angiosperm evolution (Zavada, 1982; Walker and Brenner, 1982), and recently discovered mid-Cretaceous flowers are similar to those of the Magnoliaceae (Dilcher and Crane, 1982). Magnoliaceaen-type flowers may be considered morphological analogs of the bisporangiate fructifications of the cycadeoids—so close is the gross similarity, in fact, that when first described by Wieland, cycadeoids were considered to be the missing link between the cycadopsid gymnosperms and the angiosperms (Arber and Parkin, 1907). It is now obvious that this similarity does not stem from a close relationship (e.g., Doyle, 1978). Instead, it seems likely that these similarities stem from convergence associated with adaptation to and coevolution with early, probably coleopteran, pollinators.

As in the Cycadeoidophyta, the hermaphroditic condition in the angiosperms may be viewed as an evolutionary consequence of selection for increasingly efficient pollination in monoecious or dioecious ancestral taxa.

Continuing the comparison, the attainment of the hermaphroditic condition in the angiosperms might have had similar evolutionary consequences, i.e., that since all fructifications would then have two attractants/rewards (ovules and pollen), the more expensive of the two (ovules) might be conserved. Whereas cycadeoidaleans protected their ovules by tough interseminal scales, the carpel may have evolved to protect ovules mechanically in the angiosperms, as suggested by Grant (1950). Alternatively, carpels could have initially protected ovules by making them cryptic.

It is tempting to follow this line of reasoning and to suggest that the diminutive size of angiosperm ovules at the time of pollination (directly linked to the reduced megagametophyte relative to gymnosperm ovules) is a consequence of selective pressure to reduce ovule predation at the time of pollination by minimizing the nutritional reward represented by the ovules. Of course, the displacement of ovule development until after fertilization has energetic and genetic advantages (Westoby and Rice, 1982) that would also have favored the reduced ovule.

If the carpel is an adaptation related to coevolution with insect pollinators, then two other extremely important attributes of angiosperms may be ultimately related to coevolution with insects. (1) Evolutionary modifications of the carpel for fruit dispersal have been extremely important in angiosperm success by providing a means of enhancing the stochastic isolation of small populations in new environments (Vrba, 1980; Tiffney, 1982). (2) The carpel has additional importance of a more subtle variety. The carpel places sterile sporophytic tissue between the ovules and pollen grains, a situation with two potentially important ramifications. First, contact between the pollen grain/pollen tube and sporophytic tissue present the circumstances required for the evolution of self-incompatibility as observed in modern angiosperms. It is possible that the increase in outcrossing effected by such incompatibility systems has conferred certain advantages on the angiosperms (Whitehouse, 1950) and that the hermaphroditic condition provided the selective impetus for the evolution of self-incompatibility mechanisms in taxa that were self-fertile but already had interspecific incompatibility and perhaps even the ancient S locus (Pandey, 1980). It is even possible that the angiosperms originated from an insect-pollinated monoecious lineage that already had some form of self-incompatibility and that the attainment of the hermaphroditic condition provided selective pressure for refinements of the archaic system.

At the present time, incompatibility mechanisms remain almost totally within the realm of speculation in the paleontological milieu. However, M. S. Zavada (in preparation) has recently noted a correlation between exine structure and sporophytic incompatibility. Semitectate pollen is involved in all known cases of sporophytic incompatibility. This is assumed to facilitate communication between the incompatibility substances that are sequestered in the infratectal space and the stigmatic surface. Obviously, the relationship between pollen structure and the most common form of stigmatic incompatibility represents a chance to observe the progress of the latter in angiosperm history and early angiosperm pollen has pore (lumina) size compatible with sporophytic incompatibility. While in no way proof of conjecture that efficient incompatibility might have been induced selectively by the hermaphroditic condition, available fossil evidence is certainly consistent with that possibility.

A second potentially important advantage conferred on the angiosperms by the carpel is that, in combination with insect pollination, it represents an arena for competition and selection at the gametophyte level. The significance of this phenomenon has been elegantly evaluated by Mulcahy (1979; and Chapter 7, this volume).

Self-incompatibility, gametophytic competition, efficient pollination, the reduced female gametophyte, and increased outcrossing have all been suggested as advantages enjoyed by early angiosperms (Whitehouse, 1950; Regal, 1977; Mulcahy, 1979; Stebbins, 1981), and each of these factors may be evolutionarily related to insect pollination.

A. Faithful Pollinators

Only relatively recently have the data become available to allow a consideration of the possible significance of faithful or constant pollinators in present angiosperm diversity. Stebbins (1981), for example, minimized the direct impact of insect pollination on present angiosperm diversity by noting that constant pollinators, which he regarded as those most easily associated with the promotion of diversity, occurred too late in the fossil record relative to major angiosperm diversification to have been a significant factor. New data in the areas listed below now allow a more precise evaluation of the possibility that constant pollinators played an important role in the establishment of angiosperm diversity:

1. Growing information on the timing of the evolution of various floral features and on the fossil records of families associated with constant pollinators.
2. New data on the timing of the evolution of constant pollinators.
3. New information on patterns of diversification in the angiosperms.

B. Floral Features and Families Associated with Faithful Pollinators

The growing fossil record of flowers and inflorescences has been of direct importance in understanding the evolution of various pollination mechanisms. Fossil floral structures provide the most useful evidence for determining the states of various pollination mechanisms at various times. Floral data also allow some correlation between floral structure and other plant organs representing similar taxa at particular times. These correlations aid in interpreting nonfloral fossil data in the context of pollination mechanisms.

Fossil flowers and inflorescences are best known from Eocene deposits (Crepet, 1979a; Crepet and Daghlian 1981a,b; Zavada and Crepet, 1981) and reveal specific floral characters and families associated with constant pollinators (Apoidea, bees, Lepidoptera, moths, and butterflies). Bilateral symmetry, usually associated with bee pollination, is evident in certain undescribed fossil flowers (Crepet, 1979b). Flowers with narrow tubular corollas suggesting lepidopteran pollination are also known from Eocene deposits (Crepet, 1982). Flowers and other organs of specific families associated with constant pollinators are well represented in Middle and Lower Eocene deposits. Brush flowers representing two taxa of the Mimosaceae have been described from Middle Eocene deposits (Crepet and Dilcher, 1977; Crepet, 1982). The Zingiberidae are well known for their advanced floral structure, which is frequently characterized by well-developed zygomorphy. A variety of taxa of the Zingiberidae are known from Eocene deposits based on leaves, fruits, and even a heliconioid flower (Daghlian, 1981; Koch and Friedrich, 1971; Crepet, 1982). One of the largest modern families associated with a variety of constant insect pollinators is the Euphorbiaceae (Stebbins, 1981). This family was apparently quite well established by the Middle Eocene. Flowers and fruits representing one of the most advanced euphorbioid tribes, the Hippomane, have been reported from Middle Eocene deposits (Crepet and Daghlian, 1981a; Manchester and Dilcher, 1979).

It is apparent from floral data that bees and lepidopterans, the two most important groups of constant pollinators, were well developed by the Middle Eocene. Well authenticated nonfloral evidence for other families associated with constant pollinators supports this observation (e.g., the Onagraceae, Boraginaceae, Caprifoliaceae, and Bignoniaceae; Crepet, 1982).

The advanced state of constant pollination inferred from Middle Eocene paleobotanical data is consistent with earlier coevolution between constant pollinators and various angiosperm families. It is risky to infer prehistory based on conditions at any given time, in view of the potential of extremely rapid evolution (Gould and Eldredge, 1977) and considering the magnitude of geological time. However, additional evidence confirms the involvement of constant pollinators in pre-Middle Eocene angiosperm history. Recently, Lower Eocene flowers have been discovered with seven-lobed, sympetalous, open-funneled corollas (Crepet and Daghlian, 1981b). Pollen found within the anthers is well preserved, structurally unique, and well known in the dispersed palynomorph literature (*Pistillipollenites*). Pollen has been reported from the Maestrichtian to the Eocene (Crepet and Daghlian, 1981b). These flowers are phenetically closest to the extant Gentianaceae and have a floral morphology typical of modern flowers pollinated by bees (Faegri and van der Pijl, 1971; Crepet and Daghlian, 1981b). The presence of distinctive palynomorphs similar to those found in these flowers in earlier deposits represented a unique opportunity to determine the history of this type of floral morphology given the assumption that identical, highly derived pollen would be associated with similar floral morphology at other times. In order to ascertain that the dispersed palynomorphs were indeed identical to those found in the Eocene flower, dispersed pollen was obtained from eight Paleocene–Eocene localities. Fifty grains were isolated from each sample and statistical tests were run to determine the variation in size within and among the localities. In addition, palynomorphs from each locality were sectioned from transmission electron microscopy and prepared for scanning electron microscopy (Crepet and Daghlian, in preparation). The results revealed no significant differences in size, ultrastructure, or micromorphology over time, suggesting that flowers with morphology characteristic of bee pollination existed as early as the Paleocene and possibly as early as the Maestrichtian (although the Maestrichtian palynomorphs seem identical, we were unable to obtain samples for our own analysis and evaluation).[3] Other paleobotanical data confirm that families associated with constant pollinators existed by the Maestrichtian or even earlier (e.g., Myrtaceae, Caesalpiniaceae, and Zingiberaceae; Crepet, 1982; Muller, 1981; Hickey and Peterson, 1978).

C. The History of Faithful Pollinators

Bees (Apoidea) are the preeminent faithful pollinators. The fossil record of bees has not been useful in understanding the timing of their evolution. The first bona fide bees occur in Lower Oligocene deposits (Burnham, 1978; Michener, 1979), but they are so diverse at that time that a much earlier origin seems likely. The presence of highly eusocial bees in the Lower Oligocene is also suggestive of a significantly earlier origin, especially in light of the degree of complexity of sociality in these taxa (Apini and Meliponini; Michener, 1979).

Insights into the history of bees have come from a careful analysis of their present biogeography by Michener (1979). Bees are generally not particularly

[3]Further germane evidence exists in the recent discovery of typical zygomorphic flowers of the Papilionaceae in Paleocene deposits in the southeastern United States. Inflorescences of the tribe Hippomane (Euphorbiaceae) have also been discovered in these deposits.

good dispersers, and highly social bees are extremely bad dispersers partially because of their complex swarming behavior (Michener, 1979). Although most bee distributions can be regarded as the result of slow spreading over continents, certain distributions are most easily explained by continental movements, including two subfamilies of the Colletidae, the Fideliidae, and one subgenus of the Meliponini (Michener, 1979). The short-tongued Colletidae are considered to be the most primitive bees and are strongly associated with one particular angiosperm family, the Myrtaceae (Michener, 1979). The Fideliidae are the most primitive extant long-tongued bees, and the Meliponini (stingless honeybees) are highly social and are comparable to the honeybees (Apini) in phylogenetic advancement (Michener, 1974). The Meliponini are extremely important pollinators in the modern tropics. Michener (1979) suggests that, on the basis of modern biogeography and the timing of continental movements, all of these taxa originated by the uppermost Cretaceous and perhaps even earlier.

D. Lepidoptera

Several major features of the history of the Lepidoptera may be inferred from their fossil record. The first reliable reports of Lepidoptera concern 100–130-million-year-old amber-entombed zeuglopteran moths that are apparently closely related to the extant micropterigid genus *Sabatinca* (Whalley, 1977). On the basis of the specialized nature of these fossils, Whalley proposes a Jurassic origin for the Lepidoptera. Micropterigids had mandibulate mouthparts, suggesting that the earliest lepidopterans were either predatory or pollen feeders. Micropterigid scales are also known from Cretaceous amber (Kuhne *et al.*, 1973).

The haustellate Lepidoptera, those with mouthparts specialized for nectar feeding, are first represented by the head of a ditrysian larva in 73-million-year-old Cretaceous amber (MacKay, 1977).

The record of Paleogene Lepidoptera is better than that of the Cretaceous and is consistent with the Cretaceous record. A variety of monotrysian and ditrysian moths are known from the Paleogene, and Durden and Rose (1978) have described 27 species of butterflies including species of the highly derived Papilionidae.

E. Patterns of Diversification in the Angiosperms

It is apparent that floral features, families, and pollinators associated with constant pollination mechanisms existed during the Upper Cretaceous. In order to evaluate the potential contribution of constant pollination mechanisms to angiosperm diversity, it is essential that the occurrence of this phenomenon be considered in relation to the diversification pattern of the angiosperms. Until recently, no quantitative attempts have been made to determine the diversification pattern of angiosperms. However, Niklas *et al.* (1980) and Muller (1981) have quantitatively assessed angiosperm diversification patterns on the basis of the literature and the palynological record, respectively. There are several inherent differences between these two approaches, and each has its advantages (Crepet, 1982). Both provide a view of angiosperm diversification that is to some degree unexpected. Angiosperms began their initial radiation shortly after the first angiosperm fossils appeared in the Lower Cretaceous (Niklas *et al.*, 1980); Muller, 1981). This was followed by a period of gradual increase until either the Santonian (Muller, 1981) or the lowermost Tertiary (Niklas *et al.*, 1980), when a second radiation began that was greater in magnitude than the first. The discrepancy in the suggested timing of the second major angiosperm radiation may or may not be significant, considering the inherent constraints imposed by evaluating fossil evidence and in light of differences in the data analyzed (Crepet, 1982; see Tiffney, 1981, 1982, for further discussion). Nonetheless, it is clear that a second major angiosperm radiation began at some point between the Santonian and the lowermost Tertiary.

Thus, the timing of the second and apparently major radiation of the angiosperms is consistent with other paleontological data in suggesting that pollination by constant pollinators was an important feature in the establishment of present-day angiosperm diversity.

IV. Summary and Conclusion

Modern interactions between flowering plants and their pollinators are frequently interpreted as the results of a long and intimate coevolutionary relationship. Many basic floral features have been related to the selective power of insect pollinators over evolutionary time (Grant, 1950; Stebbins, 1981). Recently available fossil data, including the timing of the appearances of certain taxa or characters, evolutionary patterns, the reproductive morphology of certain seed plants, and direct evidence of arthropod interactions with the reproductive structures of various taxa, allow an analysis of the developing relationship between seed plants and pollinators in the framework of geologic time.

As I have interpreted the fossil record, the unfolding of this relationship provides insights into evolutionary trends in the reproductive structures of certain seed plants and the patterns of their diversification. Naturally, coevolutionary relations with arthropods other than those resulting in pollination have been important in angiosperm evolution (Niklas, 1978), and other features characterizing the flowering plants may also have been important in their rise to dominance. My goal has been not to deny the importance of other features of angiosperms, but to focus on insect pollination.

Further understanding of plant–pollinator relationships on the basis of neontological data and future paleontological data may invite a reconsideration of certain of the ideas suggested above. However, I wish to emphasize that in order to understand fully the evolutionary impact of insect pollination, it is necessary to elucidate its dynamic nature—something that can best be done by including the temporal dimension.

References

Arber, E. A. N., and Parkin, J. (1907). On the origin of angiosperms. *J. Linn. Soc. London, Bot.* **38,** 29–80.

Baker, H. G., and Hurd, P. D. (1968). Intrafloral ecology. *Annu. Rev. Entomol.* **13,** 385–414.

Borror, D. J., DeLong, D. M., and Triplehorn, C. A. (1981). "An Introduction to the Study of Insects," 5th ed. Saunders, New York.

Brenner, G. J. (1976). Middle Cretaceous provinces and the early migration of angiosperms. *In* "Origin and Early Evolution of Angiosperms" (C. B. Beck, ed.), pp. 23–47. Columbia Univ. Press, New York.

Burnham, L. (1978). Survey of social insects in the fossil record. *Psyche* **85,** 85–133.

Carpenter, F. M. (1976). Geological history and the evolution of the insects. *Proc. Int. Congr. Entomol., 15th, 1976* pp. 63–70.

Chaloner, W. G. (1976). The evolution of the adaptive features in fossil exines. *In* "The Evolutionary Significance of the Exine" (I. K. Ferguson and J. Muller, eds.), pp. 1–4. Academic Press, New York.

Crepet, W. L. (1972). Investigations of North American cycadeoids: Pollination mechanisms in Cycadeoida. *Am. J. Bot.* **59,** 1048–1056.

Crepet, W. L. (1974). Investigations of North American cycadeoids: The reproductive biology of Cycadeoidea. *Palaeontographica* **148B,** 144–159.

Crepet, W. L. (1979a). Some aspects of the pollination biology of Middle Eocene angiosperms. *Rev. Palaeobot. Palynol.* **27,** 213–238.

Crepet, W. L. (1979b). Insect pollination: A paleontological perspective. *BioScience* **29,** 102–108.

Crepet, W. L. (1982). Advanced (constant) insect pollination mechanisms: Pattern of evolution and implications vis a vis angiosperm diversity. *Ann. M. Bot. Gard.* (in press).

Crepet, W. L. (1983). Punctuated Equilibria: A fossil angiosperm perspective. *Bot. Rev.* (in press).

Crepet, W. L., and Daghlian, C. P. (1981a). Euphorbioid inflorescences from the Middle Eocene Claiborne Formation. *Am. J. Bot.* **69,** 258–266.

Crepet, W. L., and Daghlian, C. P. (1981b). Lower Eocene and Paleocene Gentianaceae: Floral and palynological evidence. *Science* **214,** 75–77.

Crepet, W. L., and Dilcher, D. L. (1977). Investigations of angiosperms from the Eocene of North America: A mimosoid inflorescence. *Amer. J. Bot.* **64,** 714–725.

Crepet, W. L., and Zavada, M. S. (1983). Evolutionary implications of the ultrastructure of pollen of *Caytonanthus.* (Manuscript submitted.)

Cruden, W. C. (1977). Pollen-ovule ratios: A conservative indicator of breeding systems in flowering plants. *Evolution (Lawrence, Kans.)* **31,** 32–42.

Daghlian, C. P. (1981). A review of the fossil record of monocotyledons. *Bot. Rev.* **47,** 517–555.

Dilcher, D. L. (1979). Early angiosperm reproduction: An introductory report. *Rev. Palaeobot. Palynol.* **27,** 291–328.

Dilcher, D. L., and Crane, P. R. (1982). *Archaeanthus:* An early angiosperm from the Cenomanian of the western interior of North America. *Ann. M. Bot. Gard.* (in press).

Doyle, J. A. (1969). Cretaceous angiosperm pollen of the Atlantic Coastal Plain and its evolutionary significance. *J. Arnold Arbor., Harv. Univ.* **50,** 1–35.

Doyle, J. A. (1978). Origin of angiosperms. *Annu. Rev. Ecol. Syst.* **9,** 365–392.
Doyle, J. A., and Hickey, L. J. (1976). Pollen and leaves from the mid-Cretaceous Potomac Group and their bearing on early angiosperm evolution. *In* "The Origin and Early Evolution of the Angiosperms" C. B. Beck, ed., Columbia Univ. Press, New York.
Durden, C. J., and Rose, H. (1978). Butterflies from the Middle Eocene: The eariest occurrance of fossil Papilionoidea (Lepidoptera). *Pearce-Sellards Ser. Tex. Memor. Mus.* **29,** 1–25.
Faegri, F., and van der Pijl, L. (1971). "The Principles of Pollination Ecology." Pergamon, Oxford.
Friis, E. M. (1982). Preliminary report of Upper Cretaceous angiosperm reproductive organs from Sweden and their level of organization. *Ann. M. Bot. Gard.* (in press).
Gould, S. J., and Eldredge, N. (1977). Punctuated equilibria: The tempo and mode of evolution reconsidered. *Paleobiology* **3,** 115–151.
Grant, V. (1949). Pollinating systems as isolating mechanisms. *Evolution* (*Lawrence, Kans.*) **3,** 82–97.
Grant, V. (1950). The protection of ovulesin flowering plants. *Evolution* (*Lawrence, Kans.*) **4,** 179–201.
Hickey, L. J., and Doyle, J. A. (1977). Early Cretaceous fossil evidence for angiosperm evolution. *Bot. Rev.* **43,** 2–104.
Hickey, L. J., and Peterson, R. K. (1978). *Zingiberopsis,* a fossil genus of the ginger family from Late Cretaceous to Early Eocene sediments of Western Interior North America. *Can. J. Bot.* **56,** 1136–1152.
Janzen, D. H. (1971). Seed predation by animals. *Annu. Rev. Ecol. Syst.* **2,** 465–492.
Koch, B. E., and Friedrich, W. L. (1971). Fruchte und Samen von Spirematospermum aus der Miozänen Fasterholt-flora in Danemark. *Palaeontographica* **136B,** 1–46.
Kuhne, W. G., Kubig, L., and Schlüter, T. (1973). Eine Micropterygide (Lepidoptera, Homoneura) aus mittelcretazischen Harz Westfrankreichs. *Mitt. Deut. Entomol. Ges.* **32,** 61–64.
Levin, D. A. (1979). Pollinator foraging behavior: Genetic implications for plants. *In* "Topics in Plant Biology" (O. T. Solbrig, S. K. Jain, G. B. Johnson, and P. H. Raven, eds.), Columbia Univ. Press, New York.
MacKay, M. R. (1977). Lepidoptera in Cretaceous amber. *Science* **167,** 379–380.
Manchester, S. R., and Dilcher, D. L. (1979). A euphorbiaceous fruit from the Middle Eocene of Tennessee. *Bot. Soc. Am. Misc. Ser. Publ.* **157,** 34. (Abstr.)
Michener, C. D. (1974). The social behavior of bees: A comparative study. Harvard Univ. Press, Cambridge, Massachusetts.
Michener, C. D. (1979). Biogeography of the bees. *Ann. M. Bot. Gard.* **66,** 277–347.
Mulcahy, D. L. (1979). The rise of the angiosperms: A genecological factor. *Science* **206,** 20–23.
Muller, J. (1981). Fossil pollen records of extant angiosperms. *Bot. Rev.* **47,** 1–142.
Niklas, K. J. (1978). Coupled evolutionary rates and the fossil record. *Brittonia* **30,** 373–394.
Niklas, K. J., Tiffney, B. H., and Knoll, A. H. (1980). Apparent changes in the diversity of fossil plants. *Evol. Biol.* **12,** 1–89.
Phillips, T. L., and DiMichele, W. A. (1981). Paleoecology of Middle Pennsylvanian age coal swamps in the southern Illinois/Herrin Coal Member at Sahara Mine No. 6. *In* "Paleobotany, Paleoecology and Evolution" K. Niklas, ed., Vol. 1, pp. 231–284. Praeger, New York.
Pohl, F. (1937). Die Pollenerzeugung der Windbluter. Eine vergleichende Untersuchung mit Ausbliken auf den Bestaubungshaushalt tierblutiger Gewächse und die pollenanalytische Waldgeschichtsborschung. *Beih. Bot. Zbl.* **56,** 365–470.
Regal, P. J. (1977). Ecology and the evolution of flowering plant dominance. *Science* **196,** 622–629.
Richardson, E. S. (1980). Life at Mazon Creek. Middle and Late Pennsylvania strata on margin of Illinois Basin. *10th Annu. Field Conf., Great Lakes Sect., S.E.P.M. Univ., Illinois* pp. 217–224.
Scheihing, M. H. (1980). Reduction of wind velocity by the forest canopy and the rarity of non-

arborescent plants in the Upper Carboniferous fossil record. *Argumenta Palaeobotanica* **6,** 133–138.

Scott, A. C., and Taylor, T. N. (1983). Plant/animal interactions during the Upper Carboniferous. *Bot. Rev.* (in press).

Silander, J. A. (1983). Microevolution in clonal plants. *In* "The Population Biology of Evolution of Clonal Organisms" (J. B. C. Jackson, L. W. Buss, and R. Cook. eds.), Yale Univ. Press, New Haven, Connecticut (in press).

Stanley, R. B., and Linskins, H. F. (1974). "Pollen: Biology, Biochemistry, Management." Springer-Verlag, Berlin and New York.

Stebbins, G. L. (1981). Why are there so many species of flowering plants? *BioScience* **31,** 573–577.

Takhtajan, A. L. (1969). "Flowering Plants: Origin and Dispersal." Smithsonian, Washington.

Taylor, T. N. (1981a). Pollen and pollen organ evolution in early seed plants. *In* "Paleobotany, Paleoecology, and Evolution" (K. Niklas, ed.), Praeger. New York.

Taylor, T. N. (1981b). "Paleobotany: An Introduction to Fossil Plant Biology." McGraw-Hill, New York.

Taylor, T. N., and Millay, M. A. (1979). Pollination biology and reproduction in early seed plants. *Rev. Palaeobot. Palynol.* **27,** 329–355.

Taylor. T. N., and Scott, A. C. (1983). Interactions between plants and animals during the Carboniferous. *BioScience* (in press).

Thomas, H. H. (1925). The Caytoniales: A new group of angiospermous plants from the Jurassic rocks of Yorkshire. *Philos. Trans. R. Soc. London, Ser. B.* **213,** 299–363.

Tiffney, B. H. (1977). Dicotyledonous angiosperm flower from the Upper Cretaceous of Martha's Vineyard. Massachusetts. *Nature (London)* **265,** 136–137.

Tiffney, B. H. (1981). Diversity and major events in the evolution of land plants. *In* "Paleobotany, Paleoecology, and Evolution" (K. Niklas, ed.), Praeger, New York.

Tiffney, B. H. (1982). Seed size, dispersal syndromes, and the rise of the angiosperms: Evidence and hypothesis. *Ann. M. Bot. Gard.* (in press).

Vrba, E. S. (1980). Evolution, species, and fossils: How did life evolve? *S. Afr. J. Sci.* **76,** 61–84.

Walker, J. W., and Brenner, G. J. (1983). Fossil tetrads of the Anonaceae. *Science* (in press).

Walker, J. W., and Walker, A. G. (1982). Same grain combined light, scanning electron, and transmission electron microscopy of Lower Cretaceous angiosperm pollen. *Ann. M. Bot. Gard.* (in press).

Westoby, M. and Rice, B. (1982). Evolution of the seed plants and inclusive fitness of plant tissues. *Evolution (Lawrence, Kans.)* **36,** 713–724.

Whalley, P. (1977). Lower Cretaceous Lepidoptera. *Nature (London)* **266,** 526.

Whitehead, D. R. (1969). Wind pollination in the angiosperms: Evolutionary and environmental considerations. *Evolution, (Lawrence, Kans.)* **22,** 28–35.

Whitehouse, H. L. K. (1950). Multiple-allelomorph incompatibility of pollen and style in the evolution of the angiosperms. *Ann. Bot.* **14,** 198–216.

Zavada, M. S. (1982). Morphology, ultrastructure, and evolutionary significance of monosulcate pollen. Ph.D. Dissertation, Univ. of Connecticut, Storrs, Connecticut.

Zavada, M. S., and Crepet, W. L. (1981). Investigations of angiosperms from the Eocene of North America: Flowers of the Celtidoideae. *Am. J. Bot.* **68,** 924–933.

CHAPTER 4

Pollinator–Plant Interactions and the Evolution of Breeding Systems

ROBERT WYATT
Department of Botany
University of Georgia
Athens, Georgia

I. Introduction and Review

A. Historical Separation of Studies of Pollination Ecology and of Breeding Systems

Research in pollination biology unfortunately has proceeded historically in isolation from research in plant-breeding systems. Studies of pollinator–plant interactions were, until very recently, descriptive and static in approach. No

POLLINATION BIOLOGY

ISBN 0-12-583980-4 (hardcover)
ISBN 0-12-583982-0 (paperback)

consideration was given to the genetic bases of floral adaptations for pollination, and the implications of floral morphology and pollinator behavior for population genetics were totally overlooked. On the other hand, the study of plant-breeding systems, which evolved as a field of study only within the last 100 years, developed as an offshoot of genetics. Ecology was largely ignored in discussions of the evolution of breeding systems, and although experimental approaches were used to discover their genetic basis, breeding systems themselves were viewed statically and categorized in a purely descriptive framework. In particular, the important role of pollinators was ignored.

B. Pollination Ecology

1. Traditional Approaches: Descriptive and Static

Early studies of pollination ecology, such as those of Koelreuter (1761–1766) and Sprengel (1793), the fathers of pollination ecology (Faegri and van der Pijl, 1979), were devoted entirely to descriptions of structural adaptations of flowers to insect pollination. It is understandable that the amazingly acute observations of these workers were marred occasionally by overzealous interpretations. Darwin's (1876, 1877) research laid the basis for an experimental approach to plant reproductive biology and provided a scientific conceptual framework in which to study reciprocal adaptations of pollinators and plants. Unfortunately, as noted by East (1940, p. 450), "those who followed Darwin's lead in studying the general problems of reproductive biology in plants lacked somewhat his catholicity, his penetrating logic, and his discrimination." Unduly influenced by the Darwin-Knight law, which states that cross-pollination is absolutely essential, these workers ignored the phenomenon of self-fertilization and sought to interpret all aspects of plant–pollinator interactions in terms of adaptations for cross-pollination. Among the many botanists who devoted themselves to the study of pollination in the second half of the nineteenth century were H. Müller, F. Müller, Delpino, Hildebrand, and Knuth. Classic pollination ecology reached its zenith in Knuth's three-volume handbook of flower pollination (Knuth, 1906–1909).

During the first half of the twentieth century, studies of pollination biology were eclipsed. The field was revived, however, by the careful observations of workers such as Baker (1961) and Grant and Grant (1965), who synthesized information regarding pollinators and plants into an evolutionary framework. Their research and that of many others has recently been well summarized in overviews of pollination biology by Proctor and Yeo (1973) and Faegri and van der Pijl (1979). These authors recognize pollination syndromes, suites of characters that represent adaptations to particular types of pollinators (Table I). This approach is important because it helps to focus attention on characteristics of

TABLE I

FLORAL SYNDROMES OF ANIMAL-POLLINATED PLANTS[a]

Syndrome	Pollinators	Anthesis	Colors	Odors	Flower shapes	Flower depth	Nectar guides	Rewards
Cantharophily	Beetles	Day and night	Variable, usually dull	Strong, fruity or aminoid	Actinomorphic	Flat to bowl-shaped	None	Pollen or food bodies
Sapromyophily	Carrion and dung flies	Day and night	Purple-brown or greenish	Strong, often of decaying protein	Usually actinomorphic	None, or deep if traps involved	None	None
Myophily	Syrphids and bee flies	Day and night	Variable	Variable	Usually actinomorphic	None to moderate	None	None or pollen or nectar
Melittophily	Bees	Day and night or diurnal	Variable but no pure red	Present, usually sweet	Actinomorphic or zygomorphic	None to moderate	Present	Nectar (41.6%) and pollen; open or concealed
Sphingophily	Hawkmoths	Nocturnal or crepuscular	White or pale to green	Strong, usually sweet	Actinomorphic; held horizontal or pendant	Deep, narrow tube or spur	None	Ample nectar (22.1%); concealed
Phalaenophily	Small moths	Nocturnal or crepuscular	While or pale to green	Moderately strong, sweet	Actinomorphic; held horizontal or pendant	Moderately deep tube	None	Nectar; concealed
Psychophily	Butterflies	Day and night or diurnal	Bright red, yellow, or blue	Moderately strong, sweet	Usually actinomorphic; upright	Deep narrow tube or spur	Present	Nectar (22.8%); concealed
Ornithophily	Birds	Diurnal	Bright red	None	Actinomorphic or zygomorphic	Deep, wide tube or spur	None	Ample nectar (25.4%); concealed
Chiropterophily	Bats	Nocturnal	Dull white or green	Strong, fermented	Actinomorphic or zygomorphic	Brush- or bowl-shaped	None	Ample nectar (18.9%) and ample pollen; open

[a]Modified from Baker and Hurd (1968) and Faegri and van der Pijl (1979). Nectar concentrations are those reported by Pyke and Waser (1981).

adaptive significance and provides a framework in which to organize a welter of descriptive facts. There is also danger, however, in accepting these syndromes too literally and extrapolating probable pollinators from knowledge of blossom classes and colors. Exceptions are numerous, and as Faegri and van der Pijl (1979, p. 96) caution: "Such conclusions can only be hypotheses requiring verification by observation of actual conditions." As pollination ecologists, we must be very careful in categorizing plants as "cantharophilous," "phalaenophilous," or "chiropterophilous" and must not become so entranced by scientific jargon that we lose sight of the biologically important aspects of pollinator–plant interactions. We must also avoid the pitfall that hampered the progress of pollination ecology in the late 1800s, when "observers became less interested in what happened and more interested in what they thought ought to happen" (East, 1940, p. 451). Yet another danger inherent in a too facile acceptance of pollination syndromes is the increasingly frequent observation that even many highly specialized flowers receive visits from a range of visitors (Baker and Hurd, 1968; Stebbins, 1970). Such a situation is amply illustrated by the controversy surrounding pollination of *Aquilegia* in the Sierra Nevada of California. Chase and Raven (1975) observed hawkmoths visiting typical red-flowered *A. formosa,* which Grant (1952) had supposed was pollinated strictly by hummingbirds, thus decreasing the probability of hybridization with white-flowered, hawkmoth-pollinated *A. pubescens* (but see Grant, 1976). Even within a class of pollinators there is variation in behavior between species, and Heinrich (1976) emphasizes that such variation also exists in bumblebees between castes and individuals within castes. Only careful observations can determine if visitors are or are not pollinating flowers. Even if they are not effective pollinators, the activity of these visitors can have important effects on the reproductive biology of the plants (Baker and Hurd, 1968). The adaptive significance of certain characters of flowers, therefore, may not be directly attributable to the major pollinator of a species, but rather to selection pressures exerted by other floral visitors. For example, Raven (1972) suggests that the red color characteristic of hummingbird-pollinated plants is a mechanism to diminish visits by nectar-robbing bees, whose vision is insensitive to red wavelengths. Such considerations raise some doubt as to the generality of Stebbins's (1974, p. 62) "most effective pollinator principle," which holds that "the characteristics of the flower will be molded by those pollinators that visit it most frequently and effectively in the region where it is evolving."

2. *Modern Approaches: Experimental and Dynamic*

In spite of early suggestions by some investigators, such as Stebbins (1950) and Fryxell (1957), that studies of the pollination biology of plants should become more experimental and dynamic rather than descriptive and static, it is only very recently that new approaches have been tried. Most importantly, this research has considered pollination in terms of its genetically important effects as one component (along with seed dispersal) of gene flow in plants. Although aspects of this problem have been of interest to crop breeders for some time (Fryxell, 1957), it is only during the last 25 years that plant evolutionists have begun to examine gene flow in natural populations. Newer techniques for measuring pollen dispersal, such as radioactive isotopes (Schlising and Turpin, 1971; Reincke and Bloom, 1979), neutron activation analysis (Gaudreau and Hardin, 1974; Handel, 1976), and electrophoretic markers (Schaal, 1980; Ennos and Clegg, 1982) have been used recently in preference to earlier techniques employing fluorescent dusts or dyes (Thies, 1953; Stephens and Finkner, 1953; Simpson and Duncan, 1956; Sindu and Singh, 1961; Stockhouse, 1976).

Levin and Kerster (1974) summarized our knowledge of potential and actual gene flow in plants, relating pollen and seed dispersal patterns to the genetic structure of plant populations. They find the evidence compelling in rejection of the earlier view that gene flow is extensive and plays a major role in the cohesion of populations and population systems. Empirical evidence suggests that gene flow, especially through pollen dispersal, is extremely restricted in range (Grant, 1958, 1971; Ehrlich and Raven, 1969; Stebbins, 1970; Bradshaw, 1972; Levin, 1979b, 1981). Typically, the pattern of animal-mediated pollen dispersal is strongly leptokurtic (Levin, 1972a, 1978, 1979b, 1981; Levin and Kerster, 1974).

The accuracy with which one can extrapolate from pollinator movements to pollen dispersal patterns, however, is not easily determined (Levin and Kerster, 1974; Primack and Silander, 1975; Levin 1978, 1979a,b, 1981; Schaal, 1980; Waddington, 1981; Waser and Price, 1982a,b; Ennos and Clegg, 1982). This relationship depends on pollinator characteristics such as pollen carryover rates and directionality of successive moves between flowers (Levin, 1981). It also depends, in terms of its genetic importance, on plant characteristics such as breeding system and inflorescence architecture (Wyatt, 1982). Another new approach in pollination ecology with important implications for pollinator–plant interactions is pollination energetics (Heinrich and Raven, 1972; Heinrich, 1975, 1979). This and related topics are discussed by Waddington (Chapter 9, this volume).

C. Breeding Systems

1. Traditional Approaches: Descriptive and Static

For the purposes of this review, I shall define "breeding systems" in a very broad sense to include all aspects of sex expression in plants that affect the relative genetic contributions to the next generation of individuals within a species. Bawa and Beach (1981) have introduced the term "sexual systems" to refer to this expanded concept of breeding systems. As such, breeding systems of plants represent one of a number of factors that control the amount of recombination per generation and that directly affect the fecundity component of fitness in natural populations (Grant, 1975).

a. The Search for Outcrossing Mechanisms. Darwin (1876, 1877) was again the first investigator to appreciate the function and significance of outcrossing mechanisms in angiosperms. His demonstration that outcrossed progeny are usually more vigorous than those produced by self-fertilization touched off a virtual deluge of studies that attempted (sometimes uncritically) to interpret any and all variations in plant-breeding systems as adaptations for outcrossing (East, 1940). In its worst form, this overzealous commitment to the Darwin-Knight law led researchers not merely to overlook self-fertilization but actually to find apparent adaptations for cross-pollination in species known today to be habitual selfers (Faegri and van der Pijl, 1979). As recently as 1963, as eminent an authority as C. D. Darlington (1963, p. 41) stated that "the reproductive history of the flowering plants is therefore largely an account of different ways in which they evade self-fertilization or escape from its consequences."

Before considering the outcrossing mechanisms documented by Darwin and his followers, it seems appropriate to discuss the rather confusing terminology and classification of sexual variation in plants. As noted by Frankel and Galun (1977), we must always be cognizant of the three levels on which we may consider sex expression in angiosperms: the individual flower, the individual plant, or the group of plants. Frankel and Galun's (1977) scheme is shown in Table II. Their list is not exhaustive and one might add, for example, "trioecious" to refer to groups in which hermaphroditic plants occur in addition to androecious and gynoecious ones (see Yampolsky and Yampolsky, 1922 or Radford *et al.*, 1974, for such additional terms). Among others, Bawa and Beach (1981), drawing on the recent work of Lloyd (1979c), have proposed an alternative classification scheme. Nevertheless, Table II is sufficient to cover the great majority of cases in nature and will suffice for purposes of this discussion.

Dioecy, involving a strict separation into male and female plants, obviously enforces obligate outcrossing (Bawa, 1980). Other forms of group-level sexuality, such as monoecy, may promote outcrossing but do so with complete effectiveness only if coupled with additional mechanisms.

A widespread feature of hermaphroditic flowers is spatial separation of the stamens and pistils, which may promote outcrossing by preventing intrafloral self-pollination (Baker and Hurd, 1968; Frankel and Galun, 1977). It does not, however, prohibit geitonogamy. At the plant level of sex expression, there may be spatial separation of staminate and pistillate flowers on different parts of the plant, again reducing the likelihood of within-plant selfing. Such patterns are common in, for example, andromonoecious plants (Primack and Lloyd, 1980; Wyatt, 1982).

TABLE II

SEX EXPRESSION IN PLANTS[a]

Individual flower
Hermaphroditic—with both stamens and pistils
Staminate—with stamens only
Pistillate—with pistils only
Individual plant
Hermaphroditic—with hermaphroditic flowers only
Monoecious—with both staminate and pistillate flowers
Androecious—with staminate flowers only
Gynoecious—with pistillate flowers only
Andromonoecious—with both hermaphroditic and staminate flowers
Gynomonoecious—with both hermaphroditic and pistillate flowers
Trimonoecious—with hermaphroditic, staminate, and pistillate flowers
Groups of plants
Hermaphroditic—with hermaphroditic plants only
Monoecious—with monoecious plants only
Dioecious—with both androecious and gynoecious plants
Androdioecious—with both hermaphroditic and androecious plants
Gynodioecious—with both hermaphroditic and gynoecious plants

[a]Terminology for individual flowers, plants, and groups of plants is arranged according to the scheme of Frankel and Galun (1977).

Dichogamy involves the temporal separation of the functioning of stamens and pistils of hermaphroditic flowers and was first described in the angiosperms by Koelreuter (1761–1766). Sprengel (1793) independently discovered the phenomenon and distinguished between male-to-female (''dichogamia androgyna'') and female-to-male (''dichogamia gynandra'') cases. Hildebrand (1867) introduced the terms ''protandry'' and ''protogyny'' to refer to situations in which the stamens or the pistils, respectively, mature first. Traditionally, dichogamy has been regarded as one of the most widespread and effective mechanisms used to prevent intrafloral self-pollination (Faegri and van der Pijl, 1979). When synchrony is expressed at the level of the entire plant, dichogamy can also be effective in preventing geitonogamy. A remarkable situation occurs in the protogynous avocado. One variety has flowers (male stage) that open on the morning of the first day, then close, and reopen (female stage) in the afternoon of the following day. The second variety opens flowers (male stage) in the afternoon of the first day, then closes, and reopens (female stage) on the following morning. Outcrossing between individuals of the two varieties is thereby assured (Stout, 1924; Frankel and Galun, 1977). Walnuts are characterized by a form of heterodichogamy involving a protandrous-protogynous dimorphism controlled by a single gene (Gleeson, 1982). Mean flowering times of the complementary mating types are synchronized such that mating is almost completely disassortative.

Perhaps the most widespread and effective mechanism for promoting outcrossing is genetic self-incompatibility, the inability of a fertile hermaphroditic seed plant to produce zygotes after self-pollination (Nettancourt, 1977). At least 78 families of flowering plants are known to include one or more self-incompatible species (Pandey, 1960). Data compiled by Fryxell (1957) allowed Brewbaker and Majumder (1961) to calculate that 250 of 600 genera studied include at least one self-incompatible species. There are two broad classes of self-incompatibility systems: homomorphic, in which interfertile members of breeding groups cannot be distinguished, and heteromorphic, in which compatible plants can be identified by distinctive morphological features. Originally, genetic self-incompatibility was believed to enforce obligate outcrossing in the same manner as dioecy (East, 1940). Subsequent research, however, revealed cases of partial self-compatibility in species otherwise believed to be self-incompatible (Pandey, 1970; Nettancourt, 1977).

Homomorphic self-incompatibility may be either sporophytic, in which the incompatibility phenotype of the pollen is determined by the genotype of the pollen-producing parent, or gametophytic, in which the haploid genotype of the individual pollen grain determines its compatibility phenotype. Nettancourt (1977) summarized the work of Brewbaker (1957), Pandey (1960), Crowe (1964, 1971), Heslop-Harrison (1968, 1975), and others, noting a number of associations between these two types of incompatibility and time of gene action, site of expression, pollen cytology, and genetic determination. He concluded that the correlation of trinucleate pollen, stigmatic inhibition, and sporophytic control is very stable. In addition, all sporophytic systems operate with a dry stigma, where there can be direct physical and biochemical interactions between individual pollen grains and stigmatic papillae. In contrast, all gametophytic systems (except in the grasses) involve wet stigmatic surfaces. Gametophytic systems are controlled by multiallelic series at one, two, or, rarely, several loci. This is also true of most sporophytic systems, although systems involving two alleles per locus and several loci are known.

Two types of heteromorphic systems are also known: distyly and tristyly. In distylous taxa, one morph has flowers with long styles and short stamens (pin), whereas the other bears flowers in reciprocal positions with short styles and long stamens (thrum). The only compatible crosses are those between the two morphs. Genetically, the incompatibility reactions and morphological differences are controlled by a supergene (Lewis, 1949; Ernst, 1955; Dowrick, 1956) that behaves as a simple Mendelian factor. The thrum is usually the heterozygote in this system and the pin is the recessive homozygote (Vuilleumier, 1967), although the opposite is found in *Hypericum aegypticum* (Ornduff, 1976) and *Armeria maritima* (Baker, 1966). With either genetic mechanism, progeny should segregate in equal ratios of pins and thrums. Tristyly, which is less common than distyly, involves long-styled flowers with one set of short stamens and one set of intermediate length, short-styled flowers with one set each of long and intermediate stamens, and mid-styled flowers with short and long sets of stamens. Fertile crosses involve transfer of pollen between flowers of the different classes from stamens to stigmas at corresponding heights (Ganders, 1979). Tristyly is controlled by two loci, each with two alleles and epistatic interactions, and at equilibrium there should be equal proportions of the three morphs (Ganders, 1979). In addition to reciprocal positioning of the stamens and pistils, heterostylous plants often show additional morphological differences such as longer stigmatic papillae (Dulberger, 1974) and more numerous, but smaller, pollen grains in pins (Vuilleumier, 1967).

b. Recognition of the Trend toward Autogamy. In direct contrast to all of these mechanisms for promoting outcrossing are a large number of adaptations in flowering plants that promote selfing. These basically involve spatial or temporal changes that are the reverse of those described above, such that they tend to allow self-pollination and/or restrict pollen transfer between individuals. For example, spatial location of the stamens and pistils within a flower and their times of maturation may develop so that pollen is deposited directly from the anthers onto the receptive stigmatic surfaces. In its most extreme form, this process is represented by cleistogamy, in which flowers habitually self-pollinate without opening.

There is also a long list of taxa in which obligate outcrossing based on genetic self-incompatibility has been replaced by predominant autogamy or self-fertilization based on self-compatibility (Uphof, 1938; East, 1940; Fryxell, 1957; Stebbins, 1970, 1974; Grant, 1971). In some cases, the morphological changes associated with a shift from predominant outcrossing to predominant selfing are minor (Stebbins, 1974). The selfing *Eupatorium microstemon* is nearly indistinguishable from its outcrossing relative, *E. sinclairii* (Baker, 1967). Autogamous populations of *Lycopersicon esculentum* and *L. pimpinellifolium* differ from their outcrossing ancestors only with respect to slight rearrangements in the structure and position of anthers and stigmas (Rick, 1950; Rick *et al.*, 1978). In other taxa, major morphological reorganization may occur in association with a shift in breeding system. Ornduff (1969) has listed a number of features characterizing a general trend from predominant outcrossing to autogamy (Table III). The obvious adaptations of outcrossing plants for pollination by animal vectors are frequently lost in their autogamous derivatives. A particularly dramatic example is Rollins's (1963a) and Lloyd's (1965) studies of *Leavenworthia*. Selfing has arisen many times in independent phyletic lineages to produce a number of new autogamous species. The morphological changes accompanying the evolution of autogamy in each case are strikingly similar.

TABLE III

CHARACTERISTICS THAT OFTEN DIFFER BETWEEN OUTCROSSING SPECIES AND THEIR SELFING DERIVATIVES[a]

Characteristics of outcrossers	Characteristics of selfers
Outcrossing	Autogamy
Self-incompatible	Self-compatible
Flowers many	Flowers fewer
Pedicels long	Pedicels shorter
Sepals large	Sepals smaller
Corolla rotate	Corolla funnelform, cylindric, or closed
Petals large	Petals smaller
Petals emarginate	Petals entire
Nectaries present	Nectaries reduced or absent
Flowers scented	Flowers scentless
Nectar guides conspicuous	Nectar guides absent
Anthers long	Anthers shorter
Anther dehiscence extrorse	Anther dehiscence introrse
Anthers distant from stigma	Anthers adjacent to stigma
Pollen grains many	Pollen grains fewer
Pollen presented	Pollen not presented
Pistil long	Pistil shorter
Stamens longer or shorter than pistil	Stamens equal in length to pistil
Style exserted	Style included
Stigmatic area well defined, papillate	Stigmatic area poorly defined, less papillate
Stigma receptivity and anther dehiscence asynchronous	Sigma receptivity and anther dehiscence synchronous
Many ovules per flower	Fewer ovules per flower
Many ovules not maturing to seed	All ovules maturing to seed
Some fruits not maturing	All fruits maturing
Narrow distribution	Wide distribution

[a]Modified from Ornduff (1969).

The discovery of a wide range of selfing as well as outcrossing mechanisms in diverse angiosperm taxa led to the concept of "balanced breeding systems" in plants (Mather, 1943; Stebbins, 1950; Grant, 1975). According to these authors, the advantages of autogamous reproduction in terms of adaptation to the immediate environment provided by increased homozygosity are traded off against outcrossing, which retains genetic variability and, hence, long-term evolutionary flexibility. Indeed, many examples have been assimilated of plants that seem to epitomize just such a mixed strategy: distylous populations that consist of typical pins and thrums plus some self-compatible homostyles (Crosby, 1949); flowers in which outcrossing is promoted early in development but, failing to be pollinated, self (Stone, 1959); and plants which produce both outcrossing, chasmogamous flowers and selfing, cleistogamous ones (Uphof, 1938). Other species, however, appear to be equally successful with monomorphic breeding systems. Recent findings regarding the population structure of several predominantly selfing species, for example, have modified earlier assumptions about genetic variability in autogamous plants (Allard *et al.*, 1968; Jain, 1969, 1975, 1976). Brown (1979) reviewed possible explanations for the "heterozygosity paradox," the finding that inbreeders show more heterozygosity than expected in allozyme surveys of natural populations.

2. Modern Approaches: Experimental and Dynamic

Studies of the evolution of autogamy have been unique in presuming the possibility of changes in the breeding systems of plants and in sometimes suggesting a role for pollinators as a selective force in such shifts. For the most part, breeding systems, like pollination syndromes, have been viewed in a descriptive, static fashion. It is only within the past decade that concepts and techniques have been developed for taking a dynamic approach to the study of plant-breeding systems. Refinement of quantitative techniques for measuring outcrossing rates in natural populations and development of quantitative methods for estimating the relative contributions of plants as male and female parents have renewed interest in variation and evolution of sex expression in plants. In addition, researchers have begun to recognize that shifts in breeding systems may result from environmental changes that affect effective rates of pollination. Alternative explanations for these transitions are now being proposed and tested.

a. Quantitative Assessments of Outcrossing Rates. Earlier studies of the evolutionary significance of breeding systems in plants were hampered by the technical difficulties involved in estimating rates of outcrossing in natural populations. Fryxell (1957) reviewed these methods and detailed the problems inherent in finding suitable genetic markers, carrying out progeny testing, and analyzing appropriate experimental designs. Although some of these techniques were employed widely by plant breeders, they have seldom been used in natural populations (but see Vasek, 1964, 1965, 1967; Harding, 1970; Harding *et al.*, 1974; Ganders *et al.*, 1977; Rick *et al.*, 1978). A major breakthrough was made with the discovery of abundant and frequently codominant allozyme polymorphisms detected by gel electrophoresis. These provide an ideal source of genetic markers for use in quantifying outcrossing rates in plant populations (Brown and Allard, 1970; Jain, 1976; Clegg, 1980). This procedure has been applied with great success to a diversity of plant species (Brown and Allard, 1970; Brown *et al.*, 1975, 1978; Rick *et al.*, 1977; Phillips and Brown, 1977; Jain, 1978; Moran and Brown, 1980; Sanders and Hamrick, 1980; Schoen, 1982a,b). Recently, statistical procedures have been formulated for estimating rates of outcrossing on a multilocus basis (Brown *et al.*, 1978; Green *et al.*, 1980; Shaw *et al.*, 1981; Ritland and Jain, 1981). Shaw and Brown (1982) suggest that maximum efficiency in estimating outcrossing in predominantly selfing taxa is provided by large samples assayed for a small number of polymorphic loci. In contrast, smaller samples assayed for more loci should be preferred in predominant outcrossers.

These techniques yield more precise and accurate measures of outcrossing than were possible previously, when assessments of breeding systems were based merely on qualitative observations of morphology or phenology. For *Limnanthes,* Jain (1976, 1978) discovered that electrophoretic estimates of outcrossing were not in close agreement with Arroyo's (1975) estimates based on morphology or protandry. Schoen (1982a) found that outcrossing rates in populations of *Gilia achilleifolia* were correlated highly with both the degree of protandry and the level of spontaneous fruit set in the absence of pollinators. Two other characters, degree of stigma exsertion and flower weight, showed the expected positive associations but were not significantly correlated with outcrossing rate.

b. Quantitative Assessments of Sex Expression. Sex expression in plants traditionally has also been classified in a qualitative manner according to gross morphological variations among individuals. This lack of precision is critical, since the important aspect of sex expression for evolutionary studies involves the actual genetic contributions made by different individuals, and simple morphological categories often fail to reflect the real situation. For example, plants that appear to be hermaphroditic may, in fact, be functioning primarily (Arroyo and Raven, 1975; Lloyd and Myall, 1976) or completely (Lloyd and Horning, 1979) as males or females. It is only very recently that interest has focused on methods for quantifying the relative contributions of different sexes or incompatibility morphs as either ovule or pollen parents of seeds (Lloyd, 1975a, 1976, 1979b, 1980a,b,c). Using relative "ovule" or "pollen" fitness, an individual or a subclass of a population can be assigned a "functional gender," defined as the proportion of genes contributed by that morph to the next generation through ovules or pollen (Lloyd, 1979b). The breeding relationships among morphs within a population are therefore assigned a quantitative value, and sexuality is described along a continuum rather than as falling into sometimes arbitrary, discrete categories (Table II). This approach provides greater insight into the biology of plant sex expression and assesses gender in plants on a quantitative, genetically effective basis that is more directly related to evolutionary processes. Lloyd's formulas have been applied successfully to plants representing several sexual strategies (Lloyd, 1976, 1979b,c, 1980c; Primack and Lloyd, 1980; Barrett, 1980). Recently, Thomson and Barrett (1981b) have extended the formulation of functional gender to incorporate temporal variation.

D. Pollinator–Plant Interactions and Breeding Systems: Progress toward a Synthesis

Recognition of variation among individual plants in the extent to which they function as males versus females has rekindled an interest in evolutionary transitions between breeding systems. Darwin's (1877) suggestion that dioecy had evolved gradually from distyly in several groups was supported by the observations of Vogel (1955), Baker (1958), and Ornduff (1966). More recently, Lloyd (1979b, 1980c), Bawa (1980), and Beach and Bawa (1980) have documented additional examples. Bawa (1980) reviewed several alternative pathways leading to the evolution of dioecy, including evolution directly from hermaphroditism (Ross, 1978), via gynodioecy (Webb, 1979), via androdioecy (Lloyd, 1975b), and via monoecy (Lloyd, 1975c). It seems safe to predict that additional examples of such transitions between breeding systems will be discovered in the near future and that additional evolutionary pathways linking formerly discrete sexual categories will be proposed.

These recent discoveries have stimulated new ideas regarding the selective forces behind the evolution of breeding systems. Dioecy, which previous generations of plant evolutionists "considered to be another device which promotes outbreeding" (Whitehouse, 1959, p. 255), is now believed to evolve in response to a number of different selective factors. The inadequacy of the outcrossing hypothesis used to explain dioecy should have been recognized when the first examples of its derivation from distyly were uncovered. Since both breeding systems enforce obligate outcrossing, the selective advantages of dioecy must lie elsewhere. Lloyd (1979c) has attacked these outcrossing arguments on a broader front, arguing that the concepts of balanced breeding systems in plants are based on faulty reasoning that depends on group selection. He suggests that "the time is overdue for botanists to forsake the notion of long-term advantages of cross-fertilization as an explanation of the properties of single populations" (p. 604).

For the evolution of dioecy, a number of alternative explanations have been proposed. Bawa (1980, p. 15) states that "selection for outcrossing has been universally proposed as the principal selective force responsible for the evolution of dioecy," but he finds "little empirical support" for that view. He proposes several additional selective pressures, including such ecological factors as allocation of resources for male and female functions, sexual selection, seed dispersal, pollination, and predation. Willson (1979) outlined the steps by which intrasexual competition for mates and mate preference could lead from hermaphroditism to dioecy, noting that sexual selection and genetic advantages of outcrossing probably operated together in the evolution of dioecy. A similar outline, emphasizing the contribution of pollinators to the differential success of plants as pollen and ovule parents, has been presented by Beach (1981). Givnish (1980) discovered that animal-dispersed gymnosperms are usually dioecious, whereas wind-dispersed species are monoecious. He constructed a model to explain these correlations in terms of a disproportionate gain in seed dispersal accruing to plants with unusually large seed crops. Bawa's (1980) argument for angiosperms is directly analogous. Cox (1981) presented evidence for niche partitioning by males and females of three dioecious species and suggested how a division of reproductive labor could lead to the evolution of dioecy. For animal-pollinated taxa, Bawa (1980) has proposed that separation of the sexes may increase effective levels of pollen transfer between plants. This is so because male plants disperse more pollen than hermaphrodites, whose stigmas trap some of their own pollen, and because stigmas of females, unlike those of hermaphrodites, do not become clogged with self-pollen (Bawa and Opler, 1975). Janzen (1971) first presented the idea that separation of the sexes would effectively make plants rarer to seed predators, perhaps favoring the evolution of dioecy as an "escape in space."

This sudden wealth of ecological explanations for the evolution of dioecy has prompted a backlash from those who believe that selection for outcrossing has played a more critical role than is implied by the authors of these new arguments. Thomson and Barrett (1981a) point to a negative association between self-incompatibility and dioecy, two alternative mechanisms that enforce outcrossing, at both the generic and familial levels. Their view echoes that of Whitehouse (1950, p. 214), who believed that "it is probable that multiple-allelomorph incompatibility is absent in families composed entirely of dioecious species, since dioecism and incompatibility appear to have the same function of promoting outbreeding." Thomson and Barrett (1981a) also suggest that monoecy, gynodioecy, and other diclinous conditions are associated with self-compatibility. They further point out that Baker's (1967) observation of a high frequency of dioecious taxa on islands may be a result of selection for outcrossing following the establishment of self-compatible, hermaphroditic colonizers. Givnish (1982) has responded by attacking the logical basis of Thomson and Barrett's (1981a) argument and by generating appropriate null hypotheses against which to test their assertions. He finds that the number of families in which dioecy and self-incompatibility co-occur is not significantly different from that expected by chance. Givnish (1982, p. 861) concludes that "these concerns do not imply that selection for outcrossing is unimportant in promoting dioecy, only that certain logical pitfalls must be avoided in arguing for its importance." This is also the view of Willson (1982) and of Bawa (1982), who have recently presented additional arguments against Thomson and Barrett's (1981a) defense of the outcrossing explanation. It seems clear that genetic and ecological hypotheses used to explain the evolution of dioecy are not mutually exclusive. Our task in the future is to evaluate critically the conceptual arguments and to gather more detailed information regarding the ways in which breeding systems affect, and are affected by, the dynamics of sexual selection, pollination, dispersal, predation, and genetic population structure.

II. Current Research

With the exception of the evolution of autogamy and the evolution of dioecy from heterostyly described above (Sections I,C,1,b and I,D, respectively), there have been no attempts to relate the evolution of breeding systems to pollinator–plant interactions. Developments in this area have been hindered on both sides by a superabundance of descriptive terminology that masks underlying genetic implications and by an overly eager desire to categorize plants into a limited number of fixed types. Hopefully, this situation will change rapidly now that new quantitative approaches have been developed for assessing variation in sexual expression and its genetic effects. In this section, I shall describe three areas of current research in which application of these new ideas about sex expression in plants may prove useful. These examples are a personal selection constituting a limited sample of situations in which it is likely that pollinator–plant interactions have played a role in the evolution of breeding systems. Additional examples have been discussed recently by Bawa and Beach (1981).

A. Dichogamy in Relation to Pollinator Classes

1. Foraging Patterns of Pollinators

In contrast to the literature on pollinator movements between plants (Section I,B,2), there are few observations of pollinator foraging behavior on individual plants or on individual inflorescences. Levin and Kerster (1973) reported assortative pollination by stature in populations of *Lythrum salicaria,* and Waddington (1979) found evidence of this same phenomenon in a community of wildflowers in the Rocky Mountains. Most other reports, however, suggest that pollinators do not forage on flowers at a single height, but rather visit a number of open flowers in an inflorescence (Wyatt, 1981). One of the best-documented cases of movement patterns within inflorescences is that of *Epilobium angustifolium,* which produces a raceme of protandrous flowers that open in sequence from bottom to top (Proctor and Yeo, 1973). Bees move from the bottom of the inflorescence up on one plant, becoming dusted with pollen (Benham, 1969). This pollen is deposited on stigmas at the bottom of the next inflorescence visited, thereby minimizing self-pollination. In addition to *E. angustifolium,* Faegri and van der Pijl (1979) discuss an identical situation in *Digitalis purpurea* and concur with Proctor and Yeo's (1973) conclusion that most insects work upward. The early observations of this phenomenon in *Delphinium* by Epling and Lewis (1952) and in *Cynoglossum officinale* and *Digitalis purpurea* by Manning (1956) have been confirmed in *Digitalis* (Percival, 1965; Percival and Morgan, 1965; Best and Bierzychudek, 1982) and in Streptanthus (Rollins, 1963b). Benham (1969) also mentions earlier work on *Acanthus mollis* that demonstrated that bees preferentially forage upward on this protandrous species.

More recently, Pyke (1978) has studied in some detail bumblebee pollination of *Delphinium nelsonii, D. barbeyi, Aconitum columbianum, E. angustifolium,* and *Penstemon strictus.* On the columnar inflorescences of these species, bees tend to commence foraging at the bottom of each inflorescence, move vertically up, leave before reaching the top, and miss flowers as they move upward. Pyke (1978) interpreted these foraging patterns as optimal in the sense of resulting in the maximum net rate of energy gain to the bumblebees, and showed that these foraging patterns are reinforced by a decrease in nectar abundance as inflorescences are ascended. Similarly, production of larger nectar rewards by older, female-stage flowers has been shown in *Streptanthus* (Rollins, 1963b), *Digitalis* (Percival and Morgan, 1965; Best and Bierzychudek, 1982), *Epilobium* (Proctor and Yeo, 1973), and *Delphinium* (Waddington, 1981).

Foraging patterns within inflorescences have been less well studied in other insects, but most flies are presumed to move upward (Proctor and Yeo, 1973). Wasps are reported to climb downward in pollinating protogynous *Scrophularia nodosa* (Faegri and van der Pijl, 1979). Such movements again enhance the legitimate transfer of outcross pollen. In *Plantago major* and *P. lanceolata,* presumed to be wind pollinated (but see Clifford, 1962; Stelleman, 1978), localization of pistillate-stage flowers above staminate-stage flowers should decrease levels of ineffectual self-pollination and therefore prevent clogging of the stigma. "Trap blossoms" such as *Aristolochia* must be regarded as a special case, as they force a downward movement of their fly pollinators and typically show very well-developed protogyny (Bawa and Beach, 1981). A similar system involving a forced downward movement of beetles and other pollinators occurs in the strongly protogynous flowers of the Nymphaeaceae (Meeuse and Schneider, 1980). In these cases, dichogamy is related to pollinator behavior within rather than between flowers.

Taken as a whole, these observations suggest that protandry should be the prevalent form of dichogamy in taxa pollinated by bees or flies, whereas protogyny should prevail in taxa pollinated by wasps, beetles, or other vectors that tend to direct pollen transfer downward. To test these predictions, I scored appropriate characteristics of the pollination ecology of 43 dichogamous species whose "case histories" were described by Faegri and van der Pijl (1979), by Proctor and Yeo (1973), or by Percival (1965) (Table IV). All taxa that were unequivocally protandrous or protogynous were included. I recorded for each species or genus the most important categories of pollinators and the characteristic inflorescence type. Data regarding pollinators were gleaned from accounts in

TABLE IV

CHARACTERISTICS OF THE POLLINATION BIOLOGY OF 43 DICHOGAMOUS ANGIOSPERMS

Taxon	Pollinators	Inflorescence
Protandrous		
Arenaria uniflora (Walt.) Muhl.	Flies, bees	Cymes
Campanula spp.	Large bees	Racemes
Coronilla emerus L.	Bumblebees	Axillary heads
Epilobium angustifolium L.	Bumblebees	Racemes
Geranium spp.	Bees	Cymes
Geranium pratense L.	Bees	Cymes
Geranium pyrenaicum L.	Bees, flies	Cymes
Impatiens spp.	Bees	Solitary, racemes
Heracleum sphondylium L.	Bees, flies, beetles	Umbels
Malva spp.	Bees	Solitary, clusters
Mentha spp.	Flies, bees	Thyrse
Parnassia palustris L.	Flies, bees	Solitary
Pedicularis lanceolata Michx.	Bumblebees	Spikes
Pelargonium spp.	Bees, flies	Umbels
Polygala chamaebuxus L.	Bees	Racemes
Ruta graveolens L.	Flies, bees	Cymes
Saxifraga spp.	Flies	Cymes
Saxifraga aizoides L.	Flies	Cymes
Stellaria holostea L.	Bees	Cymes
Teucrium scorodonium L.	Bees	Racemes
Thymus spp.	Bees	Thyrse
Protogynous		
Actaea spicata L.	Beetles	Racemes
Anemone pulsatilla L.	Bees	Solitary
Arisarum vulgare Targ. Tozz.	Flies (trap)	Spadix
Aristolochia spp.	Flies (trap)	Solitary, clusters
Aristolochia clematitis L.	Flies (trap)	Clusters
Arum nigrum Schott.	Beetles, flies (trap)	Spadix
Bartsia alpina L.	Bumblebees	Racemes
Calluna vulgaris L.	Flies, bees, thrips	Racemes
Calycanthus occidentalis Hook. and Arn.	Beetles	Solitary
Euphorbia paralias L.	Flies	Umbels
Ficus spp.	Wasps	Clusters
Helleborus spp.	Bees	Cymes
Helleborus foetidus L.	Bees	Cymes
Magnolia spp.	Beetles	Solitary
Nymphaea citrina Peter	Beetles	Solitary
Parietaria spp.	Wind	Cymes
Pedicularis sceptrum-carolinum L.	Bumblebees	Spikes
Plantago major L.	Wind	Spikes
Ranunculus acris L.	Beetles, moths	Cymes
Scrophularia spp.	Wasps	Cymes
Scrophularia nodosa L.	Wasps	Cymes
Sorbus aucuparia L.	Beetles, flies, bees	Corymb

the books cited above, supplemented by lists from Knuth (1906–1909). Inflorescence types were determined by referring to taxonomic descriptions in floras. I examined the relationship between dichogamy and pollen vectors by constructing a simple contingency table and performing a chi-square analysis (Sokal and Rohlf, 1969). Bees and flies were considered as one group and were contrasted to wasps, beetles, and other vectors. I omitted from the statistical test four trap flowers and two species that are wind pollinated.

2. *Correlations between Dichogamy and Pollinator Classes*

It is immediately obvious that plants pollinated by bees and/or flies are predominantly protandrous and that plants pollinated by other vectors are exclusively protogynous (Table V). It appears, therefore, that the predicted association between dichogamy and pollinator class is supported overwhelmingly by these data.

The numbers of taxa in the two groups in Table IV do not reflect the real frequency of protandry and protogyny in the angiosperms as a whole, which are heavily dominated by protandrous species (Sprengel, 1793; Müller, 1883; Burtt, 1978; Lloyd and Yates, 1982). For example, of 235 hermaphroditic species in New Zealand, 37% are protandrous and only 8% protogynous (Thomson, 1881; Godley, 1979). Bawa and Beach (1981) and Webb (1981) argued that intrasexual

TABLE V

RELATIONSHIP BETWEEN POLLINATOR CLASSES AND DICHOGAMY IN ANGIOSPERMS

Pollinator class	Protandrous	Protogynous
Bees	15	5
Flies	6	2
Beetles	0	6
Traps (flies and beetles)	0	4
Wasps	0	3
Wind	0	2
Total	21	22

Chi-square	Protandrous	Protogynous	Row totals
Bees and flies	21	7	28
Other vectors	0	15	15
Column totals	21	22	43

$\chi^2 = 21.99$; $P < 0.005$. Omitting wind and trap blossoms, $\chi^2 = 15.61$; $P < 0.005$.

selection and mate competition should favor release of pollen before conspecific stigmas have been pollinated and exposure of stigmas only after pollen has been removed from a diverse array of genotypes. Lloyd and Yates (1982), on the other hand, explained the bias in favor of protandry in terms of selection for separation of the male and female functions of perfect flowers: Carpels mature into fruits and are therefore much more difficult to remove from the path of pollinators after they have completed their function than are the ephemeral stamens. A simpler explanation than either of these is that selection for dichogamy is more likely to lead to protandry than protogyny, because protandry merely represents the normal pattern of centripetal maturation within the flower. Protogyny, on the other hand, involves a reversal in the maturation sequence for stamens and pistils. This view conforms to Stebbins's (1970) principle of "selection along the lines of least resistance." Faegri and van der Pijl (1979) suggest that protogyny is more effective than protandry in preventing self-pollination, because pollen from other flowers is given a head start in competition to fertilize ovules. Unless all of the pollen of a protandrous species is shed before stigmas become receptive, some contamination with self-pollen is likely. Webb (1981, p. 335) agrees that "when weakly developed, protogyny is more effective than protandry in promoting outcrossing." It is possible, however, for either system to be absolutely effective in preventing intrafloral selfing if there is no overlap whatsoever between pollen release and stigma receptivity.

Thus far, we have considered only the maturation sequences of stamens and pistils within individual flowers. However, among monoecious plants protogyny is frequent and widespread (Knuth, 1906–1909). Monoecious species of wind-pollinated *Carex, Typha,* and *Sparganium,* for example, typically bear staminate flowers above the pistillate flowers and characteristically are strongly protogynous. This phenomenon is sometimes called "second-order protogyny." Unlike the situation in individual flowers where protogyny reverses the natural sequence of maturation, in monoecious taxa such as *Arum nigrum* (Table IV) with pistillate flowers clustered toward the base of the inflorescence, one would expect acropetal maturation to result in protogyny. The only change necessary in such a system is a more sharply defined separation of maturation times between the staminate and pistillate portions of the inflorescence.

Contrary to expectation, exclusion of inflorescences of noncolumnar form (i.e., those in which flowers are not arranged in a vertical array) made no difference in the outcome of the chi-square test relating dichogamy and pollen vectors (Table V). Data concerning typical patterns of maturation within noncolumnar inflorescences and foraging patterns of pollinators on such inflorescences would be most informative. In *Heracleum sphondylium* (Table IV), for example, do the protandrous flowers mature in a centripetal sequence with pollinators landing in the center of the umbel and moving toward the edge? Such a pattern does characterize the heads of the Compositae (Thomson and Plowright,

1980). Also, is there a definite sequence of maturation of primary and secondary units of the inflorescence that correlates with the foraging behavior of pollinators? Cruden and Hermann-Parker (1977), Webb (1981), and Lindsey (1982) find that such sequences do exist in the Umbelliferae. For columnar inflorescences too, we need additional direct observations of the foraging behavior of pollinators, the architecture of the inflorescence, and the form of dichogamy.

3. Alternative Explanations for Protandry and Protogyny

Since the correlation between dichogamy and pollen transfer directionality is not perfect, there obviously are other factors that affect these relationships. One of these is the pattern of maturation of the flowers in an inflorescence. Although many species of plants display acropetal maturation, other species show basipetal maturation, in which the uppermost flowers mature first (Weberling, 1965). This would reverse the predictions with respect to dichogamy, with protogyny being expected where pollen transfer is directed upward and protandry where pollen is moved downward. In species with an irregular or divergent pattern of maturation, dichogamy should not evolve. Similarly, dichogamy would not be expected in plants that are pollinated by vectors yielding no net directionality of pollen transfer such as, perhaps, hummingbirds (Epling and Lewis, 1952; Price and Waser, 1979). Bawa and Beach (1981) argue that protogyny is a very specialized form of dichogamy, usually explained by the peculiar life-history attributes of beetle pollinators. It also should be pointed out that floral mechanisms that restrict or otherwise alter the natural foraging behavior of pollinators may provide exceptions. Flies, which typically forage upward, are forced to move downward in trap blossoms such as those of *Aristolochia clematitis* and *Arisarum vulgare* (Table IV). In these taxa, protogyny rather than protandry has evolved. Theoretically, a similar situation could occur if a plant manipulated its nectar quantity and quality so that lower flowers in the inflorescence contained enhanced rewards, inducing pollinators to visit them first. Such a response, however, might be observed only after an initial phase in which bees, for example, must learn to overcome an already established behavior involving foraging upward (Waddington and Heinrich, 1979; Best and Bierzychudek, 1982).

The adaptive significance of dichogamy has been presumed since the time of Darwin (1876) to be prevention of self-pollination. This conclusion follows logically from considerations of the usual inbreeding depression that results from selfing of a typically outcrossed plant and from the expected effects of protandry and protogyny. This argument, however, applies only to self-compatible taxa and not to self-incompatible ones. Faegri and van der Pijl (1979, p. 29) explain the dilemma of self-incompatible species that are dichogamous by suggesting that dichogamy is a "phylogenetic relict" in those taxa or that the "self-incompatibility is not invariably absolute." It seems more likely that in self-incompatible species, protandry and protogyny serve to prevent coverage of the stigmatic surface with ineffectual self-pollen. Lloyd and Yates (1982) have come to this same conclusion independently as a result of their studies of *Wahlenbergia albomarginata.* Bawa and Opler (1975) also proposed stigma clogging as a factor involved in the evolution of dioecious taxa from self-incompatible progenitors (Section II,C,3). If dichogamy does represent a mechanism to prevent coverage of the stigmas with self-pollen, we might predict that it is most likely to evolve in those self-incompatible species that possess limited stigmatic surface area for receipt of pollen.

B. The Evolution of Dioecy from Heterostyly

1. Examples of the Transition

Among the best-documented examples of transitions in plant-breeding systems are those involving the evolution of dioecy from heterostyly. Darwin (1877) first recognized this possibility, and it has subsequently been proposed for *Nymphoides* (Ornduff, 1966, 1970; Pratap Reddy and Bir Bahadur, 1976; Barrett, 1980), *Byrsocarpus* (Baker, 1962), and *Cordia* (Opler *et al.,* 1975; Lloyd, 1979b) and for *Mussaenda* (Baker, 1958), *Psychotria* (Sohmer, 1977), *Coussarea* (Beach and Bawa, 1980), and other Rubiaceae (Vogel, 1955; Fosberg, 1956; Vuilleumier, 1967; Bir Bahadur, 1968; Opler *et al.,* 1975). In all of these cases, pins have evolved to become functionally female and thrums to become functionally male (Ganders, 1979; Bawa and Beach, 1981).

In all of these examples, the transition from distyly to dioecy is inferred by comparing closely related taxa. There is only one species in which distylous populations have been supposed to coexist with dioecious ones: *Mitchella repens* (Crowe, 1964). Darwin (1877) accepted Meehan's (1868) observation that some populations of this species are dioecious, although his own studies indicated a typical distylous condition. Darwin (1877) supposed that this represented, within *M. repens,* a trend toward dioecy, seen among the related genera *Nertera* and *Coprosma* (Lloyd, 1979b). Evidence for the purported tendency for pins to become functionally female and for thrums to become functionally male were lacking in Darwin's (1877) data, which showed only slightly greater seed production by pins. Similar results were obtained by Ganders (1975) and by Keegan *et al.* (1979). These studies concluded that Meehan's (1868) original observations were in error and that there is no evidence of gender specialization in *M. repens.*

The evolution of dioecy from distyly obviously must involve gender specialization such that the floral morphs no longer make equal contributions to their offspring through pollen and ovules (Willson, 1979; Lloyd, 1979c). Assessing the extent to which sexual selection is acting in distylous taxa can be done by use of Lloyd's (1979b) equations for estimating functional gender. Lloyd's (1979b)

TABLE VI

ESTIMATES OF FREQUENCIES, SEED PRODUCTION, AND FUNCTIONAL GENDER OF PINS AND THRUMS OF *MITCHELLA REPENS*[a]

Population	Long-styled plants			Short-styled plants		
	Frequency	Seed output	Gender (femaleness)	Frequency	Seed output	Gender (femaleness)
Mitchella repens (N.C.)	0.308	4.17	0.224	0.692	6.45	0.776
Mitchella repens (Md.)	0.511	6.49	0.475	0.489	6.33	0.525
Cordia dentata	0.457	17	0.443	0.543	18	0.557
Cordia pringlei	0.440	17	0.471	0.560	15	0.529
Cordia inermis	0.654	9.5	0.947	0.346	1	0.053
Cordia colococca	0.401	33	0.957	0.599	1	0.043
Cordia panamensis	0.321	19	1.000	0.679	0	0.000
Nymphoides indica	0.502	10.20	0.390	0.498	15.40	0.610

[a]Comparative data for *Cordia* are from Lloyd (1979b); for *Nymphoides*, Barrett (1980); and for *M. repens* in Maryland, Ganders (1975).

application of this method showed that the species of *Cordia* studied by Opler *et al.* (1975) clustered into two discrete groups: those that were typically distylous (*C. dentata* and *C. pringlei* in Table VI) and those that were functionally dioecious (*C. inermis, C. collococca,* and *C. panamensis* in Table VI). In the former, nearly equal contributions are being made to the next generation by each morph through pollen and ovules. In the latter, individuals contribute almost exclusively as males or as females. Barrett (1980) reported that thrums of distylous *Nymphoides indica* in Brazil produce significantly greater numbers of seeds than pins. His calculations of gender showed greater femaleness for thrums than pins (Table VI). There was some variation among individuals, and Barrett (1980) concluded that this population was not evolving toward dioecy. In *M. repens,* there appears to be variation among populations in functional gender. In North Carolina populations, thrums contribute more than 75% of the genes transmitted by ovule production (Table VI). In Maryland populations, on the other hand, the functional gender of pins is 0.48 and that of thrums is 0.52 (calculated from data of Ganders, 1975; Table VI). These values reflect a distylous system with nearly perfect balance between the morphs. It is interesting that both Barrett's (1980) and R. Wyatt and T. R. Meagher's (unpublished) studies detected deviations from equal functional gender in the direction of thrums as female parents, a trend counter to the rule that pins evolve into females. In other cases, pin plants sometimes produce more seeds than thrums (Bir Bahadur, 1970), although perhaps in most cases seed production is not significantly different between morphs (Ornduff, 1966, 1975; Barrett, 1978; Wyatt and Hellwig, 1979; Ganders, 1979).

2. Origins and Functions of Distyly

Before proceeding to a discussion of the selective forces that favor the evolution of dioecy from heterostyly, it might be well to review briefly the origin of heterostyly itself and the evolution of its morphological correlates. As noted earlier (Section I,C,1), distyly is controlled by a supergene that segregates as a simple Mendelian factor (Charlesworth and Charlesworth, 1979). Short-styled thrums represent the Ss heterozygote, and the long-styled pins represent the ss homozygote in most species (Vuilleumier, 1967; Yeo, 1975; Ganders, 1979). The converse is true, however, in *Armeria maritima* (Baker, 1966) and in *Hypericum aegypticum* (Ornduff, 1979). Explanations for this strong tendency of thrums to be heterozygous (Fig. 1) are lacking. This is an especially puzzling problem because heterostyly is generally assumed to have arisen independently many times in different phyletic lines of the angiosperms (Nettancourt, 1977).

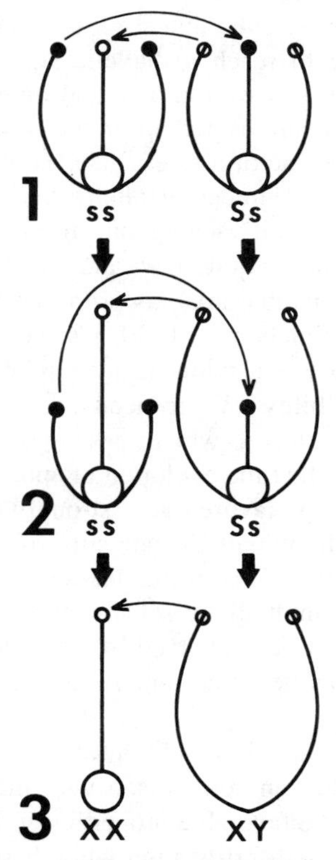

Fig. 1. Stages in the evolution of dioecy from distyly. Stage 1 represents the original monomorphic population consisting of two mating types. By stage 2, the mating types have become dimorphic for style and stamen lengths, enhancing disassortative pollination and resulting in typical distyly, with the thrums being heterozygotes. In stage 3 dioecy has become established, with the thrums representing the heterogametic (male) sex. (For a full discussion, see text.)

Ganders (1979) considered the evolution of heterostyly "an unsolved mystery." Nevertheless, drawing from the models of Charlesworth and Charlesworth (1979), he presented a "plausible scenario." He supposes that the progenitors of distylous taxa were monomorphic and self-compatible but largely outcrossed, therefore building up significant genetic loads of deleterious recessive alleles. A change in pollinators increased selfing and led to inbreeding depression. Mutation produced a new pollen type that was incompatible with the stigma type in the population, establishing functional gynodioecy. Mutation then produced a new stigma type that was compatible with the new pollen type. At this point, the population had achieved a diallelic self-incompatibility system.

It would appear that a simpler evolutionary pathway to diallelic self-incompatibility exists. If we assume that the ancestors of distylous species possessed a one-locus, multiallelic incompatibility system (Whitehouse, 1950; Brewbaker, 1957, 1959; Lewis and Crowe, 1958; Crowe, 1964; Pandey, 1958, 1960; Nettancourt, 1977; Beach and Kress, 1980; but see East, 1940; Bateman, 1952; Grant, 1975), then it is possible to reach a diallelic state merely by losing alleles (Crowe, 1964). This process is well known and might be especially important when population sizes are small, as perhaps in the case of rapid speciation. Imrie *et al.* (1972) calculated the rate of loss of S alleles by genetic drift and showed that four was the maximum number of alleles that could be maintained in a population of 32 plants originally segregating for six alleles. Furthermore, the mutation rate of new S alleles is quite low, and the evolution of new systems of incompatibility in self-compatible taxa, as proposed by Ganders (1979), is unknown (Mather, 1943; Whitehouse, 1950; Crowe, 1964). Ganders's (1979, p. 620) statement that "no heterostylous taxa are related to taxa with multiallelic sporophytic self-incompatibility" is erroneous, as both Brewbaker (1957) and Nettancourt (1977) list families in which these systems co-occur. Also, Beach and Kress (1980) point out that the evolution of sporophytic from gametophytic self-incompatibility should be favored very strongly and could have developed rapidly from ancestors with multiallelic gametophytic systems that were reduced to two alleles. Ganders (1979) argued that the ancestors of heterostylous plants could not have possessed multiallelic self-incompatibility or the less efficient diallelic system would never have evolved. It is possible, however, that such an "evolutionary choice" was never an option, as suggested by the operation of genetic drift.

Regardless of how the diallelic condition arises, once it has become established it is highly inefficient in comparison with multiallelic systems. Whitehouse (1950) emphasized the favorable properties of multiallelic incompatibility as a device for preventing self-fertilization while hardly restricting cross-fertilization. If the number of alleles is large, as it is in populations of species of *Trifolium*, the cross-fertilizing efficiency of such plants is close to 100%. In diallelic systems (and also in the case of dioecy), the average cross-fertilizing efficiency is only 50% (Whitehouse, 1950). Whitehouse (1950, p. 205) argues that multiallelic incompatibility is therefore "of much greater evolutionary significance than either dioecism or two-allelomorph incompatibility, and may indeed be the key to success of the angiosperms." This disadvantage of diallelic systems can be overcome by mechanisms that favor pollen transfer between the two mating types. Darwin (1877) proposed that the reciprocal positioning of anthers and stigmas in distylous taxa was a device to promote such disassortative pollination. He observed that pin and thrum pollen was found in different positions on bumblebees and moths visiting *Primula veris*. Similarly, Rosnov and Screbtsova (1958) and Olesen (1979) detected differential deposition of pollen on visitors to *Fagopyrum* and *Pulmonaria,* respectively. Ganders (1979) reviewed the complications involved in using stigmatic pollen load data to evaluate the importance of reciprocal positioning of anthers and stigmas in promoting disassortative pollination. Nevertheless, he concluded that "distyly does enhance disassortative pollination as Darwin hypothesized," (p. 624) and rejected the suggestion of Mather and DeWinton (1941) and Dulberger (1975a,b) that these morphological differences are merely linked to the physiological self-incompatibility system.

There are other secondary sex characteristics of distylous plants. One of the most common is a size dimorphism in pollen (Ganders, 1979). Thrum pollen is typically larger than that produced by pins, prompting Darwin (1877) to suggest that the size difference was related to the requirement for larger energy reserves in thrum pollen, which must grow down the long styles of pin flowers. Some doubt is cast on this view by Ganders's (1979) observation that there is no correspondence between the ratio of thrum to pin pollen volume and the ratio of pin to thrum style length. A strong correlation exists, however, between pollen grain number and pollen volume. Coupled with data showing that thrum stigmas capture less total pollen than pins, this observation suggests that selection may have favored increased pollen production by pins to compensate for unequal pollen flow. Ganders (1979) argues that a trade-off in size accompanied the increase in the number of pollen grains produced by pins, resulting in size dimorphism. If this is true, it would appear that the pollen size of thrums should reflect more closely the pollen size of the ancestors of distylous species. It might therefore be possible to test Ganders's (1979) hypothesis by comparing pollen sizes in distylous species and in their monomorphic progenitors.

Polymorphisms in pollen shape and ornamentation have also been detected in distylous taxa (Baker, 1966; Dulberger, 1974, 1975b). In some species these differences seem to be related to dimorphisms in stigmatic papillae, which are frequently longer in pins than in thrums. Dulberger (1975b) found that pollen adhered better in legitimate pollinations of distylous Plumbaginaceae, but Zavala (1978) detected no difference in species of *Lythrum*. It is possible that these characteristics are selected as a means of further enhancing disassortative pollination only in those taxa in which reciprocal positioning of anthers and stigmas is insufficient in itself. Baker (1966), however, suggested that these secondary morphological dimorphisms were added in sequence, with reciprocal placement of anthers and stigmas occurring last. A final dimorphism observed in a few distylous species is corolla size. Typically, corollas of thrums are smaller than those of pins, but exceptions are numerous (Wyatt and Hellwig, 1979). It is interesting to note that in 14 of 20 dioecious species studied by Bawa and Opler (1975), staminate flowers were smaller than pistillate flowers. This finding is consistent with the idea that thrums evolve into males and pins into females. Caution is urged, however, since Baker (1948) and Whitehouse (1950) detected an opposite trend, and the only true valid comparisons are those between flowers of dioecious taxa and their known distylous progenitors.

3. Alternative Explanations of Why Pins Evolve into Females

Crowe's (1964) work on the origins of dioecy seems to have been overlooked in recent discussions of heterostyly (Ganders, 1979) and of the evolution of dioecy (Lloyd, 1980c; Bawa, 1980). She noted that "the genetics of heterostyly and dioecy confirm the link implied by morphological data" (p. 441). In both, one mating type is homozygous (ss, long-styled pins and XX females) and the other is heterozygous (Ss, short-styled thrums and XY males), and the contribution of mating type alleles to the next generation is unequal (3X:1Y in dioecy and 3s:1S in heterostyly), resulting in maintenance of equal numbers of the two mating types (Fig. 1). Crowe (1964) states that only a simple change in the relation between the alleles S and s in a heterostylous species is required to create functional dioecy in a single step. If a mutation replaced dominance with individual action of alleles in Ss styles, then the heterozygous thrums would act only as males and homozygous pins only as females. Selection might then eliminate the pistils of the former thrums and the stamens of the former pins at the same or different rates, depending on the relative detriment to fitness occasioned by retaining each. Alternatively, if dominance were to give way to individual action of alleles in the pollen of heterozygous thrums, they would function as females and the homozygous pins would be males (Crowe, 1964). Therefore, when the thrum form is male (as in all cases known at present), we expect males to be heterogametic, but when the thrum form is female, the females should be heterogametic. This hypothesis can be tested if dioecious plants can be found that have evolved from ancestors with distyly of the form found in, for example, *A. maritima* (Baker, 1966) or *H. aegypticum* (Ornduff, 1979).

It is obvious that the observed tendency for pins to become female and thrums to become male when distyly evolves into dioecy may be as simple as the fact that in nearly all distylous species, the thrum is the heterogametic morph. Most authors, however, have overlooked this genetic explanation and have sought, instead, ecological explanations. Here I will briefly consider three hypotheses to explain why pins should do better as ovule parents, whereas thrums should be more effective as pollen parents: (1) asymmetric pollen flow (Baker, 1958; Willson, 1979; Beach and Bawa, 1980), (2) resource allocation (Casper and Charnov, 1982), and (3) gametophytic selection.

Baker (1958) noted that the stigmas of thrum plants of *Mussaenda chippii* and *M. tristigmatica* seldom receive pollen because the corolla tube is choked with hairs, so that thrums are functionally staminate. Most studies of pollen flow in populations of heterostylous plants have detected significant asymmetry. Typically, stigmas of pin flowers capture more pollen grains than stigmas of thrums (Ganders, 1979). Pin stigmas also have higher levels of self-pollination. Bawa (1980) attributed this to greater accessibility of the exserted floral parts: styles, in the case of pins, and anthers, in the case of thrums. Beach and Bawa (1980) believe that this lack of complementarity in pollen flow is often associated with a change from pollination by flower visitors that have long mouthparts on which pin and thrum pollen is deposited differentially to pollination by short-tongued vectors on whose bodies pollen is not localized. They note, in support of this view, that average flower sizes of dioecious species of *Cordia* and *Coussarea* are smaller than for their distylous ancestors. Furthermore, while the distylous species are pollinated by long-tongued Lepidoptera, the dioecious taxa are pollinated by short-tongued Hymenoptera.

Another factor that may be involved in promoting preferential pollen flow from long stamens to long styles is greater stigmatic surface area of pins. In *M. repens,* the stimatic surface area is greater for thrums, an observation consonant with the fact that these plants are more fecund than pins (Table VI). Contrary to the expected trend, this ecological situation could favor the transformation of thrums into females and pins into males. This lack of complemetarity in *M. repens* may be due to the extension of stigmas of pins far beyond the bearded throat of the corolla, with the anthers of thrums in a similar position (Ganders, 1975; Keegan *et al.,* 1979). Pollinators (primarily bumblebees) seeking nectar from the base of the corolla can therefore easily avoid touching those parts, thus enhancing pollen flow from pin anthers to thrum stigmas located at the level of the bearded throat. It is not clear to what extent the morphological arrangement of stigmas and anthers of *M. repens* is typical of distylous plants. It does not appear, however, that this species is evolving toward dioecy.

Alternatively, the tendency for pins to become females might be explained in terms of resource allocation. Because the pollen grains produced by pins are smaller and presumably less expensive to produce, pins may have more resources to invest in seed production. This might be especially important in taxa with nonepipetalous stamens, where resources of thrums must be invested in filaments as well as in anthers and pollen (Fig. 1). Casper and Charnov (1982) developed sex allocation models to explain specialization of pins or thrums to produce either pollen or seeds. They also predict conditions under which natural populations should deviate from a 1:1 ratio of the morphs.

A final hypothesis that could explain why pins become females involves gametophytic selection (Mulcahy, 1974, 1979). Because of the longer distance that pollen tubes must grow through the long styles of pins, the selective sieve for male gametophytes is significantly stronger. To the extent that performance of sporophytes is correlated with that of gametophytes, seeds produced on pin plants should be more vigorous, favoring further specialization of pins as seed parents.

C. The Evolution of Autogamy

1. The Evolution of Self-Compatibility

Stebbins (1974, p. 51) observed that "the evolutionary pathway from obligate outcrossing based upon self-incompatibility to predominant self-fertilization has probably been followed by more different lines of evolution in flowering plants than has any other." Numerous examples of this trend have been described and documented (Section I,C,1,b). In all cases, it appears that self-compatibility has been acquired through loss mutations of the self-incompatibility locus or loci (Lewis, 1955). Nettancourt (1977) reviewed a number of genetic analyses of mutations causing self-compatibility and concluded that the spontaneous rate ranges from 1.7 to 4.3 per million pollen grains in different clones of *Oenothera organensis,* from 0.2 to 2.3 in *Prunus avium,* and from 0.2 to 0.4 in *Nicotiana alata.* In *Chrysanthemum,* Ronald and Ascher (1975) reported self-compatibility to be a rare dominant with a complex pattern of inheritance. Lloyd (1965) found self-compatibility in *Leavenworthia* to be recessive. Rare crossover events in the supergenes controlling heterostyly versus homostyly produce simply inherited changes from obligate outcrossing to self-fertilization (Crosby, 1949; Dowrick, 1956; Mather, 1973; Ganders, 1979).

2. The Evolution of Self-Pollination

In other taxa, the evolution of autogamy may occur, so that self-compatible but predominantly outcrossing plants change such that self-pollination is assured. This may involve a breakdown in dichogamy or any other device that promotes outcrossing (Section I,C,1,a). In *Clarkia xantiana,* for example, stigma maturation, which regulates outcrossing rates, "is under relatively simple genetic control, and may be determined primarily by a single gene" (Moore and Lewis, 1965, p. 110–111). Determination of outcrossing rates in *Lycopersicon* is largely through morphological characters with a simple genetic basis (Rick, 1950), although polygenic control of certain features is postulated for some species (Rick *et al.,* 1978). In some taxa, there appears to be a large environmental component to breeding system variation (Jain, 1976).

3. Alternative Hypotheses

The state of the art with respect to the evolution of autogamy has been succinctly summarized by Jain (1976, p. 470), who noted that "we have more speculation than rigorous thought, and more theory than real data" on the questions of when, how, and why it evolved. Nevertheless, Jain (1976) listed six alternative hypotheses to explain the evolution of autogamy: (1) reproductive assurance, (2) adaptedness-adaptability balance, (3) reproductive isolation, (4) automatic selection, (5) breakdown syndrome, and (6) rapid spread of favorable recessive alleles. To this list he added Stebbin's (1950, 1957, 1958) "synthetic approach," which combines aspects of the two models most frequently invoked: reproductive assurance and greater potential for local adaptation of autogamous plants. Hypotheses 1, 3, and 4 depend on the nature of pollinator–plant interactions and therefore will be emphasized in this discussion, rather than the other hypotheses which are based on genetic phenomena.

The reproductive assurance hypothesis was first proposed by Darwin (1876), who viewed autogamy itself, however, as a rare event favored by natural selection only under highly unusual circumstances. Darlington (1939, 1963) suggested that autogamy could arise whenever environmental conditions prevented effective pollination by the normal vectors, as at the margins of the range of a species. Hagerup (1951), for example, suggested that excessive moisture in northern oceanic climates limited favorable days for flight by insects, leading to the evolution of self-pollination. Baker (1955) extended the range of such conditions selecting for autogamy to colonization by a single propagule following long-distance dispersal; only if a plant is autogamous can a new population effectively be started from a single colonizing individual. This principle has subsequently been termed "Baker's Rule." It certainly appears to be true that autogamy is characteristic of successful colonizing species (references cited by Baker and Stebbins, 1965; Stebbins, 1970). Others have suggested additional circumstances favoring autogamy as a form of reproductive assurance, such as competition for limited pollen vectors (Grant, 1958; Grant and Grant, 1965; Levin, 1972b; Grant and Flake, 1974). Lloyd (1979a) has devised models that make quantitative predictions concerning when selfing will be favored by such circumstances and what types of selfing (i.e., competing, prior, or delayed) are most likely to evolve.

Mather (1953, 1973) suggested that selection for autogamy might occur when environmental conditions allowed only a narrow range of genotypes to survive and reproduce. The gain in immediate fitness through increased local adaptation might overbalance the decrease in long-term fitness through loss of genetic flexibility (Mather and DeWinton, 1941). The genetic consequences of such a shift can be stated precisely (Jain, 1976); however, the ecological unknowns regarding how often or how much environmental variation would trigger changes in the breeding systems have not been defined adequately. Although the adaptation-adaptability hypothesis is superficially attractive, no concrete examples of its actual operation have been advanced. This is undoubtedly due in part to the difficulty of measuring environmental variation in ways meaningful to the genetics or ecology of plant populations. Lloyd (1979c, p. 604) has attacked this type of explanation on the grounds "that individual selection rather than group selection is primarily responsible for the evolution of the characters of populations."

Autogamy may also arise as a mechanism to restrict gene flow between closely related, sympatric species. The classic studies of saltational speciation in *Clarkia* (Lewis, 1962, 1966, 1973; Raven, 1964; Moore and Lewis, 1965; Vasek, 1968, 1971; Bartholomew *et al.*, 1973) incorporated selfing as an important adjunct of catastrophic selection. These observations of the effectiveness of autogamy as a breeding barrier have been amplified further by the work of Gottlieb (1973a,b, 1974) on sympatric speciation in *Clarkia*, and *Stephanomeria*. Autogamy can function to restrict gene flow and thereby allow genetic differentiation on a very local level between closely adjacent populations, as shown by Antonovics (1968) and others (Lefebvre, 1970; Dickinson and Antonovics, 1973). Levin (1975) states that autogamy may be selected whenever two taxa that produce sterile hybrids share pollinators and occur in mixed populations.

Automatic selection hypotheses include Fisher's (1941) model in which a gene for autogamy spreads, since these plants can serve as the pollen parent of seeds on other genotypes in addition to being the pollen and ovule parent of their own selfed progeny. Wells (1979) and Lloyd (1979a) proposed similar, but more general, models in which the advantage of self-fertilization does not depend on the environmental context. Crosby's (1949) example of a similar advantage enjoyed by homostylous plants of *Primula* and Bodmer's (1960a,b) demonstration that breakdown in such heteromorphic breeding systems leads inexorably to selfing suggest that when selection for outcrossing is relaxed, these systems are supplanted permanently by autogamy. The dynamics of such breakdowns have been discussed by Baker (1966), Ornduff (1972), Charlesworth (1979), Barrett (1979), and Ganders (1979). Finally, Haldane (1932) suggested that selfing might be selected for on the basis of more rapid fixation of favorable recessive genes.

4. *Examples and Possible Explanations*

As Jain (1976, p. 477) has pointed out, all hypotheses to explain the evolution of autogamy involve "implied arguments in favor of opportunistic advantages of certain rare events, presumption of the availability of genetic variants modifying breeding systems, selection or chance as alternatives in certain unspecified ways, and a general untested evolutionary principle that what can happen does happen." However, Wells (1979) and Lloyd (1979a) recently presented models with clearly stated hypotheses that make predictions that can be tested by observations or experiments. Unfortunately, a general problem with most studies of the evolution of autogamy to date is that comparisons are usually made between congeneric species that differ with respect to their rates of outcrossing. This approach implicitly assumes that all morphological, genetic, and ecological differences can be attributed directly to breeding systems alone.

The evolution of autogamy in the cedar glade endemic *Leavenworthia* constitutes one of the best-documented examples of the derivation of self-compatible and self-pollinating races and species from outcrossing progenitors (Rollins, 1963a; Lloyd, 1965; Solbrig, 1972; Solbrig and Rollins, 1977). Lloyd (1965) concluded that the evolution of self-compatibility and concomitant changes in other characters in *L. crassa* and *L. alabamica* are associated with the nature of the glades. Populations of these species that invaded "poorer" glades contained fewer plants with fewer flowers per plant. Earlier flowering at these sites also was favored, because their shallow soil caused them to dry out earlier in the season. The relative paucity of insect pollinators early in the season selected for the ability to self. According to Lloyd (1965), the advantage of having a greater percentage of their flowers pollinated at these sites more than compensated for decreases in the number and quality of seeds resulting from each self-pollination.

Arroyo (1973, 1975) also proposed paucity of pollinators as the selective force behind the evolution of autogamy in *Limnanthes floccosa*. She argued that periodic droughts resulted in reduced populations of bees, as well as plants, and that even if cross-pollination occurred, inadequate soil moisture might prevent seed maturation. Autogamy was therefore advantageous in assuring reproduction and, because such plants could flower earlier, in escaping the harmful effects of drought. She rejected the idea that autogamy resulted from selection for local adaptation, because autogamous populations showed increased chiasma frequency, a trend antithetical to (and possibly compensatory for) the effects of autogamy on genetic population structure. Unfortunately, Arroyo (1973, 1975) presented no data bearing directly on the question of pollinator scarcity and the effects of drought on seed maturation.

In *Lycopersicon pimpinellifolium,* Rick and his co-workers (Rick and Fobes, 1975; Rick *et al.*, 1977) originally contended that a lack of pollinators in certain portions of the range had led to autogamy. Pollination rates of a range of genotypes grown in a common garden, however, revealed very high pollinator activity in southern Peru, an area previously assumed to be depauperate in effective pollinators (Rick *et al.*, 1978). The selective basis for the evolution of selfing and associated changes in flower size, anther length, and stigma exsertion are therefore unclear at present. Similarly, Schoen (1982a,b) was unable to explain the greater frequency of autogamous populations of *Gilia achilleifolia* in the northern parts of its range in California. He did suggest, however, that a past northward migration had brought the species outside the range of its normal pollinators, so that autogamous variants were favored in these "small, unattractive and newly founded populations" (Schoen, 1982a, p. 358).

A final example of the evolution of autogamy that suggests a different kind of role for pollinator–plant interactions involves the species of *Arenaria* endemic to granite outcrops in the southeastern United States (R. Wyatt, unpublished). In this group, *A. alabamensis* has been considered an autogamous derivative of *A. uniflora,* a more widespread, largely outcrossing species (Weaver, 1970; McCormick *et al.*, 1971; Wyatt, 1977). It was described originally from only two localities in northeastern Alabama but has subsequently been reported from outcrops in north-central Alabama, northern South Carolina, and southwestern North Carolina (Wyatt, 1977; Wyatt and Fowler, 1977; Fig. 2). In contrast, *A. uniflora* occurs from south-central North Carolina to eastern Alabama (Fig. 2). Selfers are therefore restricted to the margins of the range of the outcrossers and therefore also to the limits of occurrence of such exposed rock habitats in the Southeast. At no site do the selfers and outcrossers occur sympatrically. Morphological differences between the two sets of populations are associated with floral features, and all are consistent with those characters typically correlated with a trend to autogamy (Ornduff, 1969; Table III).

Plants from all populations are self-compatible but differ in their rates of outcrossing due to prevention of intrafloral selfing by strong protandry in the outcrossers (R. Wyatt, unpublished). Because few flowers are open at a time and

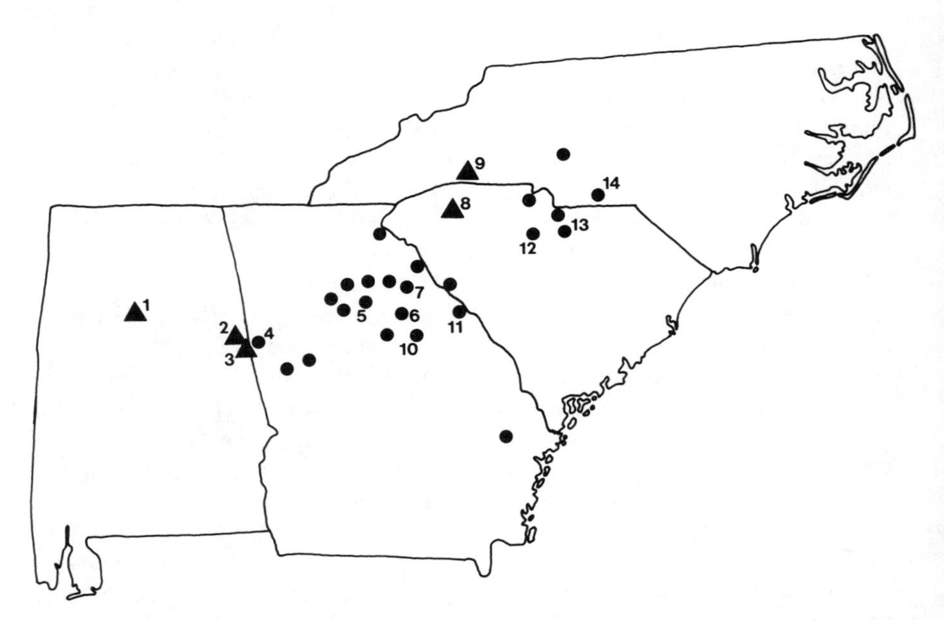

Fig. 2. Distribution of outcrossing (circles) and selfing (triangles) populations of *Arenaria uniflora* in the southeastern United States. The selfing populations have been called *A. alabamensis*. Numbers indicate field sites selected for intensive study.

stems of these exceptionally dense-growing plants are intertwined, geitonogamy is low. Limited variability is seen within populations for morphological characters, including degree of protandry. All populations show some degree of interfertility in crosses, although crosses between geographically contiguous populations are most interfertile. Despite some variation between years, transplant studies suggest that plants perform best at their native sites and outperform morphs transplanted from other sites.

A search for possible ecological and genetic factors favoring particular morphs at a given site was initiated on several fronts. To test the possibility that selfers were being favored on outcrops whose environments were relatively predictable, I attempted to correlate outcrossing rates with variability in such environmental factors as mean monthly temperature, mean monthly precipitation, and mean monthly evaporation during the growing season. The coefficient of variation was used as the estimate of variability for each factor, and no significant correlations were detected. I also collected pollinators and estimated their abundance at each site, finding no clear-cut evidence that the major pollinators (small native bees and flies) were absent from outcrops where the selfing morphs occurred. Furthermore, if any trend was present in terms of mean date of flowering, it was in the direction of earlier dates for populations of outcrossers in central Georgia. Also, the fact that outcrossers transplanted to sites at the margins of the range were able to set some seed suggests that pollinators are present, although the level of pollinator activity may be somewhat reduced compared to that of central sites.

What, then, is the explanation for the evolution of autogamy in these granite outcrop races of *Arenaria?* I believe that autogamy arose in response to competition for pollinators (Levin, 1972b) between outcrossing morphs of *A. uniflora* and a second winter annual species endemic to rock outcrops, *A. glabra,* which is pollinated by the same species of bees and flies. Originally, a correlation between mean precipitation during the growing season and the percentage of plants spontaneously setting fruit in the absence of pollinators in a greenhouse (autofertility) was puzzling (Fig. 3). Such a result is directly contrary to the results of Lloyd (1965), Arroyo (1973, 1975), and others who found autogamous plants to be associated with dry, marginal habitats. When the pattern of rainfall in the Southeast was examined, however, it became clear that *A. glabra* occurs only on relatively wet granite outcrops in eastern Alabama and northwestern North and South Carolina. Wherever this species, which produces larger, showier flowers in much greater abundance than the outcrossing morphs of *A. uniflora,* occurs with populations of the other species, it is the selfing morph (*A. alabamensis*) that is present. Most taxonomists view *A. glabra* as a very recent (perhaps Pleistocene) derivative of *A. groenlandica* (Weaver, 1970), an arctic-alpine species that extends southward at high elevations in the Appalachian Mountains. Indeed, some taxonomists still consider it merely a variety of the latter (Radford *et al.*, 1968). It is not clear, therefore, whether *A. glabra* has

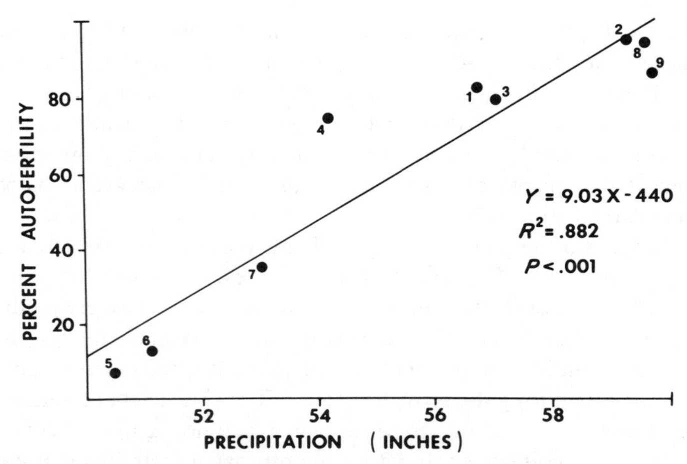

Fig. 3. Percentage of autofertility versus mean annual precipitation during the growing season (September–May). Data points are means for the nine populations of *Arenaria uniflora* numbered as in Fig. 2.

achieved its total potential range and is prevented by lack of drought tolerance from invading outcrops in central Georgia or if it might not continue to colonize new sites, driving even the central populations of *A. uniflora* to dominance by the selfing morph. Experiments are in progress to test these alternative hypotheses and to examine further the possible importance of competition for pollinators. A possible role for automatic selection must also be evaluated.

Other examples of competition for pollinator service acting as a stimulus for the evolution of autogamy may be discovered as researchers begin to look for ecological correlates of autogamy in insect-pollinated taxa. With few exceptions (Lloyd, 1965; Rick *et al.*, 1978), the hypothesis of inadequate or unpredictable visits by pollinators has been accepted uncritically without actual field data and without consideration of the composition of the local community. The fact that the areas in which most of these studies have been carried out are rich in pollinator fauna suggests that pollinators should not be scarce. Furthermore, the great diversity of self-incompatible, entomophilous species would also seem to indicate a possible role for competition for pollinator service. It is obvious that, as Levin (1972b, p. 669) suggests, "the burden of proof lies both with the advocate of pollinator paucity as well as that of competition for pollinators."

III. Summary

Only recently have attempts been made to relate the evolution of breeding systems in plants to interactions with their pollinators. Earlier explanations of variations in plant sexuality and pollination systems focused only on the genetic benefits that result from outcrossing. Breeding systems were viewed as fixed characteristics that could be ordered into discrete categories. This view has been challenged by the demonstration that plants show great variability in functional gender, their relative genetic contributions as ovule versus pollen parents or seeds. Methods for quantifying functional gender have been developed, and precise assessments of sexuality can be made along a continuum. Precise quantification of rates of outcrossing are also possible now through the use of genetic markers detected electrophoretically. These tools have rekindled interest in evolutionary transitions between breeding systems based on ecological, as well as genetic, selective factors.

It is suggested, for example, that dichogamy in the angiosperms has evolved (at least partly) in response to the directionality of pollen transfer by different vectors. Plants pollinated by bees, flies, or other agents that direct pollen transfer upward are predominantly protandrous, with anthers dehiscing before stigmas are receptive. Plants pollinated by wasps, beetles, or other vectors that direct pollen transfer downward are exclusively protogynous, with stigmas ceasing to be receptive before anthers dehisce. Such an arrangement minimizes self-pollination in self-compatible taxa and prevents coverage of the stigmatic surface with ineffectual self-pollen in self-incompatible species.

Dioecy, or separate sexes, has apparently evolved from distyly in a number of angiosperm groups. In all cases studied thus far, long-styled morphs (pins) become females and short-styled morphs (thrums) become males. A simple explanation for this tendency is that in nearly all distylous species, the thrums are heterogametic. A change from dominance to individual action of the incompatibility alleles in thrum styles would create a functionally dioecious breeding system in a single step. Alternatively, the tendency for pins to perform better as ovule parents may be due to asymmetric pollen flow (with greater pollen deposition on the long, exserted pin stigmas), resource allocation (with the energy saved by production of small pollen grains in pins being diverted to ovule parent functions), or gametophytic selection (with the long styles of pins serving as a more effective means of selecting for vigorous male gametophytes).

Autogamy, or predominant self-fertilization, has evolved in many different groups of angiosperms. Hypotheses to explain its evolution include reproductive assurance, which favors autogamy whenever the normal pollinators are missing or reduced in abundance, and automatic selection, in which a gene determining selfing spreads because such plants can serve as the pollen parent of seeds on other genotypes in addition to being the pollen and ovule parent of their own selfed seeds. Other hypotheses invoke special circumstances or, like the adaptation-adaptability balance hypothesis, depend on group selection. Few studies have been done in which these explanations have been examined critically, but reproductive assurance seems to be involved in most, if not all, cases. Competition for pollinators appears to have selected for autogamy in some taxa.

Acknowledgments

I thank David M. Lane, Thomas R. Meagher, and especially Ann Stoneburner for their help in collecting some of the data presented in this chapter. I also thank Janis Antonovics, Spencer C. H. Barrett, K. S. Bawa, David G. Lloyd, B. J. D. Meeuse, Daniel J. Schoen, and Andrew G. Stephenson for reading earlier drafts of portions of this manuscript.

Support for my studies of the evolution of autogamy in *Arenaria* was provided by NSF Grant No. 8001986.

References

Allard, R. W., Jain, S. K., and Workman, P. L. (1968). The genetics of inbreeding populations. *Adv. Genet.* **14,** 55–131.

Antonovics, J. (1968). Evolution in closely adjacent plant populations. V. Evolution of self-fertility. *Heredity* **23,** 219–238.

Arroyo, M. T. K. (1973). Chiasma frequency evidence on the evolution of autogamy in *Limnanthes floccosa* (Limnanthaceae). *Evolution (Lawrence, Kans.)* **27,** 679–688.

Arroyo, M. T. K. (1975). Electrophoretic studies of genetic variation in natural populations of allogamous *Limnanthes alba* and autogamous *Limnanthes floccosa* (Limnanthaceae). *Heredity* **35,** 153–164.

Arroyo, M. T. K., and Raven, P. H. (1975). The evolution of subdioecy in morphologically gynodioecious species of *Fuchsia* Sect. *Encliandra* (Onagraceae). *Evolution (Lawrence, Kans.)* **29,** 500–511.

Baker, H. G. (1948). Dimorphism and monomorphism in the Plumbaginaceae. I. A survey of the family. *Ann. Bot. (London)* **12,** 207–219.

Baker, H. G. (1955). Self-compatibility and establishment after "long-distance" dispersal. *Evolution (Lawrence, Kans.)* **9,** 347–348.

Baker, H. G. (1958). Studies in the reproductive biology of West African Rubiaceae. *J. West. Afr. Sci. Assoc.* **4,** 9–24.

Baker, H. G. (1961). The adaptation of flowering plants to nocturnal and crepuscular pollinators. *Q. Rev. Biol.* **36,** 64–73.

Baker, H. G. (1962). Heterostyly in the Connaraceae with special reference to *Byrsocarpus coccineus*. *Bot. Gaz. (Chicago)* **123,** 206–211.

Baker, H. G. (1966). The evolution, functioning and breakdown of heteromorphic incompatibility systems. I. The Plumbaginaceae. *Evolution (Lawrence, Kans.)* **20,** 349–368.

Baker, H. G. (1967). The evolution of weedy taxa in the *Eupatorium microstemon* species aggregate. *Taxon* **16,** 293–300.

Baker, H. G., and Hurd, P. D. (1968). Intrafloral ecology. *Annu. Rev. Entomol.* **13,** 385–414.

Baker, H. G., and Stebbins, G. L. (1965). "Genetics of Colonizing Species." Academic Press, New York.

Barrett, S. C. H. (1978). Heterostyly in a tropical weed: The reproductive biology of the *Turnera ulmifolia* complex (Turneraceae). *Can. J. Bot.* **56,** 1713–1725.

Barrett, S. C. H. (1979). The evolutionary breakdown of tristyly in *Eichhornia crassipes* (Mart.) Solms (Water Hyacinth). *Evolution (Lawrence, Kans.)* **33,** 499–510.

Barrett, S. C. H. (1980). Dimorphic incompatibility and gender in *Nymphoides indica* (Menyanthaceae). *Can. J. Bot.* **58,** 1938–1942.

Bartholomew, B., Eaton, L. C., and Raven, P. H. (1973). *Clarkia rubicunda:* a model of plant evolution in semiarid regions. *Evolution (Lawrence, Kans.)* **27,** 505–517.

Bateman, A. J. (1952). Self-incompatibility systems in angiosperms. *Heredity* **6,** 285–310.

Bawa, K. (1980). Evolution of dioecy in flowering plants. *Annu. Rev. Ecol. Syst.* **11,** 15–39.

Bawa, K. S. (1982). Outcrossing and the incidence of dioecism in island floras. *Am. Nat.* **119,** 866–871.

Bawa, K. S., and Beach, J. H. (1981). Evolution of sexual systems in flowering plants. *Ann. M. Bot. Gard.* **68,** 254–274.

Bawa, K. S., and Opler, P. A. (1975). Dioecism in tropical forest trees. *Evolution (Lawrence, Kans.)* **29,** 167–179.

Beach, J. H. (1981). Pollinator foraging and the evolution of dioecy. *Am. Nat.* **118,** 572–577.

Beach, J. H., and Bawa, K. S. (1980). Role of pollinators in the evolution of dioecy from distyly. *Evolution (Lawrence, Kans.)* **34,** 467–474.

Beach, J. H., and Kress, W. J. (1980). Sporophyte versus gametophyte: A note on the origin of self-incompatibility in flowering plants. *Syst. Bot.* **5,** 1–5.

Benham, B. R. (1969). Insect visitors to *Chamaenerion angustifolium* and their behavior in relation to pollen. *Entomologist* **102,** 211–228.

Best, L. S., and Bierzychudek, P. (1982). Pollinator foraging on foxglove (*Digitalis purpurea*): A test of a new model. *Evolution (Lawrence, Kans.)* **36,** 70–79.

Bir Bahadur (1968). Heterostyly in Rubiaceae: A review. *J. Osmania Univ.* (*Science*), *Golden Jubilee Vol.* pp. 207–238.

Bir Bahadur (1970). Homostyly and heterostyly in *Oldenlandia umbellata* L. *J. Genet.* **60,** 192–198.

Bodmer, W. F. (1960a). The genetics of homostyly in populations of *Primula vulgaris*. *Phil. R. Soc. London, Trans. Ser. B* **242,** 517–549.

Bodmer, W. F. (1960b). Discrete stochastic processes in population genetics. *J. R. Statis. Soc., Ser. B* **22,** 218–236.

Bradshaw, A. D. (1972). Some of the evolutionary consequences of being a plant. *Evol. Biol.* **5,** 25–44.

Brewbaker, J. L. (1957). Pollen cytology and incompatibility systems in plants. *J. Hered.* **48,** 271–277.

Brewbaker, J. L. (1959). Biology of the angiosperm pollen grain. *Indian J. Genet. Plant Breed.* **19,** 121–133.

Brewbaker, J. L., and Majumder, S. K. (1961). Incompatibility and the pollen grain. *Recent Adv. Bot.* **2,** 1503–1508.

Brown, A. H. D. (1979). Enzyme polymorphism in plant populations. *Theor. Popul. Biol.* **15,** 1–42.

Brown, A. H. D., and Allard, R. W. (1970). Estimation of the mating system in open-pollinated maize populations using isozyme polymorphisms. *Genetics* **66,** 133–145.

Brown, A. H. D., Matheson, A. C., and Eldridge, K. G. (1975). Estimation of the mating system of *Eucalyptus obliqua* L'Herit. by using allozyme polymorphisms. *Aust. J. Bot.* **23,** 931–949.

Brown, A. H. D., Zohary, D., and Nevo, E. (1978). Outcrossing rates and heterozygosity in natural populations of *Hordeum spontaneum* Koch in Israel. *Heredity* **41,** 49–62.

Burtt, B. L. (1978). Aspects of diversification in the capitulum. *In* "The Biology and Chemistry of the Compositae" (V. H. Heywood, J. B. Harborne, and B. L. Turner, eds.), pp. 41–59. Academic Press, New York.

Casper, B. B., and Charnov, E. L. (1982). Sex allocation in heterostylous plants. *J. Theor. Biol.* **96,** 143–149.

Charlesworth, D. (1979). The evolution and breakdown of tristyly. *Evolution (Lawrence, Kans.)* **33,** 486–498.

Charlesworth, D., and Charlesworth, B. (1979). A model for the evolution of distyly. *Am. Nat.* **114,** 467–498.

Chase, V. C., and Raven, P. H. (1975). Evolutionary and ecological relationships between *Aquilegia formosa* and *A. pubescens* (Ranunculaceae), two perennial plants. *Evolution (Lawrence, Kans.)* **29,** 474–486.

Clegg, M. T. (1980). Measuring plant mating systems. *BioScience* **30,** 814–818.

Clifford, H. T. (1962). Insect pollination of *Plantago lanceolata* L. *Nature (London)* **193,** 196.

Cox, P. A. (1981). Niche partitioning between sexes of dioecious plants. *Am. Nat.* **117,** 295–307.

Crosby, J. L. (1949). Selection of an unfavorable gene complex. *Evolution (Lawrence, Kans.)* **3,** 212–230.

Crowe, L. K. (1964). The evolution of outbreeding in plants. I. The angiosperms. *Heredity* **19,** 435–457.

Crowe, L. K. (1971). The polygenic control of outbreeding in *Borago officinalis. Heredity* **27,** 111–118.

Cruden, R. W., and Hermann-Parker, S. M. (1977). Temproal dioecism: An alternative to dioecism? *Evolution (Lawrence, Kans.)* **31,** 863–866.

Darlington, C. D. (1939). "The Evolution of Genetic Systems." Cambridge Univ. Press, London and New York.

Darlington, C. D. (1963). "Chromosome Botany and the Origins of Cultivated Plants." Allen and Unwin, London.

Darwin, C. (1876). "The Effects of Cross- and Self-fertilisation in the Vegetable Kingdom." Murray, London.

Darwin, C. (1877). "The Different Forms of Flowers on Plants of the Same Species." Murray, London.

Dickinson, H., and Antonovics, J. (1973). Theoretical considerations of sympatric divergence. *Am. Nat.* **107,** 256–274.

Dowrick, V. P. J. (1956). Heterostyly and homostyly in *Primula obconica. Heredity* **10,** 219–236.

Dulberger, R. (1974). Structural dimorphism of stigmatic papillae in distylous *Linum* species. *Am. J. Bot.* **61,** 238–243.

Dulberger, R. (1975a). S-gene action and the significance of characters in the heterostylous syndrome. *Heredity* **35,** 407–415.

Dulberger, R. (1975b). Intermorph structural differences between stigmatic papillae and pollen grains in relation to incompatibility in Plumbaginaceae. *Proc. R. Soc. London, Ser. B* **188,** 257–274.

East, E. M. (1940). The distribution of self-sterility in the flowering plants. *Proc. Am. Philos. Soc.* **82,** 449–518.

Ehrlich, P. R., and Raven, P. H. (1969). Differentiation of populations. *Science* **165,** 1228–1232.

Ennos, R. A., and Clegg, M. T. (1982). Effect of population substructuring on estimates of outcrossing rate in plant populations. *Heredity* **48,** 283–292.

Epling, C., and Lewis, H. (1952). Increase of the adaptive range of the genus *Delphinium. Evolution (Lawrence, Kans.)* **6,** 253–267.

Ernst, A. (1955). Self-fertility in monomorphic primulas. *Genetica* **27,** 91–148.

Faegri, K., and van der Pijl, L. (1979). "The Principles of Pollination Ecology." Pergamon, Oxford.

Fisher, R. A. (1941). Average excess and average effect of a gene substitution. *Ann. Eugen.* **11,** 53–63.

Fosberg, F. R. (1956). Studies in Pacific Rubiaceae. *Brittonia* **8,** 165–178.

Frankel, R., and Galun, E. (1977). "Pollination Mechanisms, Reproduction and Plant Breeding." Springer-Verlag, Berlin and New York.

Fryxell, P. A. (1957). Mode of reproduction in higher plants. *Bot. Rev.* **23,** 135–233.

Ganders, F. R. (1975). Fecundity in distylous and self-incompatible homostylous plants of *Mitchella repens* (Rubiaceae). *Evolution (Lawrence, Kans.)* **29,** 186–188.

Ganders, F. R. (1979). The biology of heterostyly. *N. Z. J. Bot.* **17,** 607–635.

Ganders, F. R., Carey, K., and Griffiths, A. J. F. (1977). Natural selection for a fruit dimorphism in *Plectritis congesta* (Valerianaceae). *Evolution (Lawrence, Kans.)* **31,** 873–881.

Gaudreau, M. M., and Hardin, J. W. (1974). The use of neutron activation analysis in pollination ecology. *Brittonia* **26,** 316–320.

Givnish, T. J. (1980). Ecological constraints on the evolution of breeding systems in seed plants: Dioecy and dispersal in gymnosperms. *Evolution (Lawrence, Kans.)* **34,** 959–972.

Givnish, T. J. (1982). Outcrossing versus ecological constraints in the evolution of dioecy. *Am. Nat.* **119,** 849–865.

Gleeson, S. K. (1982). Heterodichogamy in walnuts: Inheritance and stable ratios. *Evolution (Lawrence, Kans.)* **36,** 892–902.

Godley, E. J. (1979). Flower biology in New Zealand. *N. Z. J. Bot.* **17,** 441–466.

Gottlieb, L. D. (1973a). Enzyme differentiation and phylogeny in *Clarkia franciscana, C. rubicunda,* and *C. amoena. Evolution (Lawrence, Kans.)* **27,** 205–214.

Gottlieb, L. D. (1973b). Genetic differentiation, sympatric speciation and the origin of a diploid species of *Stephanomeria. Am. J. Bot.* **60,** 545–553.

Gottlieb, L. D. (1974). Genetic confirmation of the origin of *Clarkia lingulata. Evolution (Lawrence, Kans.)* **28,** 244–250.

Grant, V. (1952). Isolation and hybridization between *Aquilegia formosa* and *A. pubescens. Aliso* **2,** 341–360.

Grant, V. (1958). The regulation of recombination in plants. *Cold Spring Harbor Symp. Quant. Biol.* **23,** 337–363.

Grant, V. (1971). "Plant Speciation." Columbia Univ. Press, New York.

Grant, V. (1975). "Genetics of Flowering Plants." Columbia Univ. Press, New York.

Grant, V. (1976). Isolation between *Aquilegia formosa* and *A. pubescens:* a reply and reconsideration. *Evolution (Lawrence, Kans.)* **30,** 625–628.

Grant, V., and Flake, R. H. (1974). Population structure in relation to cost of selection. *Proc. Natl. Acad. Sci. U.S.A.* **17,** 1670–1671.

Grant, V., and Grant, K. A. (1965). "Flower Pollination in the Phlox Family." Columbia Univ. Press, New York.

Green, A. G., Brown, A. H. D., and Oram, R. N. (1980). Determination of outcrossing in a breeding population of *Lupinus albus* L. *Z. Pflanzenzücht* **84,** 181–191.

Hagerup, O. (1951). Pollination in the Faroes—in spite of rain and the poverty of insects. *Biol. Medd. K. Dan. Vidensk. Selsk.* **18,** 1–48.

Haldane, J. B. S. (1932). "The Causes of Evolution." Harper, London.

Handel, S. N. (1976). Restricted pollen flow of two woodland herbs determined by neutron-activation analysis. *Nature (London)* **260,** 422–423.

Harding, J. (1970). Genetics of *Lupinus.* III. The selective disadvantage of the pink flower color mutant in *Lupinus nanus. Evolution (Lawrence, Kans.)* **24,** 120–127.

Harding, J., Mankinen, C. B., and Elliott, M. H. (1974). Genetics of *Lupinus.* VII. Outcrossing, autofertility, and variability of the nanus group. *Taxon* **23,** 729–738.

Heinrich, B. (1975). Energetics of pollination. *Annu. Rev. Ecol. Syst.* **6,** 139–170.

Heinrich, B. (1976). Foraging specializations of individual bumblebees. *Ecol. Monogr.* **46,** 105–128.

Heinrich, B. (1979). "Bumblebee Economics." Harvard Univ. Press, Cambridge, Massachusetts.

Heinrich, B., and Raven, P. H. (1972). Energetics and pollination ecology. *Science* **176,** 597–602.

Heslop-Harrison, J. (1968). Ribosome sites and S gene action. *Nature (London)* **218,** 90–91.

Heslop-Harrison, J. (1975). Incompatibility and the pollen stigma interaction. *Annu. Rev. Plant Physiol.* **26,** 403–425.

Hildebrand, F. (1867). "Die Geschlechtsverteilung bei den Pflanzen." Leipzig.

Imrie, B. C., Kirkman, C. T., and Ross, D. R. (1972). Computer simulation of a sporophytic self-incompatibility breeding system. *Aust. J. Biol. Sci.* **25,** 343–349.

Jain, S. K. (1969). Comparative ecogenetics of *Avena fatua* and *A. barbata* occurring in central California. *Evol. Biol.* **3,** 73–118.

Jain, S. K. (1975). Population structure and the effects of breeding system. *In* "Plant Genetic Resources: Today and Tomorrow" (O. Frankel and J. G. Hawkes, eds.), pp. 15–36. Cambridge Univ. Press, London and New York.

Jain, S. K. (1976). The evolution of inbreeding in plants. *Annu. Rev. Ecol. Syst.* **7,** 469–495.

Jain, S. K. (1978). Breeding system in *Limnanthes alba:* Several alternative measures. *Am. J. Bot.* **65,** 272–275.

Janzen, D. H. (1971). Euglossine bees as long-distance pollinators of tropical plants. *Science* **171,** 203–205.

Keegan, C. R., Voss, R. H., and Bawa, K. S. (1979). Heterostyly in *Mitchella repens* (Rubiaceae). *Rhodora* **81,** 567–573.

Knuth, P. (1906–1909). "Handbook of Flower Pollination." Oxford Univ. Press, (Clarendon), London and New York.

Koelreuter, J. G. (1761–1766). "Vorläufige Nachricht von einigen das Geschlecht der Pflanzen betreffenden Versuchen und Beobachtungen." Leipzig.

Lefebvre, C. (1970). Self-fertility in maritime and zinc-mine populations of *Armeria maritima* (Mill.) Willd. *Evolution (Lawrence, Kans.)* **24,** 571–577.

Levin, D. A. (1972a). Pollen exchange as a function of species proximity in *Phlox. Evolution (Lawrence, Kans.)* **26,** 251–258.

Levin, D. A. (1972b). Competition for pollinator service: A stimulus for the development of autogamy. *Evolution (Lawrence, Kans.)* **26,** 668–674.

Levin, D. A. (1975). Minority cytotype exclusion in local plant populations. *Taxon* **24,** 35–43.

Levin, D. A. (1978). Pollinator behaviour and the breeding structure of plant populations. *In* "The Pollination of Flowers by Insects" (A. J. Richards, ed.), pp. 133–150. Academic Press, New York.

Levin, D. A. (1979a). The nature of plant species. *Science* **204,** 381–384.

Levin, D. A. (1979b). Pollinator foraging behavior: Genetic implications for plants. *In* "Topics in Plant Population Biology" (O. T. Solbrig, S. Jain, G. B. Johnson, and P. H. Raven, eds.), pp. 131–153. Columbia Univ. Press, New York.

Levin, D. A. (1981). Dispersal versus gene flow in plants. *Ann. M. Bot. Gard.* **68,** 233–253.

Levin, D. A., and Kerster, H. W. (1973). Assortative pollination for stature in *Lythrum salicaria. Evolution (Lawrence, Kans.)* **27,** 144–152.

Levin, D. A., and Kerster, H. W. (1974). Gene flow in seed plants. *Evol. Biol.* **7,** 139–220.

Lewis, D. (1949). Incompatibility in flowering plants. *Biol. Rev. Cambridge Philos. Soc.* **24,** 472–496.

Lewis, D. (1955). Sexual incompatibility. *Sci. Progr. (London)* **172,** 593–605.

Lewis, D., and Crowe, L. K. (1958). Unilateral interspecific incompatibility in flowering plants. *Heredity* **12,** 233–256.

Lewis, H. (1962). Catastrophic selection as a factor in speciation. *Evolution (Lawrence, Kans.)* **16,** 257–271.

Lewis, H. (1966). Speciation in flowering plants. *Science* **152,** 167–172.

Lewis, H. (1973). The origin of diploid neospecies in *Clarkia. Am. Nat.* **107,** 161–170.

Lindsey, A. H. (1982). Floral phenology patterns and breeding systems in *Thaspium* and *Zizia* (Apiaceae). *Syst. Bot.* **7,** 1–12.

Lloyd, D. G. (1965). Evolution of self-compatibility and racal differentiation in *Leavenworthia* (Cruciferae). *Contribs. Gray Herbarium, Harv. Univ.* **195,** 3–134.

Lloyd, D. G. (1975a). Breeding systems in *Cotula.* III. Dioecious populations. *New Phytol.* **74,** 109–123.

Lloyd, D. G. (1975b). The maintenance of gynodioecy and androdioecy in angiosperms. *Genetica* **45,** 325–339.

Lloyd, D. G. (1975c). Breeding systems in *Cotula.* IV. Reversion from dioecy to monoecy. *New Phytol.* **74,** 125–145.

Lloyd, D. G. (1976). The transmission of genes via pollen and ovules in gynodioecious angiosperms. *Theor. Popul. Biol.* **9,** 299–316.

Lloyd, D. G. (1979a). Some reproductive factors affecting the selection of self-fertilization in plants. *Am. Nat.* **113,** 67–79.

Lloyd, D. G. (1979b). Evolution towards dioecy in heterostylous populations. *Plant Syst. Evol.* **131,** 71–80.

Lloyd, D. G. (1979c). Parental strategies of angiosperms. *N. Z. J. Bot.***17,** 595–606.

Lloyd, D. G. (1980a). Sexual strategies in plants. III. A quantitative method for describing the gender of plants. *N. Z. J. Bot.* **18,** 103–108.

Lloyd, D. G. (1980b). Sexual strategies in plants. I. An hypothesis of serial adjustment of maternal investment during one reproductive season. *New Phytol.* **86,** 69–79.

Lloyd, D. G. (1980c). The distributions of gender in four angiosperm species illustrating two evolutionary pathways to dioecy. *Evolution (Lawrence, Kans.)* **34,** 123–134.

Lloyd, D. G., and Horning, D. S. (1979). Distribution of sex in *Coprosma pumila* on Macquarie Island, Australia. *N. Z. J. Bot.* **17,** 5–7.

Lloyd, D. G., and Myall, A. J. (1976). Sexual dimorphism in *Cirsium arvense* (L.) Scop. *Ann. Bot. (London)* **40,** 115–123.

Lloyd, D. G., and Yates, J. M. A. (1982). Intrasexual selection and the segregation of pollen and stigmas in hermaphrodite plants, exemplified by *Wahlenbergia albomarginata* (Campanulaceae). *Evolution (Lawrence, Kans.)* **36,** 903–913.

McCormick, J. F., Bozeman, J. R., and Spongberg, S. (1971). A taxonomic revision of granite outcrop species of *Minuartia* (*Arenaria*). *Brittonia* **23,** 149–160.

Manning, A. (1956). Some aspects of the foraging behaviour of bumble-bees. *Behaviour* **9,** 164–201.

Mather, K. (1943). Polygenic inheritance and natural selection. *Biol. Rev. Cambridge Philos. Soc.* **18,** 32–64.

Mather, K. (1953). The genetical structure of populations. *Symp. Soc. Exp. Biol.* **7,** 66–95.

Mather, K. (1973). "The Genetical Structure of Populations." Chapman and Hall, London.

Mather, K., and DeWinton, D. (1941). Adaptation and counter-adaptation of the breeding system in *Primula. Ann. Bot. (London)* **5,** 297–311.

Meehan, T. (1868). *Mitchella repens* L., a dioecious plant. *Proc. Acad. Nat. Sci. Philadelphia 1868* pp. 183–184.

Meeuse, B. J. D., and Schneider, E. L. (1980). *Nymphaea* revisited: A preliminary communication. *Isr. J. Bot.* **28,** 65–79.

Moore, D. M., and Lewis, H. (1965). The evolution of self-pollination in *Clarkia xantiana. Evolution (Lawrence, Kans.)* **19,** 104–114.

Moran, G. F., and Brown, A. H. D. (1980). Temporal heterogeneity in outcrossing rates in alpine ash (*Eucalyptus delegatensis* R. T. Bak.). *Theor. Appl. Genet.* **57,** 101–105.

Mulcahy, D. L. (1974). Adaptive significance of gamete competition. *In* "Fertilization in Higher Plants" (H. F. Linskens, ed.), pp. 27–30. North-Holland Publ., Amsterdam.

Mulcahy, D. L. (1979). The rise of the angiosperms: A genecological factor. *Science* **206,** 20–23.

Müller, H. (1883). "The Fertilisation of Flowers." Macmillan, New York.

Nettancourt, D. de. (1977). "Incompatibility in Angiosperms." Springer-Verlag, Berlin and New York.

Olesen, J. M. (1979). Floral morphology and pollen flow in the heterostylous species *Pulmonaria obscura* Dumort. (Boraginaceae). *New Phytol.* **82,** 757–767.

Opler, P. A., Baker, H. G., and Frankie, G. W. (1975). Reproductive biology of some Costa Rican *Cordia* species (Boraginaceae). *Biotropica* **7,** 234–247.

Ornduff, R. (1966). The origin of dioecism from heterostyly in *Nymphoides* (Menyanthaceae). *Evolution* **20,** 309–314.

Ornduff, R. (1969). Reproductive biology in relation to systematics. *Taxon* **18,** 121–133.

Ornduff, R. (1970). Cytogeography of Nymphoides (Menyanthaceae). *Taxon* **19,** 715–719.
Ornduff, R. (1972). The breakdown of trimorphic incompatibility in *Oxalis* Section *Corniculatae. Evolution (Lawrence, Kans.)* **26,** 52–65.
Ornduff, R. (1975). Heterostyly and pollen flow in *Hypericum aegypticum* (Guttiferae). *Bot. J. Linn. Soc.* **71,** 51–57.
Ornduff, R. (1976). The reproductive system of *Amsinckia grandiflora,* a distylous species. *Syst. Bot.* **1,** 57–66.
Ornduff, R. (1979). The genetics of heterostyly in *Hypericum aegypticum. Heredity* **42,** 271–272.
Pandey, K. K. (1958). Time of S-allele action. *Nature (London)* **181,** 1220–1221.
Pandey, K. K. (1960). Evolution of gametophytic and sporophytic systems of self-incompatibility in angiosperms. *Evolution (Lawrence, Kans.)* **14,** 98–115.
Pandey, K. K. (1970). Elements of the S-gene complex. VI. Mutations of the self-incompatibility gene, pseudo-compatibility and origin of new self-incompatibility alleles. *Genetica* **41,** 477–516.
Percival, M. S. (1965). "Floral Biology." Pergamon, Oxford.
Percival, M. S., and Morgan, P. (1965). Observations on the floral biology of *Digitalis* species. *New Phytol.* **64,** 1–22.
Phillips, M. A., and Brown, A. H. D. (1977). Mating system and hybridity in *Eucalyptus pauciflora. Aust. J. Biol. Sci.* **30,** 337–344.
Pratap Reddy, N., and Bir Bahadur (1976). Heterostyly in *Nymphoides indica* (L.) Griseb. *J. Indian Bot. Soc.* **53,** 133–140.
Price, M. V., and Waser, N. M. (1979). Pollen dispersal and optimal outcrossing in *Delphinium nelsoni. Nature (London)* **277,** 294–296.
Primack, R. B., and Lloyd, D. G. (1980). Sexual strategies in plants. IV. The distributions of gender in two monomorphic shrub populations. *N. Z. J. Bot.* **18,** 109–114.
Primack, R. B., and Silander, J. A. (1975). Measuring the relative importance of different pollinators to plants. *Nature (London)* **255,** 143–144.
Proctor, M. C. F., and Yeo, P. F. (1973). "The Pollination of Flowers." Collins, London.
Pyke, G. H. (1978). Optimal foraging in bumblebees and coevolution with their plants. *Oecologia* **36,** 281–293.
Pyke, G. H., and Waser, N. M. (1981). On the production of dilute nectars by hummingbird and honeyeater flowers. *Biotropica* **13,** 260–270.
Radford, A. E., Ahles, H. E., and Bell, C. R. (1968). "Manual of the Vascular Flora of the Carolinas." Univ. of North Carolina Press, Chapel Hill.
Radford, A. E., Dickison, W. C., Massey, J. R., and Bell, C. R. (1974). "Vascular Plant Systematics." Harper and Row, New York.
Raven, P. H. (1964). Catastrophic selection and edaphic endemism. *Evolution (Lawrence, Kans.)* **18,** 336–338.
Raven, P. H. (1972). Why are bird-visited flowers predominantly red? *Evolution (Lawrence, Kans.)* **26,** 674.
Reincke, D. C., and Bloom, W. L. (1979). Pollen dispersal in natural populations: a method for tracking individual pollen grains. *Syst. Bot.* **4,** 223–229.
Rick, C. M. (1950). Pollination relations of *Lycopersicon esculentum* in native and foreign regions. *Evolution (Lawrence, Kans.)* **4,** 110–122.
Rick, C. M., and Fobes, J. F. (1975). Allozyme variation in the cultivated tomato and closely related species. *Bull. Torrey Bot. Club* **102,** 376–384.
Rick, C. M., Fobes, J. F., and Holle, M. (1977). Genetic variation in *Lycopersicon pimpinellifolium:* evidence of evolutionary change in mating systems. *Plant Syst. Evol.* **127,** 139–170.
Rick, C. M., Holle, M., and Thorp, R. W. (1978). Rates of cross-pollination in *Lycopersicon*

pimpinellifolium: Impact of genetic variation in floral characters. *Plant Syst. Evol.* **129,** 31–44.

Ritland, K., and Jain, S. K. (1981). A model for the estimation of outcrossing rate and gene frequencies using *n* independent loci. *Heredity* **47,** 35–52.

Rollins, R. C. (1963a). The evolution and systematics of *Leavenworthia* (Cruciferae). *Contribs. Gray Herbarium, Harv. Univ.* **192,** 3–198.

Rollins, R. C. (1963b). Protandry in two species of *Streptanthus* (Cruciferae). *Rhodora* **65,** 45–49.

Ronald, W. G., and Ascher, D. D. (1975). Self-compatibility in garden *Chrysanthemum* L.: Occurrence, inheritance, and breeding potential. *Theor. Appl. Genet.* **46,** 45–54.

Rosov, S. A., and Screbtsova, N. D. (1958). Honeybees and selective fertilization of plants. *XVII Int. Beekeeping Congr.* **2,** 494–501.

Ross, M. D. (1978). The evolution of gynodioecy and subdioecy. *Evolution* (*Lawrence, Kans.*) **32,** 174–188.

Sanders, T. B., and Hamrick, J. L. (1980). Variation in the breeding system of *Elymus canadensis. Evolution* (*Lawrence, Kans.*) **34,** 117–122.

Schaal, B. A. (1980). Measurement of gene flow in *Lupinus texensis. Nature* (*London*) **284,** 450–451.

Schlising, R. A., and Turpin, R. A. (1971). Hummingbird dispersal of *Delphinium cardinale* pollen treated with radioactive iodine. *Am. J. Bot.* **58,** 401–406.

Schoen, D. J. (1982a). The breeding system of *Gilia achilleifolia:* Variation in floral characteristics and outcrossing rate. *Evolution* (*Lawrence, Kans.*) **36,** 352–360.

Schoen, D. J. (1982b). Genetic variation and the breeding system of *Gilia achilleifolia. Evolution* (*Lawrence, Kans.*) **36,** 361–370.

Shaw, D. V., and Brown, A. H. D. (1982). Optimum number of marker loci for estimating outcrossing in plant populations. *Theor. Appl. Genet.* **61,** 321–325.

Shaw, D. V., Kahler, A. L., and Allard, R. W. (1981). A multilocus estimator of mating system parameters in plant populations. *Proc. Natl. Acad. Sci. U.S.A.* **78,** 1298–1302.

Simpson, D. M., and Duncan, E. N. (1956). Cotton pollen dispersal by insects. *Agron. J.* **48,** 305–308.

Sindu, A. S., and Singh, S. (1961). Studies on the agents of cross-pollination of cotton. *Indian Cotton Grow. Rev.* **15,** 341–353.

Sohmer, S. (1977). *Psychotria* L. (Rubiaceae) in the Hawaiian Islands. *Lyonia* **1,** 103–186.

Sokal, R. R., and Rohlf, F. J. (1969). "Biometry." Freeman, San Francisco, California.

Solbrig, O. T. (1972). Breeding system and genetic variation in *Leavenworthia. Evolution* (*Lawrence, Kans.*) **26,** 155–160.

Solbrig, O. T., and Rollins, R. L. (1977). The evolution of autogamy in species of the mustard genus *Leavenworthia. Evolution* (*Lawrence, Kans.*) **31,** 265–281.

Sprengel, C. K. (1793). "Das Entdecke Geheimniss der Natur in Bau und in der Befruchtung der Blumen." Berlin.

Stebbins, G. L. (1950). "Variation and Evolution in Plants." Columbia Univ. Press, New York.

Stebbins, G. L. (1957). Self-fertilization and population variability in the higher plants. *Am. Nat.* **41,** 337–354.

Stebbins, G. L. (1958). Longevity, habitat, and release of genetic variability in higher plants. *Cold Spring Harbor Symp. Quant. Biol.* **23,** 365–378.

Stebbins, G. L. (1970). Adaptive radiation in angiosperms. I. Pollination mechanisms. *Annu. Rev. Ecol. Syst.* **1,** 307–326.

Stebbins, G. L. (1974). "Flowering Plants: Evolution Above the Species Level." Belknap Press, Cambridge, Massachusetts.

Stelleman, P. (1978). The possible role of insect visits in pollination of reputedly anemophilous plants, exemplified by *Plantago lanceolata,* and syrphid flies. *In* "The Pollination of Flowers by Insects" (A. J. Richards, ed.), pp. 41–46. Academic Press, New York.

Stephens, S. G., and Finkner, M. D. (1953). Natural crossing in cotton. *Agron. J.* **48,** 257–269.
Stockhouse, R. E. (1976). A new method for studying pollen dispersal using micronized fluorescent dusts. *Am. Midl. Nat.* **96,** 241–245.
Stone, D. E. (1959). A unique balanced breeding system in the vernal pool mouse-tails. *Evolution (Lawrence, Kans.)* **13,** 151–174.
Stout, A. B. (1924). The flower mechanisms of avocados with reference to pollination and production of fruit. *J. N. Y. Bot. Gard.* **25,** 1–7.
Thies, S. A. (1953). Agents concerned with natural crossing of corn. *Agron. J.* **45,** 481–484.
Thomson, G. M. (1881). On the fertilisation, etc. of New Zealand flowering plants. *Trans. N. Z. Inst.* **13,** 241–291.
Thomson, J. D., and Barrett, S. C. H. (1981a). Selection for outcrossing, sexual selection, and the evolution of dioecy in plants. *Am. Nat.* **118,** 443–449.
Thomson, J. D., and Barrett, S. C. H. (1981b). Temporal variation of gender in *Aralia spinosa* Vent. (Araliaceae). *Evolution (Lawrence, Kans.)* **35,** 1094–1107.
Thomson, J. D., and Plowright, R. C. (1980). Pollen carryover, nectar rewards, and pollinator behavior with special reference to *Diervilla lonicera. Oecologia* **46,** 68–74.
Uphof, J. C. T. (1938). Cleistogamous flowers. *Bot. Rev.* **4,** 21–49.
Vasek, F. C. (1964). The evolution of *Clarkia unguiculata* derivatives adapted to relatively xeric environments. *Evolution (Lawrence, Kans.)* **18,** 26–42.
Vasek, F. C. (1965). Outcrossing in natural populations. II. *Clarkia unguiculata. Evolution (Lawrence, Kans.)* **19,** 152–156.
Vasek, F. C. (1967). Outcrossing in natural populations. III. The Deer Creek population of *Clarkia exilis. Evolution (Lawrence, Kans.)* **21,** 241–248.
Vasek, F. C. (1968). The relationships of two ecologically marginal, sympatric *Clarkia* populations. *Am. Nat.* **102,** 25–40.
Vasek, F. C. (1971). Variation in marginal populations of *Clarkia. Ecology* **52,** 1046–1051.
Vogel, S. (1955). Über den Blutendimorphismus einiger sudafrikanischer Pflanzen. *Osterr. Bot. Z.* **102,** 25–40.
Vuilleumier, B. S. (1967). The origin and evolutionary development of heterostyly in the angiosperms. *Evolution (Lawrence, Kans.)* **21,** 210–226.
Waddington, K. D. (1979). Divergence in inflorescence height: An evolutionary response to pollinator fidelity. *Oecologia* **40,** 43–50.
Waddington, K. D. (1981). Factors influencing pollen flow in bumblebee-pollinated *Delphinium virescens.* Oikos **37,** 153–159.
Waddington, K. D., and Heinrich, B. (1979). The foraging movements of bumblebees on vertical "inflorescences": An experimental analysis. *J. Comp. Physiol.* **134,** 113–117.
Waser, N. M., and Price, M. V. (1982a). A comparison of pollen fluorescent dye carry-over by natural pollinators of *Ipomopsis aggregata* (Polemoniaceae). *Ecology* **63,** 1168–1172.
Waser, N. M., and Price, M. V. (1982b). Optimal and actual outcrossing in plants, and the nature of plant-pollinator interaction. *In* "Handbook of Experimental Pollination Ecology" (C. E. Jones and R. J. Little, eds.), Van Nostrand-Reinhold, New York.
Weaver, R. E. (1970). The Arenarias of the southeastern granitic flat-rocks. *Bull. Torrey Bot. Club* **97,** 40–52.
Webb, C. J. (1979). Breeding systems and the evolution of dioecy in New Zealand apioid Umbelliferae. *Evolution (Lawrence, Kans.)* **33,** 662–672.
Webb, C. J. (1981). Andromonoecism, protandry, and sexual selection in Umbelliferae. *N. Z. J. Bot.* **19,** 335–338.
Weberling, F. (1965). Typology of inflorescences. *J. Linn. Soc. London, Bot.* **59,** 215–221.
Wells, H. (1979). Self-fertilization: advantageous or deleterious? *Evolution (Lawrence, Kans.)* **33,** 252–255.

Whitehouse, H. L. K. (1950). Multiple-allelomorph incompatibility of pollen and style in the evolution of the angiosperms. *Ann. Bot. (London)* **14,** 199–216.
Whitehouse, H. L. K. (1959). Cross- and self-fertilization in plants. *In* ''Darwin's Biological Work'' (P. R. Bell, ed.), pp. 207–261. Cambridge Univ. Press, London and New York.
Willson, M. F. (1979). Sexual selection in plants. *Am. Nat.* **113,** 777–790.
Willson, M. F. (1982). Sexual selection and dicliny in angiosperms. *Am. Nat.* **119,** 579–583.
Wyatt, R. (1977). *Arenaria alabamensis:* A new combination for a granite outcrop endemic from North Carolina and Alabama. *Bull. Torrey Bot. Club* **104,** 243–244.
Wyatt, R. (1981). The reproductive biology of *Asclepias tuberosa:* II. Factors determining fruit-set. *New Phytol.* **86,** 375–385.
Wyatt, R. (1982). Inflorescence architecture: How flower number, arrangement, and phenology affect pollination and fruit-set. *Am. J. Bot.* **69,** 585–594.
Wyatt, R., and Fowler, N. (1977). The vascular flora and vegetation of the North Carolina granite outcrops. *Bull. Torrey Bot. Club* **104,** 245–253.
Wyatt, R., and Hellwig, R. L. (1979). Factors determining fruit-set in heterostylous bluets, *Houstonia caerulea* (Rubiaceae). *Syst. Bot.* **4,** 103–114.
Yampolsky, C., and Yampolsky, H. (1922). Distribution of sex forms in the phanerogamic flora. *Bibl. Genet.* **3,** 1–62.
Yeo, P. F. (1975). Some aspects of heterostyly. *New Phytol.* **75,** 147–153.
Zavala, M. E. (1978). Threeness and fourness in pollen of *Lythrum junceum*. *Am. Bot. Soc., Misc. Ser., Publ.* **156,** 63.

CHAPTER 5

Wind Pollination: Some Ecological and Evolutionary Perspectives

DONALD R. WHITEHEAD
Department of Biology
Indiana University
Bloomington, Indiana

I. Introduction

When one views the range of adaptations for wind pollination and the variety of conditions under which it occurs, it becomes clear that plants are, as Henry Horn (1971) has suggested, "crafty green strategists." More specifically, they appear to be especially astute environmental biophysicists. The multitude of adaptations developed to facilitate wind pollination (number of pollen grains, pollen size and sculptural characteristics, flower structure and location, timing of flowering and pollen release) are predictable from a basic understanding of the aerodynamics of particle transport and impaction on objects in the environment.

In this chapter, I will review (1) the factors controlling pollen release, transport, and capture; (2) the adaptations that maximize the efficiency of wind pollination; (3) the environmental conditions most suitable for effective wind

POLLINATION BIOLOGY

ISBN 0-12-583980-4 (hardcover)
ISBN 0-12-583982-0 (paperback)

pollination; (4) the ''ecogeography'' of wind pollination (well reviewed recently by Regal, 1982); and (5) some evolutionary aspects of wind pollination. Not surprisingly, this chapter will appear to be similar to a previous paper (Whitehead, 1969). I can justify this similarity by arguing that the relevant laws of physics have not changed much in 14 years.

II. General Considerations

In contrast to biotic pollination, which is directional and dependent on the behavior of an animal vector, wind pollination is a relatively passive process. Pollen release, transport, and capture are controlled primarily by microclimatic factors, the most important of which is wind. Thus, characteristics of the physical environment exercise considerable control over the effectiveness of wind pollination.

Wind pollination is most likely to be successful if certain idealized conditions are met. These include (1) the production of large numbers of pollen grains; (2) pollen grains with appropriate aerodynamic characteristics; (3) flower and inflorescence structure and location on the plant designed to maximize the probability of pollen's entrainment in moving air; (4) stigmatic surfaces structured and positioned to maximize collection efficiency; (5) pollen release timed within both the season and the day to maximize the possibility of pollen capture by receptive conspecifics downwind; (6) relatively close spacing of compatible plants; (7) vegetational structure that is relatively open to minimize filtration of pollen by nonstigmatic surfaces; (8) wind velocity within an acceptable range to ensure transport and minimize downwind dispersion; (9) relatively low humidity and a low probability of rainfall; and (10) unambiguous environmental cues to coordinate flowering.

These broad generalizations can be derived intuitively from a consideration of the factors controlling particle (pollen) transport and impaction.

III. Aerodynamic Considerations

At the simplest level, pollen transport involves an interaction between two variables: the settling (terminal) velocity of the pollen and wind velocity. Pollen terminal velocities range from 2 to 6 cm/sec, whereas average wind velocities within most plant communities range from 1 to 10 m/sec (Tauber, 1965; Geiger, 1966). In general, pollen dispersal will increase as wind velocity increases and as settling velocity decreases. However, there are some significant exceptions which will be discussed below.

The steady-state settling speed for spherical objects in the size range of pollen of wind-pollinated taxa (20–40 μm) can be derived from Stokes' law (Harrington, 1979) and is proportional to pollen size (diameter) and density (mass/volume). Pollen density will depend upon such variables as exine thickness and structure and water content of the pollen grain. Although the grains of many wind-pollinated plants are characterized by relatively small size and a thin exine with psilate (smooth) sculpture (Faegri and van der Pijl, 1979; Faegri and Iversen, 1975; Whitehead, 1969), many taxa have larger grains and/or a more complex exine structure, often with elaborate air spaces and features such as bladders. The latter effectively lower pollen density and thus facilitate transport in air.

The density of grains also appears to be influenced by humidity. For example, the density of *Ambrosia* kept at high humidity rises significantly, apparently because the protoplast expands to fill or occlude the air spaces within the exine. Ragweed pollen grains in unopened anthers are completely saturated and contain no air spaces, yet within seconds of dehiscence, all of the excess water can be evaporated, the density lowered, and the grains suspended in air (Harrington, 1979; Harrington and Metzger, 1963). I suspect that these phenomena can be generalized to most wind-pollinated plants.

These considerations suggest that small, light grains will disperse readily and that dispersal distances will be greater in situations characterized by higher wind velocities. Although this is undoubtedly true, it does not mean that wind pollination is most effective when these conditions are fulfilled. Pollen grains that are very small and light will be carried readily by moving air, but objects in their flight path (such as stigmatic surfaces) will have greater difficulty in capturing them. In addition, higher wind velocities are often accompanied by high turbulence, and the atmospheric pollen concentration decreases downwind from an emitting source much more rapidly under such circumstances. In short, many additional variables influence the effectiveness of wind pollination.

Clearly, if wind pollination is to be effective, the number of pollen grains reaching receptive plants from pollinating individuals upwind must be high. This will be true if the strength (number of grains released per unit time) of the emitting source is high, if the plants are closely spaced, if there are few obstacles (e.g., vegetation) between the plants, and if certain atmospheric conditions (temperature and wind profiles) prevail.

Both theoretical considerations and experimental data demonstrate that pollen concentration downwind from an elevated source drops rapidly at a rate that is inversely proportional to the square of the distance between them (Tauber, 1965; Colwell, 1951; Harrington, 1979; Raynor *et al.*, 1970, 1974; Player, 1979). The concentration gradient is sharply influenced by the turbulent intensity of the atmosphere, which is controlled by surface roughness, by wind shear, and especially by the vertical temperature profile (Mason, 1979). The ambient lapse rate influences the character and magnitude of eddies. A superadiabatic lapse rate results in a highly unstable atmosphere with high turbulence (Mason, 1979; Slade, 1968; Tauber, 1965). Under such conditions, the dispersive capacity of the atmosphere is high and the pollen concentration will decline very sharply with distance from the source. The effectiveness of wind pollination would be lower under such circumstances.

On the other hand, in atmospheres characterized by subadiabatic lapse rates or inversions, eddying is suppressed and the dispersive capacity of the atmosphere is much lower. In such situations, the pollen concentration would decrease less sharply with distance and the efficiency of wind pollination would be increased. These conditions often occur in the morning as a night-formed inversion begins to "burn off" (Mason, 1979).

Accordingly, the effectiveness of wind pollination will be greater if pollen is released when the dispersive power of the atmosphere is lower.

It is also important to consider the variables introduced by the structure of the vegetation within which pollination is occurring. It is obvious that the pollen released by one plant is not intercepted solely by receptive stigmatic surfaces of conspecifics. Trunks, branches, and leaves of *all* species in the environment will pose physical barriers to transport. A significant fraction of grains transported by air currents within a forest will be lost by impaction on various plant structures. The importance of this filtration process was first suggested by Tauber (1965, 1967) and subsequently quantified for a forest in Denmark (Tauber, 1977). In deciduous forests, leafing results in the introduction of many more surfaces which can intercept pollen, as well as a significant lowering of wind velocity.

The factors that control the impaction of pollen-sized particles are important to review, as they control not only the depletion of pollen within a forest system (by filtration and raindrop scavenging) but also the capture of pollen by stigmatic surfaces. The most important factor is collection efficiency, which is defined as the ratio of the number of particles impacting on an object to the number that would have passed through the air space had the object not been there (Tauber, 1965). Collection efficiency is proportional to wind velocity and to the density and diameter of particles and inversely proportional to the diameter of the collecting object. The trajectories of small and/or light pollen grains will diverge little from the airstream flowing around a potential collecting object. Consequently, such grains are not likely to impact. In contrast, larger and/or denser grains may have sufficiently greater forward momentum to diverge significantly from the air streamlines. Thus, the probability of impaction increases with increasing size and density. At higher wind velocities the inertia of both small and large grains will be greater, the grains will diverge more sharply from the streamlines arching around an object, and they will be more likely to be intercepted. It is especially important to emphasize that collection efficiency is inversely proportional to the diameter of the collecting object; small objects collect more efficiently than large ones. This is due to the fact that large-diameter objects have proportionately thicker boundary layers of air surrounding them. This results in a wider divergence of air streamlines around the object (and a decrease in collection efficiency). The boundary layer around both large and small objects thins somewhat at higher wind velocities, thus contributing to the increase in collection efficiency at higher wind velocities.

As an illustration of these general principles, Tauber (1965) has pointed out how the collection efficiencies for birch (21 μm) and beech (42 μm) pollen would differ under varying conditions. For example, if the collecting object has a diameter of 0.8 cm and the wind velocity is 1 m/sec, the collection efficiencies will be, respectively, 0.09 and 0.46. Thus, beech grains will be collected five times more efficiently than birch grains. If the wind velocity is increased to 2 m/sec, the collection efficiencies for both increase, but the increase is proportionately greater for birch (0.24 for birch, 0.61 for beech).

In short, small, light pollen grains are less likely to be captured than larger, denser grains; small branches (and leaves and stigmas) will have higher collection efficiencies than large ones; and the amount of filtration of pollen within the vegetation will increase with wind velocity.

Similar physical processes control the scavenging of pollen by raindrops. As anyone afflicted with hay fever realizes, rainfall is extremely effective in "washing out" particles entrained in the atmosphere. Raindrop scavenging differs from vegetational filtration in that the collecting objects (raindrops) are in motion—in fact, moving more rapidly than the particles they collect (raindrop terminal velocity is much higher than that of pollen grains). Despite this difference, the mechanisms involved in raindrop scavenging are virtually identical to those just described. Larger, dense pollen grains are captured more easily than small grains; collection efficiency is proportional to raindrop velocity and inversely proportional to droplet size. A light shower with fine droplets scavenges more efficiently than a thunderstorm with large droplets (McDonald, 1962; Harrington, 1979). It is important to emphasize that a moderate shower, regardless of droplet size, results in the virtual elimination of all pollen-sized particles in the atmosphere.

It follows from this discussion that wind pollination will be most effective if (1) compatible plants are closely spaced; (2) large quantities of pollen are released; and (3) release is timed so that (a) downwind dispersion is minimized (low air turbulence), (b) the probability of filtration within the vegetation is low (leafless season), (c) humidity is low (so that water may be lost rapidly from grains), and (d) the probability of rainfall is low.

The variety of ways in which these conditions are met can be determined by examining both plant communities in which anemophily is the dominant pollination mode and the characteristics of typical wind-pollinated plants. For example, compatible individuals are closely spaced in most temperate deciduous forest communities, boreal forests, prairies, steppes, and savannas, in most wetland habitats, and in many early successional systems. Although there is great interspecific variation in pollen output per plant, most wind-pollinated taxa release relatively large quantities. Pine, birch, alder, oak, and ragweed are classic examples of taxa which pollinate profusely. The seasonal timing of flowering is well known for many wind-pollinated taxa. Taxa in temperate deciduous forest environments tend to pollinate in early spring before leafing (aspen, birch, elm, some maples, ash) or just as leafing occurs (beech, oak, hickory, walnut) (Ogden and Lewis, 1960). For such taxa, temperature appears to be the critical factor triggering flowering (Solomon, 1979). Flowering time for wind-pollinated plants in the understory of forests or in open habitats is more variable, with some taxa flowering early in the year (e.g., alder) and others much later (e.g., ragweed). Although less is known concerning the diurnal pattern of pollen release for most taxa, ragweed has been especially well studied. Ragweed pollen release generally occurs in mid- to late morning and is correlated with high temperatures and/or low humidity (Bianchi *et al.,* 1959; Payne, 1963; Harrington and Metzger, 1963)—conditions appropriate for entrainment and effective dispersal.

Many adaptations of pollen grains, flowers, and inflorescences of wind-pollinated plants are also predictable on the basis of the aerodynamic principles just discussed. For example, small and/or light grains (e.g., chestnut, alder, aspen, birch) can be entrained in air readily and dispersed for great distances. Moreover, such grains avoid filtration by following the airstream around potential collecting objects. There is an obvious shortcoming. If small and/or light grains readily avoid impaction on twigs, branches, and leaves, they will be equally difficult for stigmatic surfaces to capture. Thus, pollen grains of wind-pollinated plants should not be too small and light, but neither should they be too large or they will not travel far and will be filtered readily by the vegetation. The pollen grains of most wind-pollinated taxa range between 20 and 40 μm, although some are smaller (chestnut) and some significantly larger (many conifers).

In addition to size and density, there are other morphological characteristics of pollen that facilitate wind pollination. Among taxa with larger grains (many conifers), there are complex air sacs signed to increase buoyancy. The sculpture of the exine is seldom ornate (usually psilate or scabrate). Wind-pollinated members of families in which most taxa are pollinated by animals generally have distinctly different pollen morphologies. The family compositae is an excellent example. The pollen grains of most composites possess a thick, complex exine and a distinctly echinate (spiny) sculpture. The spines are often large and robust. In contrast, the grains of *Ambrosia* and *Artemisia* are characterized by reduced spines and possess complex air spaces (cavea) in the exine (Wodehouse, 1935; Skvarla and Larson, 1965; McAndrews *et al.,* 1973).

Many stigmatic adaptations (to maximize collection efficiency) can be explained by aerodynamic dynamic principles as well. Faegri and van der Pijl (1979) have mentioned the need to have a large, receptive surface to ensure pollen capture. However, increasing the volume (and diameter) of stigma may actually reduce collection efficiency. Larger objects (with thicker boundary layers) have lower collection efficiencies (Tauber, 1965; Harrington, 1979). A better strategy would be to increase the stigmatic surface by subdividing it into many substructures with smaller diameters. The feathery stigmas of many members of the Gramineae are examples of this strategy. The stigmatic adaptations required will obviously depend on pollen morphology, the number of pollen grains produced, the range of wind velocities to be expected at the time of pollen release, and so on.

Lastly, as many workers have pointed out, certain floral and inflorescence morphologies would be expected and are indeed found in wind-pollinated plants. These include reduced perianth (to expose both anthers and stigmas to wind), exserted anthers, and elaborate stigmas. Pendulous, catkin-like inflorescences appear to be particularly well designed for entrainment of pollen in moving air. These are found in many wind-pollinated taxa (e.g., alder, birch, hazel, hornbeam, walnut, hickory, aspen, oak).

The location of individual flowers or inflorescences is obviously important. One would expect flowers to be located on exposed parts of the plant and as high above the ground as possible. In many circumstances, it would be beneficial to have staminate flowers or inflorescences located higher than pistillate ones. This set of adaptations is particularly evident in a variety of conifers. In both spruce and fir, staminate and ovulate cones are located near the top of the tree on the tips of exposed branches. Although there is considerable overlap, staminate cones tend to be located higher in the crown (Fowells, 1965).

IV. Geographic Aspects of Wind Pollination

The geographic distribution of wind pollination is characterized by distinctive latitudinal and altitudinal patterns. The frequency of anemophily increases with both latitude and elevation. Wind pollination is generally uncommon in tropical environments, especially in lowland rain forests (Faegri and van der Pijl, 1979; Baker, 1963), and is dominant in temperate deciduous and boreal forests (Whitehead, 1969; Regal, 1982). The latitudinal trend has been especially well documented for forests in eastern North America (Regal, 1982). The frequency of wind pollination increases with elevation in mountainous regions, both in temperate and tropical latitudes (Regal, 1982; Whitehead, 1969). The frequency of anemophily appears to be higher on remote islands (Carlquist, 1966; Whitehead, 1969; Regal, 1982). In addition, the frequency of wind pollination is generally higher in early successional systems.

As Regal (1982) has pointed out, the geographic trends are more complex in a number of extreme environments. For example, areas characterized by extreme aridity are often dominated by animal-pollinated taxa. In other areas, there are significant differences in the dominant pollination syndromes in contiguous communities. Regal (1982) points out that the tropical hammocks of southern Florida are dominated by insect-pollinated species, whereas immediately adjacent oak–pine communities are dominated by wind-pollinated taxa. A similar situation exists on the coastal plain of the Carolinas and Georgia. The characteristic shrub bogs, or pocosins, of the region (which cover 908,000 ha of the North Carolina coastal plain; Richardson *et al.*, 1981) are dominated by a variety of evergreen-leaved shrubs, the majority of which are insect pollinated (Christensen *et al.*, 1981). Pine savannas and open-structured oak and pine forests are often found immediately adjacent to pocosins (Wells, 1928). These changes in dominant pollination strategy over extremely short distances are dramatic and pose interesting questions about the ecological factors influencing reproductive systems (Regal, 1982).

There has been considerable speculation concerning the environmental variables responsible for the large-scale latitudinal and altitudinal patterns. Clearly, many biotic and abiotic variables are correlated (some positively, some negatively) with the frequency of wind pollination, and it is tempting to invoke these as explanations for the observed patterns. There is no question that certain environments are appropriate for wind pollination and others are not. Tropical rain forests and the deciduous forests of eastern North America are cases in point. In tropical rain forests, a number of factors appear to influence the lower frequency of anemophily; these include (1) high species diversity (and consequent wide spacing of conspecifics); (2) absence of a leafless season (hence, a continually high pollen filtration capacity within the forest); (3) high humidity and high rainfall probability (resulting in daily washout of the pollen entrained in the air and also that on surfaces within the forest which could be resuspended and transported further); (4) the general absence of unambiguous stimuli (e.g., day length changes) which could coordinate flowering; and (5) the abundance of potential animal vectors.

Although I feel that these conditions are not conducive to effective wind pollination, they certainly do not preclude it. In fact, it is now evident that at least a few understory trees in rain forest environments are wind pollinated. For example, Bawa and Crisp (1980) have demonstrated wind pollination in *Trophis involucrata* (Moraceae) in a lowland rain forest in Costa Rica. Not surprisingly, the individuals of these species are clumped (average distance between staminate and pistillate trees is 6.6 ± 3.3 m). In addition, the pollen grains are very small and psilate (hence, the grains are well suited for dispersal within a dense forest), the peak pollen dispersal occurs between 0930 and 1130 when the probability of rain is low, and the plants flower during the season having lowest probability of rainfall.

Trophis is not unique among rain forest trees, as anemophily has been suggested for a few other taxa, many of which are in the Moraceae (Corner, 1952; Zapata and Arroyo, 1978; Bawa and Crisp, 1980). It is interesting that pollen of a variety of "moraceous" taxa occur in the late glacial and Holocene sediments of Lake Valencia in Venezuela (Bradbury *et al.*, 1981; Leyden, 1981). This suggests that a number of these taxa may also be wind pollinated.

Temperate forests present a sharp contrast. Species diversity is lower, conspecifics are much more closely spaced (clumped distributions are common), there is a distinct leafless season, flowering of many taxa occurs before leafing, there are several reliable environmental cues that can coordinate flowering, the probability of rainfall is lower, and there are fewer potential pollinators. In short, conditions are far more appropriate for wind pollination.

Although the sharp ecological contrasts between tropical rain forests and temperate deciduous forests may not be adequate to explain the differences in dominant pollination strategies (for example, see the criticisms in Daubenmire, 1972, it is difficult for me to imagine that the correlations are merely coincidental. The combination of biological (vegetational structure, spacing of individuals, availability of pollinators) and physical factors (seasonality, humidity, rainfall probability) clearly facilitates wind pollination in one environment and hinders it in the other. However, many caveats are clearly necessary, as tree communities that appear to be comparable in terms of the frequency of wind-pollinated taxa (or individuals) may be so for quite different reasons. As Regal (1982) has pointed out, clumping of individuals (in turn related to species diversity—and species diversity to climatic factors, among other things) may well be among the most important factors facilitating wind pollination in many environments. Climatic patterns may dictate the trends in other communities.

As many individuals have pointed out, simple mechanistic explanations for the various patterns will seldom be adequate (Daubenmire, 1972; Regal, 1982). The mere fact that a community is open in structure and wind velocities are high does not mean that wind pollination will be favored; arid environments are cases in point (Regal, 1982). High precipitation alone is not sufficient to preclude wind pollination; pollen washout can obviously be compensated for in a number of ways (close spacing of individuals, timing of pollen release). For example, the conifer-cominated boreal forests that occur in many mountainous regions of North America clearly experience high precipitation, yet are dominated by wind-pollinated taxa. Here a combination of factors contributes to the high frequency of anemophily. These include (1) the close spacing of compatible individuals; (2) open canopy structure (many taxa with slender pyramidal crowns result in a "canopy" permitting wind penetration); and (3) positioning of inflorescences in exposed positions near the top of each tree. Furthermore, the temporal pattern of rainfall is often more important that the amount. There are obvious daily patterns in tropical rain forests which provide understory plants with "time windows" during which wind pollination can occur. Seasonal patterns are also critical. For example, the precipitation regime experienced by the temperate rain forests in Washington is strongly seasonal, with most of the precipitation falling in the winter months (Daubenmire, 1969). Thus, pollination can be coordinated with the dry season; this and the close spacing of conspecifics permit wind pollination to be effective in this high-rainfall regime.

Regal (1982) discusses the physical explanations for the geographic patterns but suggests that dispersion, environmental uncertainty, and seasonality are more important (and testable as hypotheses). Of these, I have already emphasized the importance of close spacing of individuals and some aspects of seasonality. Regal also emphasizes the advantage conferred upon early flowering plants in environments characterized by short growing seasons. Wind-pollinated taxa would have an advantage because they would not be dependent on an unreliable resource base (availability of pollinating insects).

Regal (1982) suggests that wind-pollinated species would be selected against in areas characterized by climatic uncertainty (unpredictability), such as extremely arid regions. The outcrossing necessary for species to survive in such extreme environments would be increased by animal pollination and decreased by anemophily.

V. Summary

In summary, wind pollination is common in many environments and extremely rare in others. The ecogeographic patterns are not random. Although different combinations of factors may contribute to the frequency of anemophily in each community type, certain conditions appear to be most significant. Wind pollination appears to be favored in vegetational types in which (1) conspecific individuals are closely spaced; (2) the probability of pollen filtration within the vegetation is low (there may be significant seasonal changes in filtration capacity—for example, in deciduous forests); (3) sharp seasonality combined with short growing seasons favors early flowering; (4) the frequency of rainfall is low and/or there are daily or seasonal time windows when the probability of rainfall is low; and (5) there are unambiguous environmental cues to coordinate flowering. Thus, an interaction of biotic and abiotic factors controls the frequency of anemophily. The rarity of wind pollination in extremely arid environments may be a function of climatic uncertainty where outcrossing systems favored by animal vectors would convey a selective advantage.

The biological adaptations and the environmental conditions that facilitate wind pollination can often be predicted from an understanding of the processes which control the transport of pollen-sized particles in the atmosphere and the impaction of particles on objects in the environment.

Acknowledgments

This chapter was written while I was on a sabbatical leave at the University of Maine in Orono. I am indebted to the Department of Botany and Plant Pathology and the Institute for Quaternary Studies for providing space and facilities for the year. I have profited greatly from informal discussions with Ronald B. Davis and George L. Jacobson. Stephen T. Jackson provided an invaluable critique of the manuscript. I am grateful to Leslie A. Real for the invitation to write this chapter and for his patience in dealing with a delinquent author.

References

Baker, H. G. (1963). Evolutionary mechanisms in pollination biology. *Science* **139,** 877–883.

Bawa, K. S., and Crisp, J. E. (1980). Wind-pollination in the understory of a rain forest in Costa Rica. *J. Ecol.* **68,** 871–876.

Bianchi, D. E., Schwemmin, J., and Wagner, W. H. (1959). Pollen release in common ragweed (*Ambrosia artemisiifolia*). *Bot. Gaz.* (*Chicago*) **4,** 235–243.

Bradbury, J. P., Leyden, B., Salgado-Labouriau, M., Lewis, Jr., M. W., Schubert, C., Binford, M. W., Frey, D. G., Whitehead, D. R., and Weibezahn, F. H. (1981). Late Quaternary environmental history of Lake Valencia, Venezuela. *Science* **214,** 1299–1305.

Carlquist, S. (1966). The biota of long-distance dispersal. IV. Genetic systems in the floras of oceanic islands. *Evolution* (*Lawrence, Kans.*) **20,** 433–455.

Christensen, H. L., Burchell, R. B., Liggett, A., and Simms, E. L. The structure and development of Pocosin vegetation. *In* "Pocosin Wetlands" (C. J. Richardson, ed.), pp. 43–61. Hutchinson Ross Pulb., Stroudsburg, Pennsylvania.

Colwell, R. N. (1951). The use of radioactive isotopes in determing spore distribution patterns. *Am. J. Bot.* **38,** 511–523.

Corner, E. J. H. (1952). "Wayside Trees of Malaysia," Vol. I. Government Printing Office, Singapore.

Daubenmire, R. (1969). Ecologic plant geography of the Pacific Northwest. *Madrono* **20,** 111–128.

Daubenmire, R. (1972). Phenological and other characteristics of tropical semi-deciduous forest in northwestern Costa Rica. *J. Ecol.* **60,** 147–170.

Faegri, K., and Iversen, J. (1975). "Textbook of Pollen Analysis," 3rd ed. Hafner, New York.

Faegri, K., and van der Pijl, L. (1979). "The Principles of Pollination Ecology," 3rd ed. Pergamon, Oxford.

Fowells, H. A. (1965). Silvics of forest trees of the United States. "Agriculture Handbook," No. 271. U.S. Department of Agriculture, Washington, D.C.

Geiger, R. (1966). "The Climate Near the Ground." Harvard Univ. Press, Cambridge, Massachusetts.

Harrington, J. B. (1979). Principles of deposition of microbiological particles. *In* "Aerobiology: The Ecological Systems Approach" (R. L. Edmonds, ed.), pp. 111–137. Dowden, Hutchinson, and Ross, Stroudsburg, Pennsylvania.

Harrington, J. B., and Metzger, K. (1963). Ragweed pollen density. *Am. J. Bot.* **50,** 532–539.

Horn, H. S. (1971). "The Adaptive Geometry of Trees." Princeton Univ. Press, Princeton, New Jersey.

Leyden, B. W. (1982). Late-Quaternary and Holocene history of the Lake Valencia basin, Venezuela. Ph.D. Dissertation, Indiana Univ., Bloomington.

McAndrews, J. H., Berti, A. A., and Norris, G. (1973). Key to the Quaternary pollen and spores of the Great Lakes region. Life Sciences Miscellaneous Publication, Royal Ontario Museum.

McDonald, J. E. (1962). Collection and washout of airborne pollens and spores by raindrops. *Science* **135,** 435–437.

Mason, C. J. (1979). Principles of atmospheric transport. *In* "Aerobiology: The Ecological Systems Approach" (R. L. Edmonds, ed.), pp. 85–95. Dowden, Hutchinson, and Ross, Stroudsburg, Pennsylvania.

Ogden, E. C., and Lewis, D. M. (1960). Airborne pollen and fungus spores of New York state. *N.Y. State Mus., Bull.* No. 378.

Payne, W. W. (1963). The morphology of the inflorescence of ragweeds (*Ambrosia-Franseria:* compositae). *Am. J. Bot.* **50,** 872–880.

Player, G. (1979). Pollination and wind dispersal of pollen in *Arceuthobium. Ecol. Monogr.* **49,** 73–87.

Raynor, G. S., Hayes, J. V., and Ogden, E. C. (1970). Experimental data on ragweed pollen dispersion and deposition from point and area sources. Report BNL 50224 (T-564), Brookhaven Natl. Lab., Upton, New York.

Raynor, C. S., Hayes, J. V., and Ogden, E. C. (1974). Mesoscale transport and dispersion of airborne pollens. *J. Appl. Meteorol.* **13,** 87–95.

Regal, P. J. (1982). Pollination by wind and animals: Ecology of geographic patterns. *Annu. Rev. Ecol. Syst.* **13,** 497–524.

Richardson, C. J., Evans, R., and Carr, D. (1981). Pocosins: An ecosystem in transition. *In* "Pocosins Wetlands" (C. J. Richardson, ed.), pp. 3–19. Hutchinson Ross Pulb., Stroudsburg, Pennsylvania.

Skvarla, J. J., and Larson, D. A. (1965). An electron microscopic study of pollen morphology in the compositae with special reference to the Ambrosiinae. *Grana Palynol.* **6,** 210–269.

Slade, D. H., ed. (1968). Meteorology and atomic energy. USAEC 56–61, Division of Technical Information NO. TID-24190.

Solomon, A. M. (1979). Sources and characteristics of airborne materials. Pollen. *In* "Aerobiology: The Ecological Systems Approach" (R. L. Edmonds, ed.), pp. 41–54. Dowden, Hutchinson, and Ross, Stroudsburg, Pennsylvania.

Tauber, H. (1965). Differential pollen dispersion and the interpretation of pollen diagrams. *Dan. Geol. Unders.,* [Afh.], Racke 2 NO. 89, pp. 1–70.

Tauber, H. (1967). Differential pollen dispersing and filtration. *In* "Quaternary Paleoecology" (E. J. Cushing and H. E. Wright, eds.), pp. 131–141. Yale Univ. Press, New Haven, Connecticut.

Tauber, H. (1977). Investigations of aerial pollen transport in a forested area. *Dan. Bot. Ark.* **32,** 1–121.

Wells, B. W. (1928). Plant communities of the coastal plain of North Carolina and their successional relations. *Ecology* **9,** 230–242.

Whitehead, D. R. (1969). Wind pollination in the Angiosperms: Evolutionary and environmental considerations. *Evolution* (*Lawrence, Kans.*) **23,** 28–35.

Wodehouse, R. P. (1935). "Pollen Grains." McGraw-Hill, New York.

Zapata, T. R., and Arroyo, M. T. K. (1978). Plant reproductive ecology of a secondary deciduous tropical forest in Venezuela. *Biotropica* **10,** 221–230.

CHAPTER 6

Male Competition, Female Choice, and Sexual Selection in Plants

ANDREW G. STEPHENSON

Department of Biology
Pennsylvania State University
University Park, Pennsylvania

ROBERT I. BERTIN

Department of Zoology
Miami University
Oxford, Ohio

POLLINATION BIOLOGY

ISBN 0-12-583980-4 (hardcover)
ISBN 0-12-583982-0 (paperback)

I. Introduction

Darwin suggested that in many animal species certain adults have greater reproductive access to members of the opposite sex. This advantage that certain individuals have in obtaining mates and producing offspring is termed "sexual selection." Two factors play an important role in sexual selection: (1) competition among members of one sex for reproductive access to the other sex and (2) preference by members of one sex for certain members of the other sex (Huxley, 1938). Because female gametes are larger (and more costly) than male gametes, and because the female function generally demands more resources per offspring than the male function, males typically compete for females and females typically exercise a choice of mates (Bateman, 1948; Trivers, 1972).

In the past century, numerous aspects of animal reproductive morphology and behavior have been ascribed to the effects of sexual selection, especially in species with polygynous or promiscuous mating systems, in which the opportunities for male competition and female choice are great (Darwin, 1871; Fisher, 1958; Trivers, 1972; Halliday, 1978; Borgia, 1979; and references therein). In fact, Darwin introduced the concept of sexual selection to account for the evolution of behavioral and morphological differences between the sexes because many of these differences seem to be handicaps to survival. More recently, zoologists have begun to attribute a much wider range of reproductive characteristics to sexual selection, including, for example, internal fertilization in insects, use of sperm plugs, prolonged copulation, and pre- and postcopulatory guarding behavior (Parker, 1970; Halliday, 1978).

Several aspects of sexual selection have been demonstrated in animals. Direct competition among males for mates occurs in many species, and in several instances the reproductive advantages that accrue to successful males have been clearly documented (Parker, 1970; LeBouef, 1974; Eberhard, 1979; Hamilton, 1979; McAlpine, 1979; Johnson, 1982; and references therein). Greater variance in the reproductive success of males compared to females, a result of the greater competition among males, has been suggested for several species on the basis of qualitative information and has been measured in other species (Bateman, 1948; Parker, 1970; LeBouef, 1974; Thornhill, 1976; Payne, 1979). Selective (preferential or nonrandom) mating by females occurs frequently (e.g., Bateman, 1948; Beach and LeBouef, 1967; Thornhill, 1976; Catchpole, 1980; Johnson, 1982), and in a few instances this selectivity has been shown to increase female fitness (Maynard Smith, 1956; Thornhill, 1976; O'Donald *et al.*, 1974; Partridge, 1980; but see Kingett *et al.*, 1981). Obviously, male competition and female choice can operate simultaneously and thus are not always separable (LeBouef, 1974; Thornhill, 1976; Kingett *et al.*, 1981). In a few cases, male choice has been suggested to have important influences on sexual selection. For example, male weevils (*Brentus anchorago*) may prefer large females, which may confer a reproductive advantage involving greater egg size (Johnson, 1982). That is, male competition is more intense for preferred females. Dewsbury (1982) and Nakatsuru and Kramer (1982) suggest that the potential number of effective matings by males of some animal species may be more limited than previously thought, lessening the probable intensity of male competition.

Until recently, the possibility of sexual selection in plants has been largely ignored. Haldane (1932) noted that stigmas often contain far more pollen than is needed to fertilize all ovules, leading to intraspecific competition among plants to pollinate their neighbors, which in turn selects for increased production and availability of pollen over an extended period. Huxley (1942) mentioned that competition among pollen grains is likely, and suggested that the competition should select for rapid growth of pollen tubes.

Bateman (1948) seems to have been the first to suggest specifically the existence of sexual selection in plants, as a result of the size difference in male and female gametes. He suggested that male competition should lead to increased production of pollen, and that the effects of such competition should be more apparent in monoecious or dioecious plants than in hermaphrodites with bisexual flowers.

Gilbert (1975) suggested that sexual selection in dioecious *Anguria* led to greater pollen and nectar production in male sporophytes, which compete for relatively rare female sporophytes. Janzen (1977), although not explicitly mentioning sexual selection, noted that selection pressures on male and female functions (pollen and fruit/seed production, respectively) differ greatly, that all pollen donors are not equally fit, and that female sporophytes should be selective in their production of offspring (seeds) with respect to pollen donors. Janzen deals primarily with processes occurring before pollination.

Willson (1979) stressed the potential importance of sexual selection in explaining several reproductive phenomena in plants, including the evolution of pollen aggregations (such as pollinia) and breeding systems. Charnov (1979) suggested that the concept of sexual selection is applicable to hermaphroditic species, including plants, and that it provides an explanation for the higher incidence of hybridization in plants than in animals as well as for the existence of double fertilization. Many recent papers consider the role of sexual selection in plants for the evolution of breeding systems, especially dioecy (Lloyd, 1979; Bawa, 1980a,b; Givnish, 1980; Bawa and Beach, 1981; Webb, 1981; Casper and Charnov, 1982).

Although there is a growing body of theory surrounding sexual selection as it applies to plants, and although sexual selection is now often invoked as an explanation for the evolution of breeding systems and reproductive strategies, there have been few attempts to examine in detail the potential mechanisms underlying intrasexual competition and intersexual choice in plants. The goal of this chapter is to describe such mechanisms of male competition and female choice and to evaluate their potential importance in plant sexual selection. In addition, we point out where further data are needed and indicate possible approaches for obtaining them.

We begin by examining the relative costs of pollen and seed production. We then review data comparing variance in the reproductive success of the male and female functions of plants. Next, we attempt to determine the form that male competition may take in higher plants. Then we describe possible mechanisms of female choice and review evidence that female choice improves the quality of the offspring produced. We conclude with a comparison of sexual selection in plants and animals.

A. Definitions and Terminology

To assess the importance of sexual selection in plants, we must have a clear definition of this process. Unfortunately, sexual selection was not explicitly defined at the time of its first usage (Darwin, 1859), and subsequent definitions have varied considerably (Huxley, 1938; Ehrman, 1972; Wade and Arnold, 1980). Here we define sexual selection as the differential reproductive success of individuals of the same sex and species that survive to reproductive age and are physically capable of reproduction. This definition differs from that of "natural selection" in that it eliminates, insofar as possible, any consideration of survival. From this perspective, intrasexual (usually male) competition and intersexual (usually female) choice are the primary causes of sexual selection. We believe that this definition is consistent with the usage of "sexual selection" as it has been applied to plants as well as with Darwin's original ideas, yet flexible enough to handle the peculiarities of hermaphroditic plants. We agree with the many recent investigators who consider sexual selection to be one aspect of natural selection and not a separate force as Darwin originally envisioned it (Halliday, 1978; Willson, 1979; and references therein).

The term "female choice" can have various connotations. Some investigators may consider the term anthropomorphic and inappropriate for biological usage. We use it merely as a shorthand for any evolved mechanism in which (1) pollen is received from some pollen donors but not others in a population or (2) nonrandom fertilization occurs following pollen deposition.

It is technically the gametophyte that is either male or female, not the sporophyte or the flower. For the purposes of this chapter, however, a male sporophyte (paternal sporophyte, pollen donor) is a plant that produces pollen, whereas a female sporophyte (maternal sporophyte, pollen recipient) is a plant that produces ovules, seeds, and fruits. Obviously, hermaphroditic plants are both male and female sporophytes, often simultaneously. A female flower is one that bears a gynoecium, and a male flower is one that bears an androecium.

Bisexual (perfect, hermaphroditic) flowers are both male and female, often simultaneously.

B. Relative Cost of Pollen and Seed Production

As the terms ''microgametophyte'' and ''megagametophyte'' imply, pollen grains are generally smaller and consist of fewer cells than ovules. Unfortunately, there is a paucity of data concerning the relative costs of micro- and megagametophytes. There is little doubt, however, that maternal investment (energy, nutrients, and risk to future reproduction; see Trivers, 1972) exceeds paternal investment per mature offspring because the maternal sporophyte supplies not only the larger gametophyte but also the resources necessary to fill the seeds and develop the fruit. For example, in *Amaryllis* the energy investment in an ovule is only about 1% of that in a mature seed (Smith and Evenson, 1978). Data from many different species show that mature fruits have from one to five orders of magnitude more dry weight and total protein than the ovary at anthesis (Dickmann and Kozlowski, 1969; Bollard, 1970; Sweet, 1973; Coombe, 1976; Lovett Doust and Harper, 1980; Stephenson, 1980; and references therein). These data suggest that the greatest period of maternal investment occurs, not surprisingly, after pollination. In contrast, there is no paternal investment during this period of maximal maternal investment. For the same resource investment, therefore, a plant can presumably make more pollen grains than seeds.

This asymmetry in parental investment strongly influences patterns of sexual selection because reproduction through the female function (fruit and seed production) is likely to be limited by resources, whereas reproduction through the male function (pollen) is likely to be limited by access to ovules (Bateman, 1948; Trivers, 1972). Consequently, there is likely to be competition among pollen-producing plants and their pollen grains for access to the ovules. Conversely, maternal sporophytes and their ovules should discriminate among potential mates.

C. Variances in Male and Female Reproductive Success

Wade (1979) and Wade and Arnold (1980) argue that the intensity of sexual selection acting on the males in a population of animals can be measured by the variance in the number of their mates. We agree with their emphasis on variance but suggest that for most organisms the appropriate reproductive attribute to measure is the number of offspring rather than mates and that the measure applies to either sex (R. I. Bertin and A. G. Stephenson, in preparation).

Classic sexual selection theory predicts that the variance in male reproductive success is greater than the variance in female success (Bateman, 1948; Trivers, 1972). In addition, we predict that the potential for different average variances in the success of the male and female functions in hermaphrodites is greater in outcrossing plants than in predominantly selfing plants (Willson, 1979, 1982). In the extreme case of a population of obligate selfers, variances in the success of male and female functions must be equal. However, only fragmentary data are available to evaluate these notions.

Gutierrez and Sprague (1959) interplanted 10 genetically identifiable stocks of *Zea mays* in a latin square design. The variance in reproductive success in terms of total F_1 progeny was 671,087 for pollen donors and 262,488 for pollen recipients. In hand pollinations among five lines of alfalfa (*Medicago sativa*), the variance in number of seeds per pollinated flower was higher for the pollen donors (4.13) than for the recipients (1.12). In field experiments, the variance in male function (seeds sired) was also higher than the variance in the female function (seeds produced) (0.135 vs. 0.039) if flowers were emasculated to prevent self-pollination, but not if stamens were left intact (Dane and Melton, 1973). This difference is interesting in that outcrossing is favored by emasculation and variance in male function is increased, as we predicted. No such differences were found, however, in another study of six *Medicago* lines (Sayers and Murphy, 1966). Hand pollinations among seven *Freesia* clones yielded variances for male and female functions in the number of seeds per pollinated flower of 3.03 and 4.72, respectively (Sparnaaij *et al.*, 1968). These studies provide little, if any, evidence that variances in male and female functions consistently differ and may not be relevant to naturally occurring species because of the unknown effects of previous artificial selection. The emasculated/ unemasculated comparison of *Medicago* is suggestive, but must be repeated for several species before it provides convincing support for our prediction. The few results from noncultivated species are inconclusive. When hand pollinations were made between all possible pairs of nine *Campsis* plants, the variance in the reproductive success of pollen donors (in terms of the number of fruits fathered) was similar to the variance in the reproductive success (number of fruits produced) of pollen recipients (R. I. Bertin, unpublished data). Hand pollinations among individuals representing seven populations of *Asclepias speciosa* produced higher variances in male function than female function (in terms of fruit production; Bookman, 1983). This result was obtained in each of two experiments. In one experiment, only one donor was used for each recipient plant and the variances in the male and female functions were 0.60 and 0.33, respectively. In the second experiment, multiple donors were used for each recipient and the variances in male and female functions were 0.27 and 0.13, respectively. Both the *Campsis* and *Asclepias* results have the shortcoming that any variance in the ability of donors to have their pollen deposited on recipient stigmas is ignored, because pollinations were made by hand.

These results suggest that the variance in the male funcion sometimes, but by no means always, exceeds that of the female function. An unequivocal test for such a difference in a natural population will be exceedingly difficult because of problems in assessing paternity. Thus, we may have to be content to accumulate fragmentary data and look for consistent patterns. The possibility remains that variance in the success of the female function is equal to or greater than that of the male function. This would imply intense female competition for mates. Such competition may be important in plants if, as we believe, the greatest opportunities for female choice occur after pollination. If so, selection may act strongly on pollen recipients to receive many more pollen grains than there are ovules in order to permit female choice.

II. Male Competition

"Competition" refers to the use of a resource by one individual that makes that resource less available to other individuals. Here the resource is the female gametophyte and the competitors are pollen-producing plants or individual pollen nuclei, depending on the stage of competition.

Male competition in plants can be conveniently divided into that affecing pollen deposition and that affecting fertilization once pollination occurs. In animals, an analogous point of separation is copulation, and in the context of sexual selection, most attention has been devoted to precopulatory competition. In plants, we believe that postpollination competition may be more important than prepollination competition (see also Charnov, 1979).

How can we tell whether male competition is occurring in plants? This question is conveniently broken down into two parts: (1) Do pollen donors compete for access to conspecific stigmas? (2) Do pollen grains on conspecific stigmas compete for access to egg cells?

A pollen donor removal experiment could unequivocally test for competition among pollen donors for access to conspecific stigmas. If only a fraction of the total pollen production by a plant is taken and properly deposited by the pollinators, and if this fraction increases after the removal of one or more pollen-producing plants from the population, competition can be inferred. In some pollinia-bearing genera (*Asclepias, Brassavola*), only a portion of the total pollen production is taken by the floral visitors (Willson and Rathcke, 1974; Wyatt, 1976, 1980; Willson and Price, 1977; Lynch, 1977; Schemske, 1980a). It is not known, however, what percentage of each individual's pollinia are properly deposited, nor have the effects of potentially competing donors been ascertained.

Even if two pollen donors have all of their pollen removed by pollinators, male competition might still be occurring. If one of the pollen donors was removed, the success of the other pollen donor might increase because more of its pollen was properly deposited. For example, the pollen might be picked up at a time that is more likely to lead to deposition on a stigma, or the pollen might not be covered by pollen from the other donor on the pollinator's body. An effective demonstration of such competition requires the ability to identify (by genetic markers) and to track the movement of pollen grains (by examining progeny) both before and after the removal of potential competitors. An increase in the offspring sired by the remaining donors would prove the existence of competition.

Competition will occur after pollen deposition if (1) pollen recipients are unable to mature fruit from all flowers receiving pollen or (2) more viable pollen is deposited on any stigma than there are ovules in the ovary. Many plant species produce many more flowers than fruits. Stephenson (1981) suggests that lack of resources rather than inadequate pollination often causes this disparity (but see Bierzychudek, 1981). Fewer data are available on the amount of viable pollen deposited on stigmas in natural plant populations, but those available indicate that plants usually have some flowers that receive more pollen grains than they have ovules even if some flowers are inadequately pollinated (Levin and Berube, 1972; Bertin, 1982a; Mulcahy *et al.*, 1982). Consequently, competition among microgametophytes could occur even on plants whose total fruit and seed production is limited by pollination. However, we would expect the intensity of male competition to be greatest in those populations in which resources limit fruit and seed production and the number of deposited pollen grains per stigma exceeds the number of ovules. Moreover, we expect that the intensity of male competition is greater in populations in which self-fertilization is unlikely. In such populations, the potential gains for effective male competition are much greater than in predominantly selfing species. This comparison is somewhat analogous to the comparison between promiscuous and monogamous mating systems in animals, with far more opportunity for very high variance in the number of offspring sired in the former system. This selfing/outcrossing comparison is useful because, if the degrees of outcrossing of plant species are known, one can evaluate the importance of adaptations hypothesized to be associated with intense male competition. Male competition has the potential for being a potent evolutionary force in plant populations.

Even though direct evidence for the existence of male competition and measurements of its intensity in natural populations are lacking, many reproductive characteristics of plants have been hypothesized to result from male competition. Next, we examine some of the proposed mechanisms whereby pollen donors and their microgametophytes may compete for access to the stigmas and megagametophytes of conspecifics.

A. Prepollination Competition

Prepollination competition differs from precopulatory competition in animals in the indirect nature of gamete transfer, mediated by an animal or abiotic factor.

Therefore, the only possibility for competition among pollen donors is to make more pollen available and/or to advertise it better to pollinators.

Plants have limited control over pollen movement because many important factors that influence movement, such as plant density and the presence of other flowering species, are largely or entirely beyond the control of an individual. Plants can, however, influence pollen pickup and movement, especially by evolutionary modifications affecting energy acquisition by pollinators (Levin and Kerster, 1969; Heinrich and Raven, 1972; Heinrich, 1975; Beach, 1981).

Plant characteristics that influence pollinator visitation are legion, and we will not review them all here. For convenience, we divide these characteristics into four categories: phenology, flower number and arrangement, proximate attractants, and floral rewards. We emphasize dioecious and monoecious species because in species with bisexual flowers it is often not possible to separate the effects of selection pressures on the male and female functions.

1. Phenology

A greater duration of flowering of male than female sporophytes could result from male competition (Willson, 1979; Bullock and Bawa, 1981). In dioecious species, male plants appear to flower as long as or longer than females (Lloyd and Webb, 1977; Bullock and Bawa, 1981). Individuals of the andromonoecious *Aesculus pavia* often, but not always, bear their male flowers over a longer period than their fewer bisexual flowers (Bertin, 1982b). In monoecious *Cupania guatemalensis,* male flowers are borne over a period that is usually three or four times greater than the female phase (Bawa, 1977). Andromonoecious Umbelliferae bear male and bisexual flowers over roughly the same period of time (Lindsey, 1982; Schlessman, 1982; and references therein).

In plants with only bisexual flowers, the duration of pollen availability and stigma receptivity can differ only as a result of dichogamy. This is rarely adequately synchronized among flowers on a plant, so that distinct periods of male and female function are produced. There are interesting exceptions (Cruden and Hermann-Parker, 1977; Lindsey, 1982). Even without synchrony, however, dichogamy combined with a long male phase would increase the number of flowers in the male phase, perhaps enchancing male competitive ability. The data, however, do not indicate a longer male than female phase. In several species, the female phase lasts longer than the male phase (Garnock-Jones, 1976; Schoen, 1977; Schemske *et al.,* 1978; Willson and Schemske, 1980; Bertin, 1982a; Lamont, 1982), although in two species of *Impatiens* the reverse is true (Schemske, 1978).

In addition to affecting the overall length of a flowering episode, competition among pollen donors might specifically favor earlier opening of male than female flowers. The fertilization ability of a pollen nucleus is often partly determined by the rate of pollen tube growth through the style. Thus, early arrival of a pollen grain on a stigma increases its chances in the race to the ovules. Furthermore, for some species, fruits are more likely to form from flowers produced earlier in the blooming season (Stephenson, 1979, 1981; and references therein; Bertin, 1982c). This would also select for early production of male flowers. Finally, early flowering of males in dioecious species may accustom pollinators to visiting these plants, increasing their potential for subsequent pollen donation (Onyekwelu and Harper, 1979). Darwin (1871) suggested that the most vigorous females in animal species are often the first to breed, thereby encouraging competition among males for these early mates. We are unaware of data on this point.

In most dioecious species, males do tend to flower earlier than females in a season (Lloyd and Webb, 1977; and references therein; Bawa and Opler, 1978; Onyekwelu and Harper, 1979; Bawa, 1980a,b; Conn, 1981; Conn and Blum, 1981; Thomson and Barrett, 1981), although *Populus tremuloides* may be an exception (Maini, 1972). Trends in monoecious species are less clear. In *Pinus ponderosa* (Roeser, 1941; Wang, 1977), *P. palustris* (Snyder *et al.*, 1977), *Rhopalostylis sapida* (Esler, 1969 cited in Godley, 1979), *Quercus* (three spp.) (Sharp and Sprague, 1967), and *Begonia* spp. (Matzke, 1938), opening of male flowers commonly precedes opening of female flowers in a season. The converse typically occurs in *Betula lutea* (Dancik and Barnes, 1972), *Cotula* spp. (Edgar, 1958 cited in Godley, 1979; Lloyd, 1972), *Juglans nigra* (Funk, 1970), *Dalechampia* spp. (Armbruster and Webster, 1979), and *Pinus contorta* (Critchfield, 1980). Male flowers of *C. guatemalensis* are usually borne before and after female flowers on particular individuals (Bawa, 1977). In *Ascarina lucida* male flowers mature earlier than female flowers when in inflorescences with two female flowers, but they flower at the same time in inflorescences with a single female flower (Moore, 1977). Individuals of many monoecious species vary greatly in the relative timing of male and female flower opening (Stout, 1928; Jong, 1976). Among andromonoecious species, the opening of bisexual flowers precedes the opening of male flowers in several cases, e.g., *Solanum* spp. (Martin, 1972; Dulberger *et al.*, 1981). In other species, such as *Aesculus* spp., no clear pattern emerges from the considerable interindividual variation (Benseler, 1975; Bertin, 1982b). Patterns also vary considerably among different species of Umbelliferae (Bell, 1971; Lindsey, 1982).

Thus, in monoecious and andromonoecious species there is no clear temporal pattern, although female (or bisexual) flowers seem to open before male flowers more often than the reverse. Of course, numerous selective pressures other than male competition are likely to act on male/female phenology. For example, female flowers opening before male flowers on a plant have a greater chance of being cross-pollinated than those opening concurrently. This behavior would, of course, not be important in dioecious species, wherein there is no possibility of self-pollination. Other factors may also be important in determining the temporal patterning of flowers. In *A. pavia,* for example, Bertin (1982b) suggested that patterns of abundance of flower- and fruit-eating insects may influence the time at which bisexual flowers are borne.

One aspect of intrafloral phenology that may be influenced by male competition is the pattern of anther dehiscence within a flower or on an entire plant. Pollen storage time (the period between anther dehiscence and pollen deposition) may affect the relative competitive ability of pollen in stylar tissue (Pfahler, 1967; Pfahler and Linskens, 1972). Male competition will favor patterns of pollen release that ensure optimum availability to the pollinator fauna. If, for example, pollen viability declines with time, selection might favor the frequent release of small quantities of pollen. This could potentially occur by either sequential anther dehiscence or the production of numerous sequentially opening flowers (Lloyd and Yates, 1982). Unfortunately, data on the rate of pollen removal from flowers in natural populations are inadequate to address these notions.

2. *Flower Number and Arrangement*

Flower number and arrangement clearly affect pollinator visitation patterns. Plants or inflorescences with many flowers often attract more floral visitors than those with fewer flowers (Free, 1966; Kendall and Smith, 1975; Willson and Price, 1977; Schaffer and Schaffer, 1979; Schemske, 1980b; Davis, 1981). In several pollinia-bearing families (Orchidaceae, Asclepiadaceae), total pollinia removal from large inflorescences is greater than that from smaller inflorescences (Willson and Rathke, 1974; Willson and Price, 1977; Willson and Bertin, 1979; Schemske, 1980a; Wyatt, 1980). In the orchid *Brassavola nodosa,* the number of pollinia removed per flower also increases as the number of flowers in the inflorescence increases (Schemske, 1980a), although this is not true in *Asclepias* (Lynch, 1977; Wyatt, 1980).

Several workers have suggested that increased production of pollen-bearing flowers may be favored by male competition (Gilbert, 1975; Willson, 1979; Bawa, 1980a). From this perspective, sexual selection may have played an important role in the evolution of the andromonoecious breeding system by selecting for increases in the male component of fitness (Lloyd, 1979; Willson, 1979; Primack and Lloyd, 1980; Bertin, 1982d). Furthermore, many species with bisexual flowers regularly produce far more flowers than can be developed into mature fruits with the available resources and, in many cases, the number of pollinated flowers exceeds the number of mature fruits. Consequently, these species are functionally andromonoecious (Stephenson, 1981). In these species, it has been suggested that the "surplus" flowers (in terms of fruit production) have evolved as a means of increasing the number of seeds that a plant sires (Willson and Rathcke, 1974; Willson and Price, 1977; Janzen *et al.,* 1980). If this were the only factor selecting for surplus flowers, one might expect strong selection for andromonoecism over hermaphroditism as a way of conserving the resources wasted in the formation of pistillate tissues (Willson, 1979; Lloyd, 1980; Stephenson, 1981). Thus, another explanation is that the surplus flowers allow plants to abort fruits selectively and thereby increase the average quality of the remaining offspring (see Section III,D).

In dioecious species, the number of flowers per inflorescence and/or the number of inflorescences per plant are almost always greater for males than females (Jong, 1976; Lloyd and Webb, 1977; Opler and Bawa, 1978; Connor, 1979; Bawa, 1980b; Webb and Lloyd, 1980; Bawa *et al.*, 1982; Lock and Hall, 1982; and references therein). The dioecious *Solanum* spp. examined by Anderson (1979) appear to be an exception. In many dioecious species the frequency of flowering of males appears to be greater than that of females, causing a greater lifetime production of flowers in males than females (Lloyd and Webb, 1977; and references therein; Webb, 1979; Webb and Lloyd, 1980; Bullock and Bawa, 1981; Bawa *et al.*, 1982). In some species, male plants begin flowering earlier in their lives (Valentine, 1974; Lloyd and Webb, 1977; and references therein).

The few studies of monoecious species also suggest that flower numbers are male biased. In *Begonia, Euphorbia, Sagittaria, Mercurialis, Mabea,* and certain orchids, male flowers outnumber flowers (Matzke, 1938; Dodson, 1962; Thomas, 1956; Ehrenfeld, 1979; Kaul, 1979; Steiner, 1981). A survey of "Gray's Manual" (Fernald, 1950) revealed more monoecious species with excess male flowers than with excess female flowers (Willson, 1979).

In monoecious *Cotula* spp., the percentage of male florets is variable (20–80%) and is positively correlated with corolla length, which is greater in species with high levels of outcrossing (Lloyd, 1972). Thus, highly outbred species have more male florets than less outbred species. Lloyd (1972) suggests that the reduced potential seed number in outbred species (due to few female flowers) may be offset by increases in quality resulting from more frequent cross-breeding. The reduced allocation to male function in the more inbred species may allow greater efficiency of seed production at the expense of some cross-pollination. Another, not necessarily incompatible, explanation of the correlation between male flower number and outcrossing involves male competition. In species with less selfing, the potential benefits of high pollen production are relatively great and intrasexual competition will favor increased maleness. As noted above, male competition will generally be greater in species exhibiting little selfing.

3. *Proximate Attractants*

Features other than flower number and arrangement may serve as proximate attractants, including olfactory and various visual cues. Several studies have shown the differential attractiveness to insects of different color morphs of the same species (Levin, 1972; Mogford, 1978; Kay, 1978; and references therein).

However, in no case is adequate evidence reported to exclude the importance of other factors in determining these differences. In fact, it seems unlikely that differential attractiveness will ever be based on proximate attractants alone because pollinators would soon cease to respond to them unless the differences also reflected differences in floral rewards (e.g., pollen and nectar).

Data are available for comparisons of flower (usually perianth) size in several diclinous species. Lloyd and Webb (1977) summarize data from several studies suggesting that male flowers of dioecious species tend to be larger than female flowers, although this is clearly not true in all dioecious species (Bawa and Opler, 1975; Godley, 1976). In gynodioecious species, bisexual flowers (the only pollen-producing flowers) are generally larger than female flowers (Baker, 1947; Godley, 1979; and references therein). Data from monoecious species are scant, which may mean that conspicuous size differences are rare. If male or bisexual flowers were consistently larger than female flowers, as suggested by Darwin (1877), this might be interpreted as a result of the greater intensity of male than female competition. The results would be analogous to the situation of animals, wherein the male sex is more adorned and larger, which is often thought to result from strong male competition (see Halliday, 1978). However, other selective factors affect flower size, and pollinator visitation can be enhanced in ways other than by increasing flower size. Most importantly, male and female flowers in zoophilous species probably must remain similar in size if individual pollinators are to visit both.

4. *Floral Rewards*

Pollinator behavior is markedly influenced by availability of floral rewards, especially nectar. Greater rewards generally cause pollinators to visit more flowers per inflorescence, to move shorter distances between successive flowers, and to increase their turning frequency so that they remain in a small area (McGregor *et al.*, 1959; Free, 1965; Gill and Wolf, 1977; Pyke, 1978, 1982; Heinrich, 1979; Waddington, 1980, 1981). Although some of these data reflect movement patterns among plants, they are probably also relevant to movement patterns within plants, especially large ones. Given this information, however, it is still difficult to predict patterns that would enhance male competitive ability (Pyke, 1981). Although more nectar (either in an inflorescence or in a plant) may increase visitation frequency and therefore pollen removal (Waddington, 1981), it may also restrict pollinator movement and increase self-pollination.

No consistent patterns of nectar production relative to sexuality emerge. In five of six dioecious species, nectar production is greater in female than male flowers (Bawa and Opler, 1975). Female flowers also produce more nectar in dioecious *Ephedra alte* (Bino and Meeuse, 1981). In other dioecious species, however, female flowers lack nectar and male flowers produce it (Bawa, 1980b; Bullock and Bawa, 1981; and references therein). In two andromonoecious species studied by Cruden (1976), the bisexual flowers produced more nectar than male flowers, but the reverse was true in *A. pavia* (Bertin, 1980).

We would expect the greatest quantities of pollen to be produced on those species experiencing intense male competition if the likelihood of paternity is influenced by the quantity of pollen produced. As noted above, competition should increase from predominantly selfing to predominantly outcrossing species. Pollen/ovule ratios are significantly greater in xenogamous than in highly autogamous species (Cruden, 1977). Cruden (1977, p. 32) states that this reflects "the likelihood of sufficient pollen grains reaching each stigma to result in maximum seed set." Our interpretation of pollen overproduction is not inconsistent with Cruden's, but it more specifically addresses the ultimate (evolutionary) cause of this pattern (also noted by Charnov, 1979).

B. Postpollination Competition

Following pollination, male competition will occur if the number of viable pollen grains deposited onto a stigma exceeds the number of ovules that the maternal sporophyte can nourish into mature seeds. Most phenomena involving differential germination of pollen, growth of pollen tubes, fusion with egg nuclei, and ovule and ovary development are potentially influenced by both the pollen and the maternal sporophyte. This makes it difficult to discriminate male competition and female choice, a difficulty by no means unique to plants. For convenience, we will consider most of the phenomena that may reflect both male competition and female choice in Section III. Here we describe aspects of pollen germination and pollen tube growth that seem to be strongly influenced by male attributes.

In vitro studies are useful in assessing the relative germination ability and pollen tube growth of pollen grains from different males, independent of any effects of maternal tissue. Several studies show significant differences in the average germinability of pollen grains from different clones or individuals *in vitro* (*Z. mays:* Pfahler, 1974; *M. sativa:* Straley and Melton, 1970; Dane and Melton, 1973; *Pinus radiata:* Matheson, 1980; *Costus guanaiensis:* Schemske and Fenster, 1983). The genetic basis of at least some differences in germinability is shown by differences in the proportion or rate of germination of pollen with different genotypes from a single donor (e.g., *Z. mays:* Sprague, 1933; *Datura stramonium:* Buchholz and Blakeslee, 1927).

In vitro studies have also shown significant differences in the rate or length of pollen tube growth from pollens of different donors. Most of these studies did not eliminate effects of germination times, so some of the results may be partly due to different times of germination. Pollen tubes elongation rates in *M. sativa* vary with clone (Barnes and Cleveland, 1963a; Straley and Melton, 1970). Pollen tube growth rates in *Z. mays* varied among individuals and were associated with pollen germination ability (Mulcahy, 1971; Pfahler, 1974; Ottaviano *et al.*, 1975; Sari-Gorla *et al.*, 1975). Pollen tube growth rates *in vitro* also differ among pollens from different individuals of *A. speciosa* (Bookman, 1983) and *C. guanaiensis* (Schemske and Fenster, 1983).

Several lines of evidence indicate that some aspects of male gametophyte performance are heritable. Studies with *Z. mays* and *Phaseolus limensis* illustrate that differences of single alleles can influence the outcome of pollen competition (Schwartz, 1950; Bemis, 1959; Jimenez and Nelson, 1965; and references in Mulcahy, 1979). In *Z. mays* one can select for increases in pollen tube growth rate, clearly indicating the genetic basis of this trait (Mulcahy, 1974; Johnson and Mulcahy, 1978). The average length of pollen tubes of heterozygous F_1 *Z. mays* grown *in vitro* for 2h were intermediate between those of the inbred parental lines and were also more variable (Ottaviano *et al.*, 1975; Sari-Gorla *et al.*, 1975). This is expected if pollen tube growth rates have a strong genetic component, because segregation should produce F_1 characteristics that are more variable than, but on average intermediate between, those of the inbred parental lines. Taken together these results provide clear evidence of genetic control of pollen tube growth rates in *Zea,* but additional work with other species is desirable.

The extent to which these studies can be generalized to patterns in nature is unclear. Almost all work has involved cultivated species, which may exhibit more variability in microgametophyte performance than individuals in natural populations, because cultivars may not have been continuously selected for high male performance, as might occur in nature. The remaining additive genetic variance in pollen germination and tube growth in natural populations may be very limited as a result of past selection. Bookman's (1983) and Schemske and Fenster's (1983) results suggest, however, that some variance does remain in natural populations.

Even if these objections are overcome, we are still concerned with the applicability of *in vitro* results to an *in vivo* process. Pollen from different donors could respond differently to various germination media. Barnes and Cleveland (1963a,b), Sari-Gorla *et al.* (1975), and Bookman (1983), working with *M. sativa, Z. mays,* and *A. speciosa,* respectively, found a reasonably good correlation between *in vitro* and *in vivo* results, but Pfahler and Linskens (1972), Dane and Melton (1973), and Kumar and Sarkar (cited in Sari-Gorla *et al.*, 1975) did not.

C. Differential Male Reproductive Success

In several cases, differences in male reproductive success have been demonstrated among two or more genetically distinct lines, without a clear demonstration of the relative importance of prepollination and postpollination phenomena.

Gutierrez and Sprague (1959) noted different male fecundities among lines of *Z. mays*. Dates of tasseling and silking and the weight of the shed pollen were unrelated to male fecundity, but plant height, the length of the pollen-shedding period, and the number of plants shedding pollen were related to fecundity. Hoff (1962) grew 11 races of *Euoenothera* together and found that some races were more effective at fathering seeds on other plants (male outcrossing) than others. The *parviflora* line had low male outcrossing and also low pollen production; the latter may have been partly responsible for the former. Horovitz and Harding (1972) grew the wild type and several flower color mutants of *Lupinus nanus* together and found higher rates of male outcrossing for the wild type than for the mutants, suggesting differential pollinator visitation. However, as the authors point out, these results might not be generalizable to natural conditions because of the high levels of mutants that they used in the experimental populations.

D. Male Competition: Summary

Plants may compete in terms of male reproductive output either before or after pollination. Prepollination competition seems probable in many plant populations, but an unequivocal demonstration of competition, which would involve monitoring male success before and after removal of certain donors, has not been made. The requirements for postpollination competition have been demonstrated in numerous instances. Additional work is needed to demonstrate the effects of such competition in natural plant populations.

Several predictions about the effects of male competition on phenology, flower number and arrangement, and proximate attractants can be made. In other cases (e.g., patterns of nectar amount and distribution), our knowledge of how a reproductive characteristic affects male success is insufficient to allow predictions. Where predictions are possible, they seem to be borne out by data from dioecious or monoecious species (e.g., staminate plants having more flowers and blooming longer than pistillate plants). This suggests that male competition is stronger than female competition in most instances, which we would expect from basic considerations of the expenditure per offspring of male and female sporophytes.

Exceptions to most of these generalizations exist. In some cases (e.g., female or bisexual flowers opening before male flowers in monoecious species), the trend is not what we would expect if male competition were the dominant selective force. Clearly, other selective pressures affect plant reproductive biology. The above trend might, for example, reflect selection for better survival of offspring, if we assume that pistillate flowers opening before staminate flowers are more likely to be cross-pollinated, and that offspring resulting from cross-pollinations perform better than those arising from selfing. Because this involves survival, it is not a part of sexual selection as we define it.

It is unlikely that all patterns that are consistent with the concept of male competition are in fact maintained evolutionarily by male competition because other selective pressures can produce similar effects. Consequently, unless a male reproductive advantage is demonstrated for a particular trait (e.g., more flowers per inflorescence or a longer duration of flowering), we cannot be certain that it is associated with male competition.

Postpollination competition is not only better documented but is more amenable to analysis than prepollination competition. More work is needed here on the relative ability of pollen grains from different individuals in natural populations to germinate and grow on artificial media, and the extent to which these results are related to patterns *in vivo*.

III. Female Choice

In plants the pollen-producing parent invests little in offspring (embryo, zygote, seed, seedling) beyond supplying haploid sets of chromosomes. Consequently, the only benefit that the maternal sporophyte derives from mating is genes (Trivers, 1972; Halliday, 1978). If pollen grains produced by various donors differ in genetic quality (from the perspective of the maternal sporophyte), and if reproduction is not pollen limited, then the maternal sporophyte is under selective pressure to allow only those pollen grains with the highest genetic quality to sire her seed crop (Janzen, 1977; Charnov, 1979). Differential acceptance of pollen could influence the variance in reproductive success among pollen donors and hence the intensity of sexual selection.

As noted earlier, the term "female choice" merely refers to the consistent, nonrandom reception or utilization of pollen from some donors in the population by a given maternal sporophyte. Specifically, we ask: (1) which mechanisms might provide the maternal sporophyte with some degree of control over the parental parentage of her seed crop; (2) do these mechanisms select for certain pollen genotypes and handicap other pollen genotypes; and (3) what effect does female choice have on the performance of the resultant offspring? If female sporophytes have no control over the parental parentage of their seed crop, then female choice is not an important cause of sexual selection. If female choice does occur, we would expect it to enhance the performance of the resulting offspring.

Theoretically, female choice could operate at any of several times. The maternal sporophyte could influence (1) which pollen donors deposit pollen on her stigmas, (2) which of the deposited pollen grains fertilize the ovules, (3) which of the fertilized ovules develop into mature seeds and, concurrently, (4) which of the juvenile fruits develop to maturity. This is analogous to suggesting that females of some animal species may influence who their mates are, which sperm fertilize their eggs, or which embryos are likely to develop fully. Obviously, there are important differences between plants and animals relevant to female choice. For example, each female flower often receives pollen simultaneously from several pollen donors, whereas in animals, sperm is usually received sequentially if multiple mating occurs and always if fertilization is internal. In addition, the female reproductive function of an individual plant is often divided into several more of less independent units rather than into a single clutch (Lloyd, 1979; Stephenson, 1981).

A. Selective Pollination

Can the maternal sporophyte influence which plants in the population deposit pollen on her stigmas? Clearly, many characteristics of the flowers restrict self-pollination, such as the position of the stamens relative to the stigmas in bisexual flowers, temporal separation of the sexes within a bisexual flower (dichogamy), and the production and disposition of unisexual flowers (monoecy). The role of these characteristics in attracting pollen from certain "foreign" (non-self) donors and excluding other foreign donors is less clear.

Several mechanisms may influence which foreign plants donate pollen. These include flower number, the temporal and spatial arrangement of the flowers, the quantity and quality of the nectar rewards, the time of anthesis, and the duration of stigmatic receptivity (Janzen, 1977). To this list we could add the length of the style, position of the stigma, flower color, changes in flower color, and any other plant characteristics to which a pollinator might respond. For example, a large floral display might attract pollinators from a greater distance than a smaller display, and perhaps these pollinators would transport pollen from less closely related sporophytes. Unfortunately, there are not studies on the effects of these floral characteristics in determining which plants donate pollen. There are, however, data which suggest that some of these characteristics influence the number of flowers that receive pollen. For example, it is well established that large inflorescences on some species attract more floral visitors than smaller inflorescences, and that the flowers in large inflorescences are more likely to be pollinated than those in smaller inflorescences (Willson and Rathcke, 1974; Schaffer and Schaffer, 1979; Stephenson, 1979; and Schemske, 1980a). Until the notion that floral characteristics promote selective pollination has been investigated, we cannot distinguish the roles that male competition and female choice may have played in their evolution and maintenance.

Even if floral characteristics can influence to some degree which plants donate pollen, many factors associated with pollination will not be under the direct control of the maternal sporophyte—including size of the pollinator populations, spatial distribution of the plants in the population, and variations in flowering times within a population due to environmental conditions. Each of these factors may also influence the direction and amount of pollen flow (Augspurger, 1980).

B. Selective Fertilization

Plants possess a variety of mechanisms that may influence which of the pollen grains deposited on the receptive surface of a flower fertilize the eggs within that flower. Perhaps the most thoroughly studied and widely known of these mechanisms are the genetic incompatibility systems (see reviews by Hesplop-Harrison, 1975; Lewis, 1976; Nettancourt, 1977). These systems fall into two broad classifications: sporophytic incompatibility, in which the maternal sporophytic tissue (stigma, style, ovary) rejects pollen based on the genotype of the pollen-producing parent, and gametophytic incompatibility, in which the maternal sporophyte rejects pollen based on the genotype of the pollen itself. In both types of incompatibility, the alleles at one or more loci determine the phenotype of both the pollen and the style. When pollen shares a phenotype with the style, it will be rejected and fail to effect fertilization. Plants with sporophytic incompatibility most commonly produce trinucleate pollen, and rejection occurs on a dry stigmatic surface. Plants with gametophytic incompatibility most commonly produce binucleate pollen, but rejection occurs not in the copious secretions of the stigma but rather in the style or ovary (Brewbaker, 1957; Heslop-Harrison, 1975; Nettancourt, 1977). There are, however, several exceptions to these generalizations (Nettancourt, 1977; Heslop-Harrison, 1982).

Incompatibility systems have been traditionally viewed as mechanisms that regulate self-fertilization. Some authors have suggested, however, that incompatibility may be a more general regulator of offspring quality, with reduction of selfing being one component (Fisher, 1965; Lewis, 1979; Bertin, 1982c). Several lines of evidence are consistent with this view. First, incompatibility may establish a complex set of partial and complete cross-compatibilities within a population. The nature and extent of the cross-compatibilities are governed by many factors, such as the number of loci that form the genetic basis of the system and the number of alleles at each locus within the population, as well as dominance, epistasis, and linkage patterns (see Nettancourt, 1977). Second, many incompatibility systems are not based on rigid identification and rejection of like pollen but are more variable and phenotypically plastic than originally thought (Bertin, 1982c; Willson, 1982). In many species, self-incompatibility is only partial (Cooper and Brink, 1940; Pandey, 1968; Crowe, 1971; Dane and Melton, 1973; Straley and Melton, 1970; Lundquist *et al.*, 1973; and references therein). That is, some seeds are commonly produced from self-pollinations. These studies also show that differences in the degree of self-incompatibility often exist among individuals in a population. Moreover, incompatibility systems may change with the age of the style and pollen, and finally, the degree of incompatibility can be modified by environmental conditions (Pandey, 1968; Ascher and Peloquin, 1966; Straley and Melton, 1970; Dane and Melton, 1973; and references therein).

These studies suggest that complete self-incompatibility may be merely one end of a continuum that grades smoothly into complete self-compatibility. It may be naive to divide higher plants into two groups: the self-incompatible and the self-compatible (Willson, 1982). This implies that incompatibility systems may provide plants with some degree of control over the paternal parents of their seed crop. Detailed and controlled experiments are needed to determine how the various degrees of incompatibility and the genetic bases of the incompatibility systems influence the number and identity of pollen-producing parents that contribute to a plant's seed crop, and how the number and timing of cross-pollinations influence the degree of incompatibility.

When seed production following self- and cross-pollination is similar, a plant is usually considered to be self-compatible. There is considerable evidence that nonrandom fertilization occurs even in self-compatible species. For example, when equal amounts of self and foreign (outcross) pollen were placed on the stigma of *Cheiranthus cheiri* (in a pollen mixture experiment), the foreign pollen produced significantly more fertilizations than the self-pollen (Bateman, 1956). Similar results have been reported for many other species (Squillace and Bingham, 1958; Pfahler, 1965, 1967; Weller and Ornduff, 1977; Ockendon and Currah, 1978; Matheson, 1980). In self-compatible individuals of *Brassica oleracea,* foreign pollen achieved significantly more fertilizations than self-pollen even when the self-pollen was applied to the stigma up to 6 h before the foreign pollen (Ockendon and Currah, 1978). This suggests that differences in the timing of self- and cross-pollen deposition can be overcome in some species that receive multiple visits by pollinators. In *Z. mays,* pollen mixture experiments revealed that different foreign pollen donors differed in their ability to effect fertilization, but that all foreign donors in the study did better than self-pollen (Pfahler, 1965). On the other hand, selective fertilization by self-pollen in pollen mixture experiments is also known to occur, especially in highly inbred lines (Jones, 1920). On some individuals of *Pinus monticola,* whether self- or cross-pollen produced the greater number of fertilizations depended upon which foreign donor was used in the pollen mixture (Squillace and Bingham, 1958).

These differences in the ability of pollen grains to effect fertilization reflect either differences in their rates of germination or differences in the speed of pollen tube growth (see Section II,B). It is well known that the genotype of the pollen often has a demonstrable effect on such differences (see Section II,B), but it is unlikely that the maternal sporophyte has no effect on the outcome of competitive interactions among microgametophytes. This maternal effect can be demonstrated when the outcomes of competitive interactions among pollen tubes change dramatically from one maternal sporophyte to another. The most obvious cases are those which show selective fertilization by foreign pollen in competition with self-pollen (see Section III,B). Pollen mixture experiments in which pollen from two or more foreign donors is deposited simultaneously on the same stigmatic surface indicate that selective fertilization occurs and, furthermore, that different foreign donors are sometimes selected by different maternal sporophytes (Pfahler, 1967). These studies suggest that fertilization is influenced by complex pollen–pollen and pollen–style interactions, which are determined by the genotypes of both the pollens and the maternal sporophyte and perhaps also be the genotype of the paternal parent (Pfahler, 1967; Jennings and Topham, 1971). Studies that separate the roles of microgametophytic competition and female choice or that explore the physiological mechanisms of female choice are needed. We suspect that both maternal and paternal influences will be found in most cases.

What benefit does the maternal sporophyte derive from pollen tube competition and selective fertilization? In animals, it is often argued that females mating with competitively superior males (in terms of obtaining mates) produce sons that are also competitively superior (assuming that the trait is heritable). Consequently, these sons acquire more mates and leave more progeny (see Halliday, 1978, for review). An analogous argument could apply to plants, but no data are available to evaluate it. In addition to this possibility, a recent paper by Mulcahy (1979) reviews the evidence which indicates that pollen tube competition correlates with the ability of the resultant seeds to germinate, grow, and survive. He suggests that the correlation between pre- and postzygotic qualities results from a suite of genes that are expressed in both the gametophytic and sporophytic stages of the life cycle.

Modification of sporophyte vigor by pollen competition has been demonstrated by two types of studies: experiments in which the quantity of pollen that is deposited on the stigmatic surface is varied (Mulcahy *et al.*, 1975, 1978; Ter-Avanesian, 1978; and references therein) and experiments in which the distance between the site of pollen deposition and the ovary is varied (Correns, 1914, 1928; Mulcahy and Mulcahy, 1975). In the former, excessive pollen deposition provides an opportunity for great pollen competition and, consequently, there exists great potential for female choice compared to pollinations using fewer pollen grains. In the latter studies, there is greater opportunity for fast-growing pollen tubes to surpass slower-growing tubes when the distance between the site of pollen deposition and the ovary is increased. Consequently, the opportunity for selection on the microgametophytes will increase as the distance increases between the locus of pollen deposition and the locus of fertilization. Both types of studies reveal that seeds resulting from intense pollen competition produce seedlings that are significantly more vigorous than seedlings produced from less intense competition (Mulcahy and Mulcahy, 1975; Mulcahy *et al.*, 1975). In addition, Mulcahy *et al.* (1978) showed that intense gametophytic selection in the F_1 generation can also modify the F_2 generation in *Petunia hybrida*.

The quantity of pollen that is deposited on the stigma also influences the proportion of the pollen grains that germinate. The percentage of germination tends to increase with the amount of pollen (Brewbaker and Majumder, 1961; Brewbaker and Kwack, 1963; Hornby and Charles, 1966; Jennings and Topham, 1971). Both *in vitro* and *in vivo* studies of this phenomenon reveal that pollen germination is conditioned by the secretion of growth substances provided partly by the pollen grains themselves and partly by the styles of the maternal parent and show that there are interactions between the two sources (Brewbaker and Majumder, 1961; Jennings and Topham, 1971). Density-dependent pollen germination further accentuates differences among flowers in the amount of pollen tube competition. On those flowers with large quantities of pollen on their stigmas, a high percentage of the grains germinate, intense pollen competition ensues, and a high percentage of the ovules are fertilized. In contrast, on those flowers with low pollen densities on their stigmas, there are low levels of pollen germination and minimal pollen competition, and some ovules may not be fertilized. The possible implications of these patterns are discussed below.

C. Zygotic Female Choice: Seed Abortion

Just as more pollen is often deposited on stylar tissue than is represented in fertilized ovules, there are often more fertilized ovules than mature seeds. The possibility exists, therefore, that seeds are selectively aborted. Both extrinsic and intrinsic factors are responsible for the loss of immature seeds. Extrinsic factors include damage by inclement weather, insects, and other seed predators (see reviews by Janzen, 1971; Sweet, 1973; Stephenson, 1981). However, seed losses due to extrinsic factors are unlikely to be related to male competition or female choice unless it can be shown that some zygotic genotypes are more likely to be damaged than others.

In some plants with several ovules per ovary, a fixed proportion of the ovules will invariably abort whether or not fertilization occurs. Casper and Wiens (1981) reported that only one of the four ovules in an ovary of *Cryptantha flava* normally develops into a mature seed. Additional seeds cannot be obtained even when pollination is ensured. They also note that ovule abortion is independent of ovule position. Seeds are equally likely to mature from each of the four ovule positions. Casper and Wiens speculate that this abortion system may provide an opportunity for selection at the zygote level either by competition among embryos or by direct control by the maternal sporophyte. Furthermore, many other species are known to have fixed rates of random ovule abortion (with respect to position) (Casper and Wiens, 1981). Whether or not ovule abortion in these species is also random with respect to zygotic genotype will remain unknown until genetic markers are used to analyze the abortion system.

In many species, especially legumes, certain ovule positions are more likely to produce mature seed (Cooper *et al.*, 1937; Linck, 1961; Sayers and Murphy,

1966; Horovitz *et al.*, 1976; Schaal, 1980). For example, seed set in *L. nanus* declines from the peduncular to the stylar end of the pod, whereas several species of *Medicago* have a higher seed set in the even-numbered positions (Horovitz *et al.*, 1976). In both *L. nanus* and *M. sativa,* controlled pollinations using genetic markers have revealed that foreign and self-pollen differed with respect to the positions of the ovules they fertilized (Horovitz *et al.*, 1976). Cooper and Brink (1940) showed that self-pollen reaches different ovules than foreign pollen in *M. sativa* and that the seeds from self-fertilizations are more likely to abort. These studies suggest that the growth characteristics of the pollen tubes, when coupled with ovule abortion, may influence the paternity of the seeds within a fruit. Clearly, additional studies are necessary before much confidence can be placed in this suggestion.

Among species in which the rate of seed abortion is not fixed or due to position effects, it appears that self-fertilization often increases the number of seeds that abort. For example, many gymnosperms have greater seed production following cross-pollination than following self-pollination (Squillace and Bingham, 1958; Sarvas, 1962, 1968; Johnsson, 1976; Rehfeldt, 1978; Matheson, 1980). Detailed examinations of zygote development reveal that the low seed set in self-pollinated gymnosperms is probably due to homozygosity of lethal or defective genes in the embryo and endosperm (Sarvas, 1962, 1968; Johnsson, 1976). In addition to reducing the seed yield, self-fertilization reduces the performance of the seedlings (Squillace and Bingham, 1958; Rehfeldt, 1978; Matheson, 1980). Self-fertilized seed of *Pseudotsuga menziesii* produced seedlings that were 32% shorter and 17% lighter after 4 yr than cross-fertilized seed (Rehfeldt, 1978).

Increased rates of seed abortion following self-pollination may also be due to postzygotic polygenic incompatibility systems (Crowe, 1971). In *Borago officinalis,* many loci govern the incompatibility system and, unlike typical incompatibility systems, rejection occurs following fertilization. However, this effect is variable. For example, heterozygous individuals are far less self-incompatible than highly inbred individuals. Consequently, the degree of self-incompatibility varies among individuals in a population and between generations in an inbred line. Nettancourt (1977) points out that this system eliminates the negative effects of inbreeding rather than inbreeding itself. He also points out that such a complex system may often go undetected and, therefore, may be more common than limited reports would indicate. It should be noted that polygenic postzygotic incompatibility systems are, for the most part, indistinguishable from abortion due to homozygosity of lethal genes which are expressed early in the development of the embryo or endosperm. Finally, differential seed set following cross- and self-fertilization is likely to lead to fruits that vary in seed number on the same maternal sporophyte. Fruits that result from outcrossing would contain more seeds than fruits from self-pollinated flowers.

D. Zygotic Female Choice: Fruit Abortion

Many species consistently abort a sizable portion of their immature fruit crops. These species apparently initiate more fruits than can be developed to maturity given the resources available for reproduction. Consequently, fruit abortion can be viewed as a means of adjusting a plant's fruit crop to available resources (Lloyd, 1980; Stephenson, 1981). However, all immature fruits are not equally likely to abort. Depending on the species, juvenile fruits may selectively abort on the basis of the number of developing seeds, pollen source, order of pollination, or some combination of these factors (see Stephenson, 1981).

Evidence of selective abortion based on seed number comes from several investigators employing a variety of techniques (Heinicke, 1917; Murneek, 1954; Wright, 1956; Martin *et al.*, 1961; Sarvas, 1962, 1968; Quinlan and Preston, 1968; Lee, 1980; Bertin, 1982a). For example, when the number of foreign pollen grains that are placed onto the stigmas of flowers is varied, the probability that the resulting fruit will abort varies inversely with the number of pollen grains used in pollination. Other investigators have counted the number of fertilized ovules in developing and aborted fruits and have shown that, at the time of abscission, aborted fruits contain fewer fertilized ovules than developing fruits. In some species, infructescences from which some fruits are thinned prior to any fruit abortion have little or no additional fruit abortion. If the mature fruit on thinned infructescences and on infructescences allowed to abort naturally are compared, fewer seeds per fruit are found on the thinned infructescences. This is expected only if plants abort those fruits with a below-average number of seeds. These studies show that in some species fruits with low seed numbers are selectively aborted.

Some variation in seed number among developing fruits is probably related to variations in the number of pollen grains deposited by the pollinators. This variation in seed number would be further enhanced if low pollen density also decreases the proportion of pollen grains that germinate (see Section III,B). Lee and Bazzaz (1982) have suggested that, by aborting fruits with low seed numbers, a plant may be eliminating those fruits in which there was little pollen competition for access to the ovules. Given the known effects of pollen competition on the quality of the resultant offspring, selective abortion of fruits with low seed numbers may be viewed as a means of increasing the average quality of seed crops. An alternative, but not mutually exclusive, explanation is that if the resource expenditure in pericarp on a per seed basis decreases as the number of seeds per fruit increases, it is more resource efficient to mature those fruits with the most seeds (Lee and Bazzaz, 1982).

There are also examples of selective fruit abortion based on the pollen source. In some species, fruits from self-pollinated flowers are more likely to abort than fruits from cross-pollinated flowers (Heinicke, 1917; Murneek, 1954; Sarvas, 1962, 1968). This is not surprising given that self-pollinated flowers are more likely to have fewer seeds than cross-pollinated flowers because of pre- and postzygotic incompatibility systems and the increased opportunity for the expression of recessive lethals. Other mechanisms which produce low seed numbers following self-pollination may also be important. For example, the low seed numbers in self-pollinated apples appear to result from the slower growth of self-pollen tubes, compared to foreign pollen tubes, and rapid ovule degeneration when nitrogen is scarce. On plants growing in soils fertilized with nitrogen, ovules are longer-lived and more seeds develop from self-pollination (Williams, 1965; Hill-Cottingham and Williams, 1967). In most species where selective fruit abortion of self-pollinated flowers has been demonstrated (including apples), the aborted fruits have fewer seeds. In addition to eliminating those fruits that result from low pollen competition, selective abscission of juvenile fruits based on seed number may also eliminate fruits resulting from self-pollinations. The advantages of outcrossing are well known and have often been reviewed (Darwin, 1877; Solbrig, 1976; Maynard Smith, 1978).

Two recent studies suggest that plants can also discriminate among fruits produced by different foreign pollen donors (Bertin, 1982c; Bookman, 1983). In *Campsis radicans,* many immature fruits abort (Bertin, 1980a). Hand pollinations of nearly 2500 flowers on nine plants revealed that the probability that an outcrossed flower will produce a mature fruit depends partly on the pollen donor. Individuals that were favored as donors by one recipient were not necessarily favored by other recipients. The pollen donors favored by particular recipients were usually those whose pollinations resulted in fruit with relatively many large seeds (Bertin, 1982c). In *A. speciosa,* which also aborts a large percentage of its fruits, pollen donors differed significantly in their ability to sire fruits following hand pollination. The pollen donors that produce the most mature fruits on a particular recipient also produce fruits with significantly more seeds and significantly heavier seeds. Moreover, the seeds from fruits of these favored donors produced seedlings that had a significantly greater dry weight after 1 mo (Bookman, 1983).

Janzen (1977) suggested that pollen deposition from multiple donors might have a different effect on fruit production and seed quality than deposition of pollen from a single donor. This could operate at either of two levels: the number of plants that contribute pollen to form the seeds in a given fruit, or the number of plants that contribute pollen to a plant's total fruit/seed crop. The latter type has received some attention. In the orchid, *Encyclia cordigera,* total fruit production is significantly affected by the number of foreign pollen donors (Janzen *et al.*, 1980). However, Janzen *et al.* indicate that these results are equivocal. An important point that deserves closer study is whether a diversity of pollen donors (either within or among fruits on an individual) increases the probability of fruit maturation simply because the chance of including one good donor is high, or whether there is something about the diversity itself that is favored.

That fruits are generally more likely to mature from the first rather than later pollinated flowers has been demonstrated in several species (Van Steveninck, 1957; Tamas *et al.*, 1979; Lee, 1980; Stephenson, 1980; Wyatt, 1980; Bertin, 1982c). Most species with this pattern of maturation have terminal inflorescences with acropetal development (basal to terminal). Consequently, fruits from basal flowers enjoy a spatial as well as a temporal advantage because leaf and root assimilates must pass the lower, more mature fruits en route to higher, younger fruits farther along the inflorescence. When resources are limited, younger fruits are abscised first. Temporal reduction in fruit production has also been noted in species that produce sequential inflorescences along a stem. This pattern of selective maturation tends to minimize the resources wasted by fruit abortion because those fruits that have received the greatest resource investment are preserved (Stephenson, 1981).

If the order of pollination absolutely determines the probability that an initiated fruit will mature or abort, then this pattern of selective maturation is unlikely to be related to the genotypes of the zygotes within the fruit, unless certain pollen-producing plants are more likely to pollinate the first flowers in an inflorescence. However, spatial and temporal advantages are not absolute. Some early pollinated flowers sometimes fail to mature, whereas later ones produce mature fruits (Stephenson, 1979, 1980; Lee, 1980; Wyatt, 1980; Bertin, 1982c). Consequently, other factors must be influencing the pattern of fruit maturation. In a preliminary study of this phenomenon, Bertin (1982c) observed that in *C. radicans* those pollen donors that were favored during the first half of the sequence of hand pollinations were even more favored during the second half of the sequence relative to the nonfavored donors (see above). This suggests that plants become more selective with respect to pollen donors as the number of cross-pollinations increases. Clearly, this possibility merits further investigation.

E. Mechanisms of Female Choice: Summary

This section has reviewed the mechanisms by which the maternal sporophyte may exercise some control over the paternity of her seed crop. It is suggested that floral characteristics and the timing of the floral display of a plant may influence which plants in a population donate pollen (Janzen, 1977). There are, however, too few data to allow proper evaluation of these characteristics from the perspective of female choice. In all probability, events not under the direct control of the maternal sporophyte lead to temporal and spatial variability in the number of pollinated flowers, in the number of pollen grains per stigma, and in the paternal parentage of the donated pollen.

Considerable evidence indicates that a plant's mature seed crop does not represent a random sample of the pollen genotypes deposited onto its stigmas. We know that the various incompatibility systems and differences in the growth rates of pollen tubes with different genotypes lead to nonrandom fertilization. We also know that microgametophytic competition can influence the performance of the resultant offspring and that seed abortion is nonrandom with respect to the genotypes of the developing zygotes. Fruit abortion is nonrandom and is related to the number of seeds per fruit, the paternal parentage of the seeds within a fruit, or the time of pollination. Each of these patterns of abortion may affect the quality and the parentage of the resultant seed crop. Consequently, selective fertilization, seed abortion, and fruit abortion shield a plant from some of the random events associated with pollination (see Mulcahy, 1979).

Traditionally, the different potential mechanisms of female choice have been viewed as separate, isolated phenomena. For example, those who study incompatibility systems rarely consider postzygotic elimination of selfing (Nettancourt, 1977). We suggest that incompatibility systems, pollen competition, and selective seed or fruit abortion represent an integrated genetic sieve, favoring those pollen and zygote genotypes likely to make the greatest contributions to the fitness of the maternal sporophyte and simultaneously allowing a plant to adjust its fruit and seed crop to match the resources available for reproduction (Bertin, 1982c). These mechanisms may operate either separately or in concert.

Pre- and postfertilization mechanisms of female choice may provide different advantages and disadvantages to the maternal sporophyte. In general, the earlier the selection the fewer the resources that are wasted (Lloyd, 1979; Bertin, 1982c). In this regard, sporophytic incompatibility may be the least expensive because rejection occurs on the stigmatic surface. Later forms of rejection progressively waste stylar tissue, ovules, enlarging seeds, or enlarging ovaries. On the other hand, pollen rejection in sporophytic incompatibility systems is based on a rather limited number of alleles possessed by the pollen-producing parent. Later forms of rejection are based on the genotype of the pollen itself, the vigor of the pollen tubes relative to other pollen tubes in the style, the genetic congruity of the microgametophyte and megagametophyte, and characteristics of the fruit relative to other fruits competing for the same resources. In short, some of the later forms of rejection may be based on characteristics more closely related to offspring quality compared to the somewhat arbitrary selection against one or a few alleles in sporophytic incompatibility systems. Furthermore, selection for pollen tube vigor and selective fruit abortion is very flexible. For example, the probability that a given pollen tube will effect fertilization depends not only on the number of competing pollen tubes but also on their genotypes. Fruit abortion allows plants to respond to uncertainties associated with the number of pollinated flowers, fruit and seed predation, and fluctuations in the amount of resources available for reproduction, such as those caused by leaf herbivory (Stephenson, 1981).

The quantity and quality of data supporting these mechanisms of female choice vary greatly. Few data were collected in order to examine directly female choice in plants. Consequently, many of the most important quesions were not addressed. In addition, the examples are drawn predominantly from cultivated species, and their relevance to wild species is uncertain.

Although many plants can exert some control over the paternal parentage of their seed crops, this does not automatically imply that female choice causes intense sexual selection among males. In the extreme case in which each maternal sporophyte accepts pollen from a different donor for her seed crop, female choice would be strong, but its effects on sexual selection among pollen donors would be minimal. It is possible that much female choice in plants is simply positive or negative assortative mating with regard to all or part of the genotype (e.g., self-incompatibility alleles). To distinguish assortative mating from female choice, it is necessary to demonstrate a consistent preference among maternal sporophytes for particular pollen donors. If assortative mating is important, we expect preferred pollens of one maternal sporophyte not necessarily to be the preferred pollens of other sporophytes. Because genetic relatedness is reciprocal, we expect that among hermaphrodites a preference by sporophyte A for the pollen of sporophyte B will be associated with a preference by sporophyte B for the pollen of sporophyte A. This was found among nine individuals of *C. radicans* (Bertin, 1982c), but not among individuals from several populations of *A. speciosa* (Bookman, 1983). In Bookman's study, certain donors tended to be consistently successful (or preferred), suggesting that sexual selection among males may be more important in *Asclepias* than in *Campsis*. Because the *Asclepias* donors came from different populations, however, these results must be treated with caution. More research that distinguishes between assortative mating and consistent preferences by maternal sporophytes is needed before we can ascribe a major role to female choice in determining the intensity of sexual selection in plants.

IV. Discussion

A. Sexual Selection in Plants and Animals

Patterns of sexual selection in plants differ in several ways from those in animals. These differences are due largely to the fact that plants are sedentary, have limited sensory abilities, and have gamete transfer usually mediated by an external agent. Thus, sporophytes have limited influence over where their pollen is deposited or from which other sporophytes pollen is received (Lloyd, 1979).

That such control is not totally lacking is suggested by the known abilities of pollinators to respond to patterns of floral display or energy availability (Heinrich and Raven, 1972; Janzen, 1977; Beach, 1981; and references therein). However, as noted earlier, many important determinants of pollinator movement patterns are above the level of the individual plant and therefore may not be amenable to natural selection. The general effect of such limited control over pollen movement is to increase the potential importance of postpollination competition and choice relative to prepollination competition and choice. Thus, postpollination phenomena in plants are probably much more important than postcopulation phenomena in animals, although more data are needed.

Partly as a consequence of postpollination competition and choice, the highly visible products of sexual selection in animals (certain secondary sexual structures and behaviors) are poorly developed in plants. Instead, many adaptations related to male competition or female choice occur on or in the pistil. Any communication involved is undoubtedly chemical, and is therefore apparent only with chemical methods or by examining its consequences in offspring genotypes.

As a result of the importance of postpollination female choice in some plants and its demonstrated effect of increasing components of offspring quality (Mulcahy *et al.*, 1975, 1978; Bertin, 1982c; Bookman, 1983), selection will exist for female sporophytes to receive many pollen grains in order to maximize the possibility for such choice. In particular, selection may favor adaptations that lead to receipt of many more pollen grains than there are ovules to be fertilized. If true, this would have important consequences for the study of plant reproductive biology. First, it would mean that sufficiency of pollination is not easily determined (specifically it cannot be determined by counting the number of pollen grains on stigmas relative to the number of ovules in ovaries). A knowledge of pollination sufficiency can be important in developing adaptive explanations for breeding systems and floral displays. Second, it indicates that competition among female sporophytes (for pollen) may be much more intense than competition among female animals for mates. Female animals, unlike plants, have senses that allow them to be selective during the precopulation stage, lessening the need for intrasexual female competition. Given this tendency in plants, however, competition among female sporophytes might approach or even exceed that among pollen donors. This could be reflected in variances in female reproductive success equal to or greater than those in males.

In animals, the mating system is generally thought to influence the intensity of sexual selection by affecting the amount of parental investment given to each offspring (see Trivers, 1972; Halliday, 1978). In plants, mating systems such as polygyny, monogamy, or polyandry do not exist. All maternal sporophytes that outcross to some degree may be considered to be promiscuous in terms of which and how many sporophytes deposit pollen. Plants do, however, differ greatly in their degree of self-fertilization. We have argued that the degree of self- and cross-fertilization influences the intensity of sexual selection by affecting the variance in the reproductive output of the two sexual functions of hermaphrodites (Willson, 1979, 1982). When self-fertilization predominates, there can be little difference in the variance between the male and female functions. As the proportion of flowers that are outcrossed increases, there is an increasing possibility for greater variance in the reproductive output of one of the sexual functions.

A major difference between plants and those animals in which sexual selection has been considered is that plants are mostly bisexual and animals are mostly unisexual. That sexual selection can operate among hermaphrodites has been suggested by Bateman (1948), Charnov (1979), and Willson (1979).

Interpretations of reproductive traits of hermaphrodites in terms of sexual selection overlap with explanations based on the theory of optimal resource allocation. Evolutionary modifications in one sexual function of a hermaphrodite could involve (1) allocation of additional resources from nonreproductive functions (e.g., less leaf tissue, more pollen), (2) reallocation of resources within a sexual function (e.g., more but smaller pollen grains), or (3) increased investment in one sexual function at the expense of the other (e.g., fewer ovules, more pollen). The last possibility is the same as natural selection favoring an optimal allocation (in terms of fitness) of resources between sexual functions (Ross, 1982). Thus, optimal resource allocation explanations (Charnov *et al.*, 1976; Maynard Smith, 1978) and sexual selection explanations (Charnov, 1979; Willson, 1979; Bawa, 1980a) of reproductive phenomena such as breeding systems overlap. The same, of course, is true in hermaphroditic animals.

B. Sexual Selection and the Evolution of Breeding Systems

Recent papers have suggested that sexual selection may have contributed to the evolution of various plant breeding systems, including andromonoecy, heterostyly, and especially dioecy (Willson, 1979, 1982; Bawa, 1980a; Givnish, 1980, 1982; Beach, 1981; Casper and Charnov, 1982).

A comprehensive review of these ideas is beyond the scope and spatial limitations of this chapter. However, the potential roles of sexual selection in the evolution of breeding systems have been discussed in part, by Robert Wyatt in Chapter 5 of this volume.

V. Conclusion

The theory of sexual selection provides a new perspective on the evolution of plant reproductive traits. It has the potential for explaining and uniting many previously unrelated aspects of reproduction, such as flowering patterns, pollen tube growth rates, and patterns of fruit and seed abortion. However, uncritical enthusiasm for the role of sexual selection in shaping breeding systems and reproductive strategies is not warranted for several reasons. First, sexual selection is unlikely to be an important force in the evolution of highly autogamous species or those species in which seed production is consistently limited by pollination. In these species, there would be little opportunity for male competition. In addition, relevant data on male competition and female choice prior to pollen deposition are almost totally missing for xenogamous species. Postpollination data are available, but are fragmentary or based on cultivated species. Finally, there are no complete and critical demonstrations of male competition or female choice for any natural populations.

Although we caution against unbridled enthusiasm for the role of sexual selection in the evolution of plant reproductive characteristics, we also suggest that investigators would be remiss to reject totally the importance of sexual selection in plants. There is little question that in xenogamous species the maternal sporophyte makes a greater resource investment per offspring than the paternal sporophyte. Furthermore, resources rather than pollination limit reproduction in many species. Consequently, the resource investment of the maternal sporophyte can be viewed as a limiting resource for which pollen donors may potentially compete. It is known that the number and distribution of flowers and the quantity and quality of the floral rewards can affect both the attraction and the movements of floral visitors. It has been shown that the growth rates of pollen tubes are influenced by their genotypes, and that there is a correlation between the competitive ability of microgametophytes and the quality of the resultant offspring. Finally, a plant may exercise some control over the paternal parentage of its seed crop by influencing the germination and growth of pollen and by preferentially maturing fruits and seeds. Consequently, plants possess known mechanisms by which male competition and female choice can occur.

Although the available data will not permit us to state the importance of sexual selection in the evolution of plant reproductive characteristics, we can say that the application of sexual selection theory to plants will benefit the way we think about plant reproduction. First, it focuses equal attention on the male and female functions of hermaphroditic species. Until recently, it has been tacitly assumed that the forces which influence fruit and seed production are sufficient to understand plant reproduction; consequently, the forces that shape the other half of the genetic contribution to future generations have been ignored. Second, because sexual selection can operate both before and after pollen deposition, it forces us to view plant reproduction as a highly integrated series of events rather than a sequence of isolated or independent phenomena. Traditionally, pollination biology, pollen–style interactions, and fruit and seed development have been separate areas of inquiry. In terms of which pollen donors father a seed crop, however, the events associated with each of these phases of reproduction can be viewed as complementary and alternative means to the same end. Finally, sexual selection provides a critical alternative to the exclusive consideration of arguments based on survival selection for the evolution of breeding systems and plant reproductive strategies. With the recent explosion of interest in plant reproductive biology, we can undoubtedly look forward to the data that will allow us to separate the fact from the fiction in this chapter.

Acknowledgments

We thank N. Burley, D. Janzen, D. Schemske, M. Willson, R. Wyatt, and especially J. Kurland for comments and criticisms on a previous draft, and S. Bookman and T. Lee for insightful discussions of the ideas presented here. The junior author expresses special thanks to D. Schemske and M. Willson for discussions on topics in this chapter prior to its conception. Many thoughts presented here appear to have been developed concurrently and often independently by them. During the preparation of this chapter, we were supported by NSF grants DEB 8105198 (A.G.S.) and DEB 8206465 (R.I.B.)

References

Anderson, G. J. (1979). Dioecious *Solanum* species of hermaphroditic origin is an example of a broad convergence. *Nature (London)* **282,** 836–838.

Armbruster, W. S., and Webster, G. L. (1979). Pollination of two species of *Dalechampia* (Euphorbiaceae) in Mexico by euglossine bees. *Biotropica* **11,** 278–283.

Ascher, P. D., and Peloquin, S. J. (1966). Effect of floral aging on the growth of compatible and incompatible pollen tubes in *Lilium longiflorum. Am. J. Bot.* **53,** 99–102.

Augspurger, C. K. (1980). Mass-flowering of a tropical shrub (*Hybanthus prunifolius*): Influence on pollinator attraction and movement. *Evolution (Lawrence, Kans.)* **34,** 475–488.

Baker, H. G. (1947). Biological flora of the British Isles. *Melandrium* (Roehling em.) Fries. *J. Ecol.* **35,** 271–292.

Barnes, D. K., and Cleveland, R. W. (1963a). Pollen tube growth of diploid alfalfa *in vitro. Crop Sci.* **3,** 291–295.

Barnes, D. K., and Cleveland, R. W. (1963b). Genetic evidence for nonrandom fertilization in alfalfa as influenced by differential pollen tube growth. *Crop Sci.* **3,** 295–297.

Bateman, A. J. (1948). Intra-sexual selection in *Drosophila. Heredity* **2,** 349–368.

Bateman, A. J. (1956). Cryptic self-incompatibility in the wall flower: *Cheiranthus cheiri* L. *Heredity* **10,** 257–261.

Bawa, K. S. (1977). The reproductive biology of *Cupania guatemalensis* Radlk. (Sapindaceae). *Evolution (Lawrence, Kans.)* **31,** 52–63.

Bawa, K. S. (1980a). Evolution of dioecy in flowering plants. *Annu. Rev. Ecol. Syst.* **11,** 15–39.

Bawa, K. S. (1980b). Mimicry of male by female flowers and intrasexual competition for pollinators in *Jacaratia dolichaula* (D. Smith) Woodson (Caricaceae). *Evolution (Lawrence, Kans.)* **34,** 467–474.

Bawa, K. S., and Beach, J. H. (1981). Evolution of sexual systems in flowering plants. *Ann. M. Bot. Gard.* **68,** 254–274.

Bawa, K. S., and Opler, P. A. (1975). Dioecism in tropical forest trees. *Evolution (Lawrence, Kans.)* **29,** 167–179.

Bawa, K. S., and Opler, P. A. (1978). Why are pistillate inflorescences of *Simarouba glauca* eaten less than staminate inflorescences? *Evolution (Lawrence, Kans.)* **32,** 673–676.

Bawa, K. S., Keegan, C. R., and Voss, R. H. (1982). Sexual dimorphism in *Aralia nudicaulis* L. (Araliaceae). *Evolution (Lawrence, Kans.)* **36,** 171–378.

Beach, F. A., and LeBoeuf, B. J. (1967). Coital behaviour in dogs. I. Preferential mating in the bitch. *Anim. Behav.* **15,** 546–558.

Beach, J. H. (1981). Pollinator foraging and the evolution of dioecy. *Am. Nat.* **118,** 572–577.

Bell, C. R. (1971). Breeding systems and floral biology of the Umbelliferae or evidence for specialization in unspecialized flowers. *In* "The Biology and Chemistry of the Umbelliferae" (V. H. Heywood, ed.), pp. 93–107. Academic Press, New York.

Bemis, W. P. (1959). Selective fertilization in lima beans. *Genetics* **44,** 555–562.

Benseler, R. W. (1975). Floral biology of California buckeye. *Madrono* **23,** 41–53.

Bertin, R. I. (1980). "The Reproductive Biologies of Some Hummingbird-Pollinated Plants." Ph.D. Thesis, Univ. of Illinois, Urbana-Champaign.

Bertin, R. I. (1982a). Floral biology, hummingbird pollination and fruit production of trumpet creeper (*Campsis radicans,* Bignoniaceae). *Am. J. Bot.* **69,** 122–134.

Bertin, R. I. (1982b). The ecology of sex expression in red buckeye. *Ecology* **63,** 445–456.

Bertin, R. I. (1982c). Paternity and fruit production in trumpet creeper (*Campsis radicans*). *Am. Nat.* **119,** 694–709.

Bertin, R. I. (1982d). The evolution and maintenance of andromonoecy. *Evol. Theor.* **6,** 25–32.

Bierzychudek, P. (1981). Pollinator limitation of plant reproductive effort. *Am. Nat.* **117,** 838–840.

Bino, R. J., and Meeuse, A. D. J. (1981). Entomophily in dioecious species of *Ephedra:* A preliminary report. *Acta Bot. Neerl.* **30,** 151–153.

Bollard, E. G. (1970). The physiology and nutrition of developing fruits. *In* "The Biochemistry of Fruits and their Products" (A. C. Hulme, ed.), pp. 387–427. Academic Press, New York.

Bookman, S. S. (1983). A demonstration of sexual selection in plants (*Asclepias speciosa* Torr.). *Evolution* (*Lawrence, Kans.*) (in press).

Borgia, G. (1979). Sexual selection and the evolution of mating systems. *In* "Sexual Selection and Reproductive Competition in Insects" (M. S. Blum and N. A. Blum, eds.), pp. 19–80. Academic Press, New York.

Brewbaker, J. L. (1957). Pollen cytology and self-incompatibility systems in plants. *J. Hered.* **48,** 271–277.

Brewbaker, J. L., and Kwack, B. H. (1963). The essential role of calcium ion in pollen tube germination and pollen tube growth. *Am. J. Bot.* **50,** 859–865.

Brewbaker, J. L., and Majumder, S. K. (1961). Cultural studies of the pollen population effect and the self-incompatibility inhibition. *Am. J. Bot.* **48,** 457–464.

Buchholz, J. T., and Blakeslee, A. F. (1927). Abnormalities in pollen tube growth in *Datura* due to the gene "tricarpel." *Proc. Natl. Acad. Sci. U.S.A.* **13,** 242–249.

Bullock, S. H., and Bawa, K. S. (1981). Sexual dimorphism and the annual flowering pattern in *Jacaratia dolichaula* (D. Smith) Woodson (Caricaceae) in a Costa Rica rain forest. *Ecology* **62,** 1494–1504.

Casper, B. B., and Charnov, E. L. (1982). Sex allocation in heterostylous plants. *J. Theor. Biol.* **96,** 143–150.

Casper, B. B., and Wiens, D. (1981). Fixed rates of random ovule abortion in *Cryptantha flava* (Boraginaceae) and its possible relation to seed dispersal. *Ecology* **62,** 866–869.

Catchpole, C. K. (1980). Sexual selection and the evolution of complex songs among European warblers of the genus *Acrocephalus*. *Behaviour* **74,** 149–166.

Charnov, E. L. (1979). Simultaneous hermaphroditism and sexual selection. *Proc. Natl. Acad. Sci. U.S.A.* **76,** 2480–2484.

Charnov, E. L., Maynard Smith, J., and Bull, J. J. (1976). Why be an hermaphrodite? *Nature* (*London*) **263,** 125–126.

Conn, J. S. (1981). Phenological differentiation between the sexes of *Rumex hastatulus* niche partitioning or different optimal reproductive strategies? *Bull. Torrey Bot. Club* **108,** 374–378.

Conn, J. S., and Blum, U. (1981). Differentiation between the sexes of *Rumex hastatulus* in net energy allocation, flowering and height. *Bull. Torrey Bot. Club* **108,** 446–455.

Connor, H. E. (1979). Breeding systems in the grasses: a survey. *N. Z. J. Bot.* **17,** 547–574.
Coombe, B. G. (1976). The development of fleshy fruits. *Annu. Rev. Plant Physiol.* **27,** 507–528.
Cooper, D. C., and Brink, R. A. (1940). Partial self-incompatibility and the collapse of fertile ovules as factors affecting seed formation in alfalfa. *J. Agric. Res.* **60,** 453–472.
Cooper, D. C., Brink, R. A., and Albrecht, H. R. (1937). Embryo mortality in relation to seed formation in alfalfa (*Medicago sativa*) *Am. J. Bot.* **24,** 203–213.
Correns, C. (1914). "Die Bestimmung und Vererbung des Gesclechtes nach neven Versuchen mit hoheren Pflanzen." Berlin, Borntrager.
Correns, C. (1928). Bestimmung, Vererbung and Verteilung des Geschlechtes bei den hoheren pflanzen. *Handb. Vererbungsw.* **2,** 1–138.
Critchfield, W. B. (1980). Genetics of lodgepole pine. *U.S. Dept. Agric. For. Serv. Res. Pap. WO-37* p. 57.
Crowe, L. K. (1971). The polygenic control of outbreeding in *Borago Officinalis. Heredity* **27,** 111–118.
Cruden, R. W. (1976). Fecundity as a function of nectar production and pollen-ovule ratios. *In* "Tropical Trees: Variation, Breeding and Conservation" (J. Burley and B. T. Styles, eds.), pp. 171–178. Academic Press, New York.
Cruden, R. W. (1977). Pollen-ovule ratios: A conservative indicator of breeding systems in flowering plants. *Evolution* (*Lawrence, Kans.*) **31,** 32–46.
Cruden, R. W., and Hermann-Parker, S. M. (1977). Temporal dioecism: An alternative to dioecism? *Evolution* (*Lawrence, Kans.*) **31,** 863–866.
Dancik, B. P., and Barnes, B. V. (1972). Natural variation and hybridization of yellow birch and bog birch in southeastern Michigan. *Silvae Genet.* **21,** 1–9.
Dane, F., and Melton, B. (1973). Effect of temperature on self- and cross-compatibility and *in vitro* pollen growth characteristics in alfalfa. *Crop Sci.* **13,** 587–591.
Darwin, C. (1859). "On the Origin of Species." Murray, London.
Darwin, C. (1871). "The Descent of Man and Selection in Relation to Sex." Murray, London.
Darwin, C. R. (1877). "The different Forms of Flowers on Plants of the Same Species." Murray, London.
Davis, M. A. (1981). The effect of pollinators, predators, and energy constraints on the floral ecology and evolution of *Trillium erectum. Oecologia* **48,** 400–406.
Dewsbury, D. A. (1982). Ejaculate cost and male choice. *Am. Nat.* **119,** 601–610.
Dickmann, D. I., and Kozlowski, T. T. (1969). Seasonal growth patterns of ovulate strobili of *Pinus resinosa* in central Wisconsin. *Can. J. Bot.* **47,** 839–848.
Dodson, C. H. (1962). Pollination and variation in the subtribe Catasetinae (orchidaceae). *Ann. M. Bot. Gard.* **49,** 35–56.
Dulberger, R., Levy, A., and Palevitch, D. (1981). Andromonoecy in *Solanum* marginatum. *Bot. Gaz.* (*Chicago*) **142,** 259–266.
Eberhard, W. G. (1979). The function of horns in *Podischnus agenor* (Dynastinae) and other beetles. *In* "Sexual Selection and Reproductive Competition in Insects" (M. S. Blum and N. A. Blum, eds.), pp. 231–258. Academic Press, New York.
Ehrenfeld, J. G. (1979). Pollination of three species of *Euphorbia* subgenus *Chamaesyce,* with special reference to bees. *Am. Midl. Nat.* **101,** 87–98.
Ehrman, L. (1972). Genetics and sexual selection. *In* "Sexual Selection and the Descent of Man 1871–1971" (B. Campbell, ed.), pp. 180–203. Aldine, Chicago.
Esler, A. E. (1969). Leaf fall and flowering of nikau. *Wellington Bot. Soc. Bull.* **36,** 19–22.
Fernald, M. L. (1950). "Gray's Manual of Botany." Van Nostrand-Reinold, Princeton, New Jersey.
Fisher, R. A. (1958). "The genetical Theory of Natural Selection." Dover, New York.
Fisher, R. A. (1965). "The Theory of Inbreeding." Academic Press, New York.
Free, J. B. (1965). The ability of bumblebees and honeybees to pollinate red clover. *J. Appl. Ecol.* **2,** 289–294.

Free, J. B. (1966). The foraging area of honeybees in an orchard of standard apple trees. *J. Appl. Ecol.* **3,** 261–268.

Funk, D. T. (1970). Genetics of black walnut (*Juglans nigra*). *U.S. Dept. Agric. For. Serv. Res. Pap. WO-10* p. 13.

Garnock-Jones, P. J. (1976). Breeding systems and pollination in New Zealand *Parahebe* (Scrophulariaceae). *N. Z. J. Bot.* **14,** 291–298.

Gilbert, L. E. (1975). Ecological consequences of a coevolved mutalism between butterflies and plants. "Coevolution of Animals and Plants" (L. E. Gilbert and P. H. Raven, eds.), pp. 210–240. Univ. Texas Press, Austin.

Gill, F. B., and Wolf, L. L. (1977). Nonrandom foraging by sunbirds in a patchy environment. *Ecology* **58,** 1284–1296.

Givnish, T. J. (1980). Ecological constraints on the evolution of breeding systems in seed plants: Dioecy and dispersal in gymnosperms. *Evolution (Lawrence, Kans.)* **34,** 959–972.

Givnish, T. J. (1982). Outcrossing versus ecological constraints in the evolution of dioecy. *Am. Nat.* **119,** 849–865.

Godley, E. J. (1976). Sex ratio in *Clematis gentianoides* DC. *N. Z. J. Bot.* **14,** 299–306.

Godley, E. J. (1979). Flower biology in New Zealand. *N. Z. J. Bot.* **17,** 441–466.

Guitierrez, M. G., and Sprague, G. F. (1959). Randomness of mating in isolated polycross plantings of maize. *Genetics* **44,** 1075–1082.

Haldane, J. B. S. (1932). "The Causes of Evolution." Harper, New York.

Halliday, T. R. (1978). Sexual selection and mate choice. *In* "Behavioural Ecology: An Evolutionary Approach" (J. R. Krebs and N. B. Davies, eds.), pp. 180–213. Blackwell, Oxford.

Hamilton, W. D. (1979). Wingless and fighting males in fig wasps and other insects. *In* "Sexual Selection and Reproductive Competition in Insects" (M. S. Blum and N. A. Blum, eds.), pp. 167–220. Academic Press, New York.

Heinicke, A. J. (1917). Factors influencing the abscission of flowers and partially developed fruits of the apple (Pyrus Malus L.) *Bull. N.Y. Agric. Exp. Stn. (Ithaca)* **393,** 43–114.

Heinrich, B. (1975). Energetics of pollination. *Annu. Rev. Ecol. Syst.* **6,** 139–170.

Heinrich, B. (1979). Resource heterogeneity and patterns of movement in foraging bumblebees. *Oecologia* **40,** 235–245.

Heinrich, B., and Raven, P. H. (1972). Energetics and pollination ecology. *Science* **176,** 597–602.

Heslop-Harrison, J. (1975). Incompatibility and the pollen stigma interaction. *Annu. Rev. Plant Physiol.* **26,** 403–425.

Heslop-Harrison, J. (1982). Pollen-stigma interaction and cross-compatibility in the grasses. *Science* **215,** 1358–1364.

Hill-Cottingham, D. G., and Williams, R. R. (1967). Effect of time of application of fertilizer nitrogen on the growth, flower development, and fruit set of maiden apple trees, var. Lord Lambourne, and on the distribution of total nitrogen within the trees. *J. Hortic. Sci.* **42,** 319–38.

Hoff, V. J. (1962). An analysis of outcrossing in certain complex-heterozygous Evoenotheras. I. Frequency of outcrossing. *Am. J. Bot.* **49,** 715–721.

Hornby, C. A., and Charles, W. B. (1966). Pollen germination as affected by variety and number of pollen grains. *Tomato Growers Co-Operative* **16,** 11.

Horovitz, A., and Harding, J. (1972). The concept of male outcrossing in hermaphrodite higher plants. *Heredity* **29,** 223–236.

Horovitz, A., Meiri, L., and Beiles, A. (1976). Effects of ovule position in fabaceous flowers on seed set and outcrossing rates. *Bot. Gaz. (Chicago)* **137,** 250–254.

Huxley, J. S. (1938). The present standing of the theory of sexual selection. *In* "Evolution: Essays on Aspects of Evolutionary Biology Presented to Prof. E. S. Godrich on his Seventieth Birthday (G. R. de Beer, ed.), pp. 11–42. Oxford Univ. Press (Clarendon), London and New York.

Huxley, J. (1942). "Evolution, the Modern Synthesis." Harper, New York.

Janzen, D. H. (1971). Seed predation by animals. *Annu. Rev. Ecol. Syst.* **2,** 465–492.
Janzen, D. H. (1977). A note on optimal mate selection by plants. *Am. Nat.* **111,** 365–371.
Janzen, D. H., DeVries, P., Gladstone, D. E., Higgins, M. L., and Lewinsohn, T. M. (1980). Self- and cross-pollination of *Encyclia cordigera* (Orchidaceae) in Santa Rosa National Park, Costa Rica. *Biotropica* **12,** 72–74.
Jennings, D. L., and Topham, P. B. (1971). Some consequences of respberry pollen dilution for its germination and or fruit development. *New Phytol.* **70,** 371–380.
Jimenez, J. R., and Nelson, O. E. (1965). A new fourth chromosome gametophyte locus in maize. *J. Hered.* **56,** 259–263.
Johnson, C. M., and Mulcahy, D. L. (1978). Male gametophyte in maize. II. Pollen vigor in inbred plants. *Theor. Appl. Genet.* **51,** 211–215.
Johnson, L. K. (1982). Sexual selection in a brentid weevil. *Evolution (Lawrence, Kans.)* **36,** 251–262.
Johnsson, H. (1976). Contributions to the genetics of empty grains in the seed of pine (*Pinus sylvestris*). *Silvae Genet.* **25,** 10–15.
Jones, D. F. (1920). Selective fertilization in pollen mixtures. *Biol. Bull. (Woods Hole, Mass.)* **38,** 251–289.
Jong, P. C. de. (1976). Flowering and sex expression in *Acer* L. Mededelingen Landbouwhogeschool Wageningen.
Kaul, R. B. (1979). Inflorescence architecture and flower sex ratios in *Sagittaria brevirostra* (Alismataceae). *Am. J. Bot.* **66,** 1062–1066.
Kay, Q. O. N. (1978). The role of preferential and assortative pollination in the maintenance of flower colour polymorphisms. *Linn. Soc. Symp. Ser. No. 6,* pp. 175–190.
Kendall, D. A., and Smith, B. D. (1975). The foraging behaviour of honeybees on ornamental *Malus* spp. used as pollinizers in apple orchards. *J. Appl. Ecol.* **12,** 465–471.
Kingett, P. D., Lambert, D. M., and Telford, S. R. (1981). Does mate choice occur in *Drosophila melanogaster? Nature (London)* **293,** 492.
Lamont, B. (1982). The reproductive biology of *Grevillea leucopteris* (Proteaceae), including reference to its glandular hairs and colonizing potential. *Flora* **172,** 1–20.
LeBouef, B. J. (1974). Male-male competition and reproductive success in elephant seals. *Am. Zool.* **14,** 163–176.
Lee, T. D. (1980). "Demographic analysis of the reproductive process in an annual legume, *Cassia fasciculata.*" Ph.D. Dissertation, Univ. of Illinois, Urbana.
Lee, T. D., and Bazzaz, F. A. (1982). Regulation of fruit maturation pattern in an annual legume, *Cassia fasciculata. Ecology* **63,** 1374–1388.
Levin, D. A. (1972). The adaptedness of corolla-color variants in experimental and natural populations of *Phlox drummondii. Am. Nat.* **106,** 57–70.
Levin, D. A., and Berube, D. E. (1972). *Phlox* and *Colias:* The efficiency of a pollination system. *Evolution (Lawrence, Kans.)* **26,** 242–250.
Levin, D. A., and Kerster, H. W. (1969). The dependence of bee-mediated pollen and gene dispersal upon plant density. *Evolution (Lawrence, Kans.)* **23,** 560–571.
Lewis, D. (1976). Incompatibility in flowering plants. *In* "Receptors and Recognition Series A, Vol. 2" (P. Cuatrecasas and M. F. Greaves, eds.), pp. 141–178. Chapman and Hall, London.
Lewis, D. (1979). "Sexual Incompatibility in Plants." Arnold, London.
Linck, A. J. (1961). The morphological development of the fruit of *Pisum sativium* var. *Alaska. Phytomorphology* **2,** 79–84.
Lindsey, A. H. (1982). Floral phenology patterns and breeding systems in *Thaspium* and *Zizia* (Apiaceae). *Syst. Bot.* **7,** 1–12.
Lloyd, D. G. (1972). Breeding systems in *Cotula* L. (Compositae, Anthemideae). II. Monoecious populations. *New Phytol.* **71,** 1195–1201.

Lloyd, D. G. (1979). Parental strategies of angiosperms. *N. Z. J. Bot.* **17,** 595–606.

Lloyd, D. G. (1980). The distribution of gender in four angiosperm species illustrating two evolutionary pathways to diocy. *Evolution (Lawrence, Kans.)* **34,** 123–134.

Lloyd, D. G., and Webb, C. J. (1977). Secondary sex characters in plants. *Bot. Rev.* **43,** 177–216.

Lloyd, D. G., and Yates, J. M. A. (1982). Intrasexual selection and the segregation of pollen by stigmas in hermaphroditic plants, exemplified by *Wahlenbergia albomarginata* (Campanulaceae). *Evolution (Lawrence, Kans.)* **36,** 903–913.

Lock, J. M., and Hall, J. B. (1982). Floral biology of *Mallotus oppositifolius* (Euphorbiaceae). *Biotropica* **14,** 153–155.

Lovett Doust, J., and Harper, J. L. (1980). The resource costs of gender and maternal support in an andromonoecious umbellifer, *Smyrnium olusatrum* L. *New Phytol.* **85,** 251–264.

Lundquist, A. U., Osterbye, U., Lorsen, K., and Linde-Laursen, I. (1973). Complex self incompatibility systems in *Ranunculus acris* L. and *Beta vulgaris* L. Hereditas **74,** 161–168.

Lynch, S. P. (1977). The floral ecology of *Asclepias solanoana* Woods. *Madrono* **24,** 159–177.

McAlpine, D. K. (1979). Agonistic behavior in *Achias australis* (Diptera, Platystomatidae) and the significance of eyestalks. *In* "Sexual Selection and Reproductive Competition in Insects" (M. S. Blum and N. A. Blum, eds.), pp. 221–230. Academic Press, New York.

McGregor, S. E., Alcorn, S. M., Kurtz, E. G., Jr., and Butler, G. D., Jr. (1959). Bee visits to saguaro flowers. *J. Econ. Entomol.* **52,** 1002–1004.

Maini, J. S. (1972). Silvics and ecology in Canada. *U.S. Dept Agric. For. Serv. Gen. Tech. Rep. NC-1* pp. 67–73.

Martin, F. W. (1972). Sterile styles in *Solanum mammosum. Phyton* **29,** 127–134.

Martin, D., Lewis, T. L., and Cerny, J. (1961). Jonathan spot-three factors related to incidence: fruit size, breakdown and seed numbers. *Aust. J. Agric. Res.* **12,** 1039–1049.

Matheson, A. C. (1980). Unexpectedly high frequencies of outcrossed seedlings among offspring from mixtures of self and cross pollen in *Pinus radiata* D. Don. *Aust. For. Res.* **10,** 21–27.

Matzke, E. B. (1938). Inflorescence patterns and sexual expression in *Begonia semperflorens. Am. J. Bot.* **25,** 465–478.

Maynard Smith, J. (1956). Fertility, mating behaviour and sexual selection in *Drosophila subobscura. J. Genet.* **54,** 261–279.

Maynard Smith, J. (1978). "The Evolution of Sex." Cambridge Univ. Press, London and New York.

Mogford, D. J. (1978). Pollination and flower colour polymorphism, with special reference to *Cirsium palustre. Linn. Soc. Symp. Ser. No. 6,* pp. 191–199.

Moore, L. B. (1977). The flowers of *Ascarina lucida* Hook. f. (Chloranthaceae). *N. Z. J. Bot.* **15,** 491–494.

Mulcahy, D. L. (1971). A correlation between gametophytic and sporophytic characteristics in *Zea mays* L. *Science* **171,** 1155–1156.

Mulcahy, D. L. (1974). Adaptive significance of gametic competition. *In* "Fertilization in Higher Plants" (H. F. Linskens, ed.), pp. 27–30. North-Holland Publ., Amsterdam.

Mulcahy, D. L. (1979). The rise of the angiosperms: A genecological factor. *Science* **206,** 20–23.

Mulcahy, D. L., and Mulcahy, G. B. (1975). The influence of gametophytic competition on sporophytic quality in *Dianthus chinensis. Theor. Appl. Genet.* **46,** 277–280.

Mulcahy, D. L., Mulcahy, G. B., and Ottaviano, E. (1975). Sporophytic expression of gametophytic competition in *Petunia hybrida. In* "Gamete Competition in Plants and Animals" (D. L. Mulcahy, ed.), pp. 227–232. North-Holland Publ., Amsterdam.

Mulcahy, D. L., Mulcahy, G. B., and Ottaviano, E. (1978). Further evidence that gametophytic selection modifies the genetic quality of the sporophyte. *Soc. Bot. Fr. Actualities Bot. nl-2* pp. 57–60.

Mulcahy, D. L., Curtis, P. S., and Snow, A. A. (1982). Pollen competition in a natural population of

Geranium maculatum. *In* "Handbook of Experimental Pollination Biology" (C. E. Jones and R. J. Little, eds.), pp. 101–123. Academic Press, New York.

Murneek, A. E. (1954). The embryo and endosperm in relation to fruit development, with special reference to the apple, *Malus sylvestris*. *Proc. Amer. Soc. Hortic. Sci.* **64,** 573–582.

Nakatsuru, K., and Kramer, D. L. (1982). Is sperm cheap? Limited male fertility and female choice in the lemon tetra (Pisces, Characidae). *Science* **216,** 753–755.

Nettancourt, D. de. (1977). "Incompatibility in angiosperms," Springer-Verlag, Berlin and New York.

Ockendon, D. J., and Currah, L. (1978). Time of cross- and self-pollination affects the amount of self-seed set by partially self-incompatible plants of *Brassica oleracea*. *Theor. Appl. Genet.* **52,** 233–237.

O'Donald, P., Wedd, N. S., and Davis, J. W. F. (1974). Mating preferences and sexual selection in the Arctic skua. *Heredity* **33,** 1–16.

Onyekwelu, S. S., and Harper, J. L. (1979). Sex ratio and niche differentiation in spinach (*Spinacia oleracea*). *Nature (London)* **282,** 609–611.

Opler, P. A., and Bawa, K. S. (1978). Sex ratios in tropical forest trees. *Evolution (Lawrence, Kans.)* **32,** 812–821.

Ottaviano, E., Sari Gorla, M., and Mulcahy, D. L. (1975). Genetic and intergametophytic influences on pollen tube growth. *In* "Gamete Competition in Plants and Animals" (D. L. Mulcahy, ed.), pp. 125–130. North-Holland Publ., Amsterdam.

Pandey, K. K. (1968). Colchicine-induced changes in the self-incompatibility behavior of *N. zotiana*. *Genetica* **39,** 257–271.

Parker, G. A. (1970). Sperm competition and its evolutionary effect on copula duration in the fly *Scatophaga stercoraria*. *J. Insect Physiol.* **16,** 1301–1328.

Partridge, L. (1980). Mate choice increases a component of offspring fitness in fruit flies. *Nature (London)* **283,** 290–291.

Payne, R. B. (1979). Sexual selection and intersexual differences in variance of breeding success. *Am. Nat.* **114,** 447–452.

Pfahler, P. L. (1965). Fertilization ability of maize pollen grains I. pollen sources. *Genetics* **52,** 513–520.

Pfahler, P. L. (1967). Fertilization ability of maize pollen grains. II. Pollen genotype, female sporophyte and pollen storage interactions. *Genetics* **57,** 513–521.

Prahler, P. L. (1974). Pollen genotype studies in maize (*Zea mays* L.). *In* "Fertilization in Higher Plants" (H. F. Linskens, ed.), pp. 3–14. North-Holland Publ., Amsterdam.

Pfahler, P. L., and Linskens, H. F. (1972). *In vitro* germination and pollen tube growth of maize (*Zea mays* L.) pollen. VI. Combined effects of storage and the alleles at the waxy (wx), sugary (su_1) and shrunken (sh_2) loci. *Theor. Appl. Genet.* **42,** 136–140.

Primack, R. B., and Lloyd, D. G. (1980). Andromonoecy in the New Zealand montane shrub manuka, *Leptospermum scoparium* (Myrtaceae). *Am. J. Bot.* **67,** 361–368.

Pyke, G. H. (1978). Optimal foraging: Movement patterns of bumblebees between inflorescences. *Theor. Pop. Biol.* **13,** 72–98.

Pyke, G. H. (1981). Optimal foraging in nectar-feeding animals and coevolution with their plants. *In* "Foraging Behavior: Ecological, Ethological, and Psychological Approaches" (A. C. Kamil and T. D. Sargent, eds.), pp. 19–38. Garland STPM Press, New York.

Pyke, G. H. (1982). Foraging in bumblebees: Rule of departure from an inflorescence. *Can. J. Zool.* **60,** 417–428.

Quinlan, J. D., and Preston, A. P. (1968). Effects of thinning blossoms and fruitlets on growth and cropping of sunset apple. *J. Hortic. Sci.* **43,** 373–381.

Rehfeldt, G. E. (1978). The genetic structure of a population of Douglas-fir (Pseudotsuga menziesii var. glauca) as reflected by its wind-pollinated progenies. *Silvae Genetica* **27,** 49–52.

Roeser, J., Jr. (1941). Some aspects of flower and cone production in ponderosa pine. *J. For.* **39,** 534–536.

Ross, M. D. (1982). Five evolutionary pathways to subdioecy. *Am. Nat.* **119,** 297–318.

Sari-Gorla, M., Ottaviano, E., and Faini, D. (1975). Genetic variability of gametophyte growth rate in maize. *Theor. Appl. Genet.* **46,** 289–294.

Sarvas, R. (1962). Investigations on the flowering and seed crop of *Pinus sylvestris. Commun. Inst. For. Fenn.* **53,** 1–198.

Sarvas, R. (1968). Investigations on the flowering and seed crop of *Picea abies. Commun. Inst. For. Fenn.* **67,** 1–84.

Sayers, E. R., and Murphy, R. P. (1966). Seed set in alfalfa as related to pollen tube growth, fertilization frequency, and post-fertilization ovule abortion. *Crop Sci.* **6,** 365–368.

Schaal, B. A. (1980). Reproductive capacity and seed size in *Lupinus texensis. Am. J. Bot.* **67,** 703–709.

Schaffer, W. M., and Schaffer, M. V. (1979). The adaptive significance of variations in reproductive habit in the Agavaceae II: Pollinator foraging behavior and selection for increased reproductive expenditure. *Ecology* **60,** 1051–1069.

Schemske, D. W. (1978). Evolution of reproductive characteristics in *Impatiens* (Balsaminaceae): the significance of cleistogamy and chasmogamy. *Ecology* **59,** 596–613.

Schemske, D. W. (1980a). Evolution of floral display in the orchid *Brassavola nodosa. Evolution (Lawrence, Kans.)* **34,** 489–493.

Schemske, D. W. (1980b). Floral ecology and hummingbird pollination of *Combretum farinosum* in Costa Rica. *Biotropica* **12,** 169–181.

Schemske, D. W., and Fenster, C. (1983). Pollen-grain interactions in a neotropical *Costus:* effects of clump size and competitors. *In* "Pollen Biology: Basic and Applied Aspects" (D. L. Mulcahy and E. Ottaviano, Eds.), Elsevier, New York.

Schemske, D. W., Willson, M. F., Melampy, M. N., Miller, L. J., Verner, L., Schemske, K. M., and Best, L. B. (1978). Flowering phenology of some spring woodland herbs. *Ecology* **59,** 351–366.

Schlessman, M. A. (1982). Expression of andromonoecy and pollination of tuberous *Lomatiums* (Umbelliferae). *Syst. Bot.* **7,** 134–149.

Schoen, D. J. (1977). Floral biology of *Diervilla lonicera* (Caprifoliaceae). *Bull. Torrey Bot. Club* **104,** 234–240.

Schwartz, D. (1950). The analysis of a case of cross-sterility in maize. *Proc. Natl. Acad. Sci. U.S.A.* **36,** 719–724.

Sharp, W. M., and Sprague, V. G. (1967). Flowering and fruiting in the white oaks. Pistillate flowering, acorn development, weather, and yields. *Ecology* **48,** 243–251.

Smith, C. A., and Evenson, W. E. (1978). Energy distribution in reproductive structures of *Amaryllis. Am. J. Bot.* **65,** 714–716.

Snyder, E. B., Dinus, R. J., and Derr, H. J. (1977). Genetics of longleaf pine. *U.S. Dept. Agric. For. Serv. Res. Pap. WO-33* p. 24.

Solbrig, O. T. (1976). On the relative advantages of cross- and self-fertilization. *Ann. M. Bot. Gard.* **63,** 262–276.

Sparnaaij, L. D., Kho, Y. O., and Baer, J. (1968). Investigations on seed production in tetraploid freesias. *Euphytica* **17,** 289–297.

Sprague, G. F. (1933). Pollen tube establishment and the deficiency of waxy seeds in certain maize crosses. *Proc. Natl. Acad. Sci. U.S.A.* **19,** 838–841.

Squillace, A. E., and Bingham, R. T. (1958). Selective fertilization in *Pinus monticola* Dougl. I. Preliminary Results. *Silvae Gen.* **7,** 188–196.

Steiner, K. E. (1981). Nectarivory and potential pollination by a neotropical marsupial. *Ann. M. Bot. Gard.* **68,** 505–513.

Stephenson, A. G. (1979). An evolutionary examination of the floral display of *Catalpa speciosa* (Bignoniaceae). *Evolution (Lawrence, Kans.)* **33,** 1200–1209.

Stephenson, A. G. (1980). Fruit set, herbivory, fruit reduction, and the fruiting strategy of *Catalpa speciosa* (Bignoniaceae). *Ecology* **61,** 57–64.

Stephenson, A. G. (1981). Flower and fruit abortion: Proximate causes and ultimate functions. *Annu. Rev. Ecol. Syst.* **12,** 253–279.

Stout, A. B. (1928). Dichogamy in flowering plants. *Bull Torrey Bot. Club* **55,** 141–153.

Straley, C., and Melton, B. (1970). Effect of temperature on self-fertility and *in vitro* pollen growth characteristic of selected alfalfa clones. *Crop Sci.* **10,** 326–329.

Sweet, G. B. (1973). Shedding of reproductive structures in forest trees. *In* ''Shedding of Plant Parts'' (T. T. Kozlowski, ed.), pp. 341–382. Academic Press, New York.

Tamas, I. A., Wallace, D. H., Ludford, P. M., and Ozbun, J. L. (1979). Effects of older fruits on abortion and abcisic acid concentration of younger fruits in *Phasedus vulgaris* L. *Plant Physiol.* **64,** 620–622.

Ter-Avanesian, D. V. (1978). The effect of varying the number of pollen grains used in fertilization. *Theor. Appl. Genet.* **52,** 77–79.

Thomas, R. G. (1956). Effects of temperature and length of day on the sex expression of monoecious and dioecious angiosperms. *Nature (London)* **178,** 552–553.

Thomson, J. D., and Barrett, S. C. H. (1981). Temporal variation of gender in *Aralia hispida* Vent. (Araliaceae). *Evolution (Lawrence, Kans.)* **35,** 1094–1107.

Thornhill, R. (1976). Sexual selection and nuptial feeding behavior in *Bittacus apicalis* (Insecta: Mecoptera). *Am. Nat.* **110,** 529–548.

Trivers, R. L. (1972). Parental investment and sexual selection. *In* ''Sexual Selection and the Descent of Man, 1871–1971'' (B. Campbell, ed.), pp. 136–179. Aldine, Chicago.

Valentine, F. A. (1974). Genetic control of sex ratio, earliness, and frequency of flowering in *Populus tremuloides*. *Proc. 22nd N.E. For. Tree Improve.* pp. 111–129.

Van Steveninck, R. F. M. (1957). Factors affecting the abscission of reproductive organs in yellow lupins (*Lupinus luteus* L.) I. The effect of different patterns of flower removal. *J. Exp. Bot.* **8,** 373–381.

Waddington, K. D. (1980). Flight patterns of foraging bees relative to density of artificial flowers and distribution of nectar. *Oecologia* **44,** 199–204.

Waddington, K. D. (1981). Factors influencing pollen flow in bumblebee-pollinated *Delphinium virescens*. *Oikos* **37,** 153–159.

Wade, M. J. (1979). Sexual selection and variance in reproductive success. *Am. Nat.* **114,** 742–747.

Wade, M. J., and Arnold, S. J. (1980). The intensity of sexual selection in relation to male sexual behavior, female choice, and sperm precedence. *Anim. Behav.* **28,** 446–461.

Wang, C.- W. (1977). Genetics of ponderosa pine. *U.S. Dept. Agric. For. Serv. Res. Pap. WO-34* p. 24.

Webb, C. J. (1979). Breeding systems and the evolution of dioecy in New Zealand apioid Umbelliferae. *Evolution (Lawrence, Kans.)* **33,** 662–672.

Webb, C. J. (1981). Andromonoecism, protandry, and sexual selection in Umbelliferae. *N. Z. J. Bot.* **19,** 335–338.

Webb, C. J., and Lloyd, D. G. (1980). Sex ratios in New Zealand apioid Umbelliferae. *N. Z. J. Bot.* **18,** 121–126.

Weller, S. G., and Ornduff, R. (1977). Cryptic self-incompatibility in *Amsinckia grandiflora*. *Evolution (Lawrence, Kans.)* **31,** 47–51.

Williams, R. R. (1965). The effect of summer nitrogen applications on the quality of apple blossom. *J. Hortic. Sci.* **40,** 31.

Willson, M. F. (1979). Sexual selection in plants. *Am. Nat.* **113,** 777–790.

Willson, M. F. (1982). Sexual selection and dicliny in angiosperms. *Am. Nat.* **119,** 579–583.

Willson, M. F., and Bertin, R. I. (1979). Flower-visitors, nectar production, and inflorescence size of *Asclepias syriaca*. *Can. J. Bot.* **57,** 1380–1388.

Willson, M., and Price, P. W. (1977). The evolution of inflorescence size in *Asclepias* (Asclepiadaceae). *Evolution (Lawrence, Kans.)* **31,** 495–511.

Willson, M. F., and Rathcke, B. J. (1974). Adaptative design of the floral display in *Ascelpias syriaca* L. *Am. Midl. Nat.* **92,** 47–57.

Willson, M. F., and Schemske, D. W. (1980). Pollinator limitation, fruit production, and floral display in pawpaw (*Asimina triloba*). *Bull. Torrey Bot. Club* **107,** 401–408.

Wright, S. T. C. (1956). Studies of fruit development in relation to plant hormones. III. Auxins in relation to fruit morphogenesis and fruit drop in the black currant *Ribes nigrum*. *J. Hortic. Sci.* **31,** 196–211.

Wyatt, R. (1976). Pollination and fruit set in *Asclepias:* A reappraisal. *Am. J. Bot.* **63,** 845–851.

Wyatt, R. (1980). The reproductive biology of *Asclepias tuberosa:* I. Flower number, arrangement, and fruit set. *New Phytol.* **85,** 119–131.

CHAPTER 7

Models of Pollen Tube Competition in *Geranium maculatum**

DAVID MULCAHY
Botany Department
University of Massachusetts
Amherst, Massachusetts

A number of references indicate that there is a significant correlation between gametophytic and sporophytic qualities (Ter-Avanesian, 1949, 1978; Ottaviano *et al.*, 1975; Mulcahy and Mulcahy, 1975, 1983; McKenna *et al.*, 1983). Gametes from pollen tubes which grow rapidly give rise, in *Petunia hybrida* and *Dianthus chinensis,* to plants which germinate sooner and grow faster than do other plants. In *Zea mays,* the resultant plants yield more, and in *D. chinensis* they exhibit increased competitive ability.

The basis for this correlation between gametophytic and sporophytic qualities is very likely a reflection of the fact that a substantial fraction (60% in *Lycopersicon esculentum*) of the structural genes which are expressed in the sporophyte are also expressed in the gametophyte (Tanksley *et al.*, 1981). As a result, whenever pollen competition causes changes in gene frequencies, these changes

*Supported by NSF Grant #DEB 7903685.

POLLINATION BIOLOGY

ISBN 0-12-583980-4 (hardcover)
ISBN 0-12-583982-0 (paperback)

may be expressed not only in the gametophytic portion of the life cycle but also in the sporophyte.

This dual expression of significant portions of the genome in both portions of the life cycle, termed "genetic overlap," suggests that pollen tube competition, intensified by the haploid nature of the gametophytic genome and possibly also by great excesses of pollen grains, could be a significant factor in adaptive processes (Mulcahy, 1979). To a large extent, however, this significance depends on the frequency and intensity of pollen tube competition in populations.

In some species, seed production is apparently limited by the availability of pollen (Bierzychudek, 1981). However, a recent study of pollen competition in *Geranium maculatum* indicated that in that species a significant degree of pollen competition was occurring in the natural population under study (Mulcahy *et al.*, 1983). The average growth rate of "successful" pollen tubes, i.e., those that reached an unfertilized ovule, was estimated to be 34–41% greater than the average growth rate in a population of randomly chosen pollen tubes. In this chapter, the data from that study are used to construct a general model of pollen tube competition. With the resultant model it is possible to examine the effect of several floral characteristics upon pollen tube competition. The result, although far from conclusive, indicates which data ought to be obtained in future studies.

I. The Model of Pollen Tube Competition

Six variables are considered in the model:

1. Time between pollinator visits
2. Average number of pollen grains left in one visit
3. Length of the style
4. Pollen tube growth rate
5. Variance in growth rate
6. Number of available ovules

Geranium maculatum is insect pollinated and, because the corolla abscisses shortly after stigmatic surfaces become receptive, the earliest and the latest pollen deposits are separated from each other by no more than 5 h. Furthermore, pollen was deposited on stigmas at a rate of approximately 11 grains/30 min. The rate at which pollinators arrive, and thus the amount of pollen deposited during an average visit, are not yet known. Thus, we have modeled pollen arrival as a series of discrete deposits of varying sizes.

Some pollen tubes were observed to reach the ovary within 30 min of germination. Their actual growth rate through the 4 mm of stylar tissue is therefore at least 0.133 mm/min, fairly typical for trinucleate pollen (Hoekstra, 1983). The distribution of pollen tube growth rates is shown in Table I.

In this species, each pistil contains 10 ovules, although only five seeds are produced. In the model, however, ovule number is one of the variables.

TABLE I

DISTRIBUTION OF OBSERVED POLLEN TUBE GROWTH RATES

Time to reach ovary (min)	Pollen tube growth rate (mm/min)	Observed frequency[a]
30	0.133	0.062
60	0.066	0.401
90	0.044	0.195
120	0.033	0.136
150	0.027	0.000

[a]The sum of observed frequencies is less than 100 because approximately 20.6% of the pollen tubes fall between the stigma and the ovary.

Knowing the assumed times of pollination, actual pollen tube growth rates, and specified style lengths, it is possible to calculate when pollen tubes for each category of pollen tube growth rate in each pollen deposit would reach the ovary. These tubes can then be sorted according to increasing time of arrival at the ovary. Then, starting with the first pollen tubes to reach the ovary, and moving on to later arrivals, it can be calculated which pollen tubes would be successful in fertilizing a specified number of ovules. At that point, it is possible to determine both the average pollen tube growth rate of successful pollen tubes and also the diversity of pollen deposits which contributed successful pollen tubes. (This diversity is expressed in the model as the Shannon–Wiener diversity index.)

The model was written in Basic on an Apple II+ computer (48K) and is presented below in outline form.

II. Program Outline

INPUT OR SET AS VARIABLES, OR, WHERE APPROPRIATE, SET AS THE START OF A LOOP:

Minutes between pollinator visits.
Number of pollen grains left by each pollinator visit.
Number of categories of pollen tube growth rate.
Percentage of pollen grains in each category of pollen tube growth rate.
Time required for each category to reach the ovary (in min).
Observed style length (in mm).
Hypothetical style length (in mm).
Number of ovules available for fertilization.

CALCULATE:

Pollen tube growth rate (in mm/min) for each category.

For each pollinator visit, the number of pollen grains in each category of pollen tube growth rate and, for each pollinator visit, the time at which each category of pollen tube growth rate will reach the ovary. (Consider time of pollination, pollen tube growth rate, and the hypothetical length of the sytle.)

SORT:

Time of fertilizations, from earliest to latest.

FERTILIZATION SUBROUTINE:

REMOVE one from that category of pollen tubes which is the first to reach the ovary and which also contains a pollen tube.

RECORD growth rate of the pollen tube removed.

CALCULATE average pollen tube growth rate or Shannon– Wiener diversity index.

REMOVE one from the number of ovules available for fertilization.

ADD one seed to those present.

End of fertilization subroutine

RETURN to FERTILIZATION SUBROUTINE if more ovules are available for fertilization. OTHERWISE, PRINT average pollen tube growth rate or Shannon–Wiener diversity index.

END or CONTINUE LOOP.

Copies of the program are available on request. If a blank soft sector, single density, 5.25-in disk is sent, it will be returned with the program copied.

III. General Effects of Changing Single Floral Variables

An important fact in the biology of *G. maculatum* is that the growth rates of the fastest and slowest pollen tubes may differ from each other by as much as a factor of 4. Consequently, if pollen deposits are not separated from each other by too much time, it may be possible for the fastest pollen tubes of a late pollen deposit to overtake and pass slower tubes of earlier deposits. This fact, in itself, generates most of the interesting aspects of the model presented here. For example, increasing the time between pollinator visits will decrease pollen tube competition by allowing more time for slow pollen tubes of early pollen deposits to reach ovules before fast pollen tubes of later deposits can do so. Similarly, decreasing the number of pollen grains left by each pollinator visit or increasing the number of ovules available for fertilization will also decrease pollen tube competition. On the other hand, increasing style length will intensify pollen tube competition by allowing more time for faster tubes of late pollinations to overtake slower tubes of early deposits.

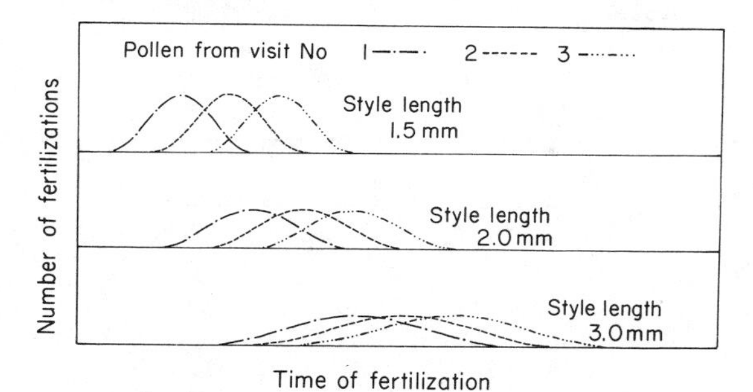

Fig. 1. Influence of style length upon pollen tube competition and diversity of pollen sources.

TABLE II

EFFECTS OF CHANGING FOUR SEPARATE POLLINATION VARIABLES

Changing variable	Direction of change in variable	Pollen tube growth rate	Shannon–Wiener diversity index
1. Time between pollinator visits	Increase	Decrease	Decrease
2. Grains per pollinator visit	Increase	Increase	Decrease
3. Ovules per flower	Increase	Decrease	Increase
4. Style length	Increase	Increase	Increase

Changes in these same variables will also have an effect upon the diversity of pollen deposits which contribute to the next generation. Decreasing the time between pollinator visits, or increasing style length, will increase the diversity of pollen sources. This stylar effect, illustrated in Fig. 1, results from more frequent successes by the fastest pollen tubes of increasing numbers of different pollen deposits as the style length is increased. Increasing the number of available ovules or decreasing the number of pollen grains deposited in a single pollinator visit will increase the diversity of successful pollen sources. These effects are summarized in Table II.

IV. Effects of Changing Two and Three Floral Variables

Changes in either ovule number or style length will modify the average growth rate of successful pollen tubes. In Table III, the combined effects of these changes are shown. As may be anticipated from the single-variable effects, pollen tube growth rate is increased by increasing style length or decreasing

TABLE III

INFLUENCE OF CHANGES IN OVULE NUMBER AND STYLE LENGTH UPON THE AVERAGE GROWTH RATE OF POLLEN TUBES WHICH REACH AN UNFERTILIZED OVULE[a]

Ovule number										
1	6	11	16	21	26	31	36	41	46	
133	74	71	65	68	67	66	67	66	66	Style = 1
133	88	74	71	73	69	69	68	67	68	Style = 3
133	88	84	80	73	74	73	70	71	70	Style = 5
133	88	84	80	78	78	74	73	72	71	Style = 7
133	100	90	87	83	78	78	76	74	74	Style = 9
133	100	90	87	83	82	78	77	76	75	Style = 11
133	111	96	91	88	85	81	80	79	77	Style = 13
133	111	96	91	88	85	84	80	79	77	Style = 15
133	122	103	95	92	89	84	83	80	78	Style = 17
133	122	103	95	92	89	87	83	80	78	Style = 19

[a]Growth rates are in mm/min ×1000. Each pollinator visit results in deposition of 10 pollen grains on the stigma. Compare with Table IV.

ovule number. Consequently, the highest pollen tube growth rate occurs in the lower left corner of Table III, and the lowest pollen tube growth rate occurs in the upper right corner. This approach becomes more interesting when a third independent variable, the quantity of pollen left in an individual pollinator visit, is considered. A comparison of Tables III and IV shows that if the amount of pollen

TABLE IV

INFLUENCE OF CHANGES IN OVULE NUMBER AND STYLE LENGTH UPON THE AVERAGE GROWTH RATE OF POLLEN TUBES WHICH REACH AN UNFERTILIZED OVULE[a]

Ovule number										
1	6	11	16	21	26	31	36	41	46	
133	133	103	91	85	82	79	77	76	75	Style = 1
133	133	103	91	85	82	79	77	76	75	Style = 3
133	133	133	116	104	97	92	88	86	84	Style = 5
133	133	133	116	104	97	92	88	86	84	Style = 7
133	133	133	133	123	112	105	99	95	92	Style = 9
133	133	133	133	123	112	105	99	95	92	Style = 11
133	133	133	133	133	128	118	111	105	101	Style = 13
133	133	133	133	133	128	118	111	105	101	Style = 15
133	133	133	133	133	133	131	122	115	110	Style = 17
133	133	133	133	133	133	131	122	115	110	Style = 19

[a]Growth rates are in mm/min ×1000. Each pollinator visit results in deposition of 100 pollen grains on the stigma.

left in each visit is increased from 10 to 100 grains, plants with short styles and many eggs will experience only a 13.6% (66 to 75) increase in pollen tube growth rate. For the same change in pollen loads, plants with the longest styles and many eggs, however, show a 41.0% increase (78 to 110). This suggests that if a high pollen tube growth rate is advantageous, the selective advantage of increasing pollen loads should be much greater in long-styled plants than in short-styled ones.

Changing the same three variables produces a more complex pattern of changes in the Shannon–Wiener diversity index. Starting with 10 pollen grains in each pollinator visit, Table V shows a major peak where both egg number and style length are highest. There is also, with long styles and low egg numbers, a secondary peak. This secondary peak in diversity is the result of fertilizations by the fastest pollen tubes of several different pollen loads. As the number of pollen grains per visit increases, the secondary peak simultaneously shifts its center to a higher egg number and increases slightly in height. Eventually, with 40 grains per visit, it becomes the primary peak and, although its height does not increase further, the peak slowly shifts to higher egg numbers. (See Tables VI and VII.)

These data suggest that if there is selective pressure to maximize the diversity of pollen sources, egg number should be high and, if possible, style length long. (See the lower right corner of Table V; $n = 10$.) This, however, is likely to greatly reduce the average pollen tube growth rate. (See Table III.) This conflict can be avoided, in the case of long-styled species, if stigma loads are intermediate (30–60 pollen grains per pollinator visit in this model). That combination

TABLE V

INFLUENCE OF CHANGES IN OVULE NUMBER AND STYLE LENGTH UPON THE DIVERSITY OF POLLEN DEPOSITS CONTRIBUTING POLLEN TUBES WHICH REACH AN UNFERTILIZED OVULE[a]

Ovule number										
1	6	11	16	21	26	31	36	41	46	
0	0	84	99	155	185	199	228	243	257	Style = 1
0	65	94	127	176	191	213	230	248	266	Style = 3
0	65	134	174	176	212	233	242	264	275	Style = 5
0	65	134	174	196	229	238	253	266	280	Style = 7
0	125	167	204	225	229	256	271	276	294	Style = 9
0	125	167	204	225	245	256	275	286	296	Style = 11
0	179	197	225	250	260	269	288	302	305	Style = 13
0	179	197	225	250	260	282	288	302	305	Style = 15
0	225	222	242	263	283	282	301	307	312	Style = 17
0	225	222	242	263	283	295	301	307	312	Style = 19

[a]Diversity is expressed as the Shannon–Wiener diversity index ×100. Each pollinator visit results in deposition of 10 pollen grains on the stigma. Compare with Tables VI and VII.

TABLE VI

INFLUENCE OF CHANGES IN OVULE NUMBER AND STYLE LENGTH UPON THE DIVERSITY OF POLLEN DEPOSITS CONTRIBUTING POLLEN TUBES WHICH REACH AN UNFERTILIZED OVULE[a]

Ovule number										
1	6	11	16	21	26	31	36	41	46	
0	0	0	0	0	0	0	58	80	91	Style = 1
0	0	0	0	45	39	63	85	94	119	Style = 3
0	91	68	54	72	114	124	125	122	119	Style = 5
0	91	68	54	72	114	124	125	146	141	Style = 7
0	158	130	106	89	135	148	149	159	183	Style = 9
0	158	130	106	89	135	148	149	159	183	Style = 11
0	158	185	154	131	150	170	172	168	195	Style = 13
0	158	185	154	131	150	170	172	168	195	Style = 15
0	158	229	199	172	150	190	193	190	204	Style = 17
0	158	229	199	172	150	190	193	190	204	Style = 19

[a]Diversity is expressed as the Shannon–Wiener diversity index ×100. Each pollinator visit results in deposition of 40 pollen grains on the stigma.

would allow both high average pollen tube growth rate and a high Shannon–Wiener diversity index. (See Table VI.)

The model thus suggests that, over a range of pollen loads, a maximal pollen tube growth rate can be achieved by low egg numbers. Furthermore, with low numbers of pollen grains per pollinator visit, the Shannon–Wiener diversity

TABLE VII

INFLUENCE OF CHANGES IN OVULE NUMBER AND STYLE LENGTH UPON THE DIVERSITY OF POLLEN DEPOSITS CONTRIBUTING POLLEN TUBES WHICH REACH AN UNFERTILIZED OVULE[a]

Ovule number										
1	6	11	16	21	26	31	36	41	46	
0	0	0	0	0	0	0	0	0	0	Style = 1
0	0	0	0	0	0	0	0	0	0	Style = 3
0	0	99	95	86	77	70	65	60	55	Style = 5
0	0	99	95	86	77	70	65	60	55	Style = 7
0	0	99	156	155	145	134	125	116	108	Style = 9
0	0	99	156	155	145	134	125	116	108	Style = 11
0	0	99	156	195	198	190	179	168	158	Style = 13
0	0	99	156	195	198	190	179	168	158	Style = 15
0	0	99	156	195	223	231	225	214	204	Style = 17
0	0	99	156	195	223	231	225	214	204	Style = 19

[a]Diversity is expressed as the Shannon–Wiener diversity index ×100. Each pollinator visit results in deposition of 100 pollen grains on the stigma.

index will be highest with high egg numbers. Moderate increases in the number of pollen grains per pollinator visit will allow both the pollen tube growth rate and the Shannon–Wiener diversity index to reach high values with low egg numbers.

V. Additional Variables

Three other variables must be considered in the model: time between pollinator visits, variance in pollen tube growth rates, and average pollen tube growth rate. Available data allow no more than a circumscription of critical points, but it is already possible to see that these three variables interact with each other very strongly. For example, competition between different pollen deposits cannot occur if a second deposit is made too late for the fastest pollen tubes of the second deposit to surpass the slowest pollen tubes of the first. As the time between deposits is shortened (see Fig. 2), pollen tube competition will increase since there is increased opportunity for the fast-growing pollen tubes of late deposits to surpass slow ones from early deposits. Ultimately, decreasing the time between pollinator visits has the same effect as increasing the size of individual pollen deposits. This effect (and limit) is approached geometrically since shortening the time between pollinator visits places not only the next but ultimately all pollen deposits within striking distance of an early deposit.

Within a range defined by the number of pollen grains left at each pollinator

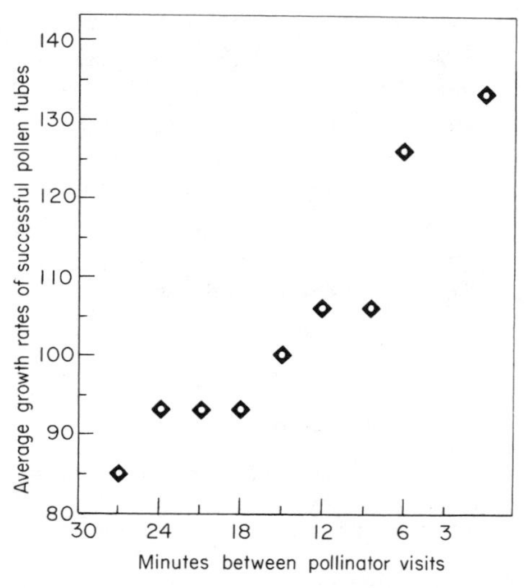

Fig. 2. Influence of time between pollinator visits and rate of successful pollen tubes.

visit and the time between visits, pollen tube competition will be directly proportional to the variance in pollen tube growth rates. Greater variance in pollen tube growth rates will increase the probability that fast pollen tubes from a later deposit will be able to overtake slower tubes of an earlier deposit.

The average pollen tube growth rate will also influence the intensity of pollen tube competition since increasing growth rates will have the same effect as increasing the time between separate pollen deposits. Thus, increased pollen tube growth rates will actually decrease pollen selection.

The average pollen tube growth rate will also modify the diversity of pollen sources. If the variance in pollen tube growth rate is independent of the average growth rate, then increasing the average will decrease the diversity of pollen sources. If, as is generally the case, variance is directly proportional to the average, diversity will be less affected.

VI. The Model and Reality

Given the assumptions of the model presented above, it is not surprising that modifying floral characteristics can influence both the pollen tube growth rate and the Shannon–Wiener diversity index. However, it is necessary to ask if, in natural populations, floral characteristics exhibit sufficient variation to have significant effects. This question may be answered by considering data published by Cruden and Miller-Ward (1981). By statistical convention, 99.9% of the variation about a mean is included in a range of ±3 SD. Thus, a population of 1000 plants should contain all the variants encompassed by this range. Applying this consideration to Cruden's data on *Epilobium angustifolium,* we see that ovule number should vary from 292 to 596 per flower and style length from 9.3 to 19.5 mm. If we assume that pollinators are depositing from 100 to 1000 pollen grains per visit (see Primack and Silander, 1978), then according to the model, average growth rates of successful pollen tubes should vary from 0.065 to 0.133 mm/min and the Shannon–Wiener diversity index should vary from 1.72 to 5.93. These figures, although extremely tenuous, indicate that observed variation in natural populations could have significant effects on both pollen tube growth rates and diversity of pollen sources.

References

Bierzychudek, P. (1981). Pollinator limitation of plant reproductive effort. *Am. Nat.* **117,** 838–840.

Cruden, R. W., and Miller-Ward, S. (1981). Pollen-ovule ratio, pollen size, and the ratio of stigmatic area to the pollen-bearing area of the polinator: An hypothesis. *Evolution (Lawrence, Kans.)* **35,** 964–975.

Hoekstra, F. (1983). Physiological evolution in angiosperm pollen: Possible role of pollen vigor. *In* ''Pollen: Biology and Implications for Plant Breeding'' (D. Mulcahy and E. Ottaviano, eds.), pp. 35–41. Elsevier Amsterdam.

McKenna, M., and Mulcahy, D. (1983). Ecological aspects of gametophytic competition in *Dianthus chinensis*. *In* ''Pollen: Biology and Implications for Plant Breeding'' (D. Mulcahy and E. Ottaviano, eds.), pp. 419–424. Elsevier, Amsterdam.

Mulcahy, D. L. (1979). The rise of the angiosperms: A genecological factor. *Science* **206,** 20–23.

Mulcahy, D. L., and Mulcahy, G. B. (1975). The influence of gametophytic competition on sporophytic quality in *Dianthus chinensis*. *Theor. Appl. Gen.* **46,** 277–280.

Mulcahy, D. L., and Mulcahy, G. B. (1983). Pollen selection: An overview. *In* ''Pollen: Biology and Implications for Plant Breeding'' (D. L. Mulcahy and E. Ottaviano, eds.), pp. 15–17. Elsevier, Amsterdam.

Mulcahy, D. L., Curtis, P., and Snow, A. (1983). Pollen competition in a natural population. *In* ''Handbook of Experimental Pollination Biology'' (C. E. Jones and R. J. Little, eds.), pp. 330–337. Van Nostrand-Reinhold, New York.

Ottaviano, E., Sari-Gorla, M., and Mulcahy, D. L. (1975). Genetic and intergametophytic influences on pollen tube growth. *In* ''Gametic Competition in Plants and Animals'' (D. L. Mulcahy, ed.), pp. 125–134. Elsevier, Amsterdam.

Primack, R. B., and Silander, J. A. (1978). Pollination intensity and seed set in the evening primrose *Oenothera frutilosa*. *Am. Midl. Nat.* **100,** 213–217.

Tanksley, S., Zamir, D., and Rick, C. M. (1981). Evidence for extensive overlap of sporophytic and gametophytic gene expression in *Lycopersicon esculantum*. *Science* **213,** 453–455.

Ter-Avanesian, D. V. (1949). The influence of the number of pollen grains used in pollination. *Tr. Prikl. Bot., Genet. Sel.* **28,** 119–133.

Ter-Avanesian, D. V. (1978). The effect of varying the number of pollen grains used in fertilization. *Theor. Appl. Genet.* **52,** 77–79.

CHAPTER 8

Pollination Ecology, Plant Population Structure, and Gene Flow

STEVEN N. HANDEL
Department of Biology
Yale University
New Haven, Connecticut

I. Pollination Dynamics in Plant Populations

Where does pollen go? This simple inquiry has important consequences for investigators in many branches of biology. The early history of flowering plants, the morphological structure of the flower, sexual reproduction and genetic recombination, and plant–animal interactions are all tightly involved with investigations on the movement of pollen grains (Schmid, 1975; Baker, 1979). Similar-

POLLINATION BIOLOGY

ISBN 0-12-583980-4 (hardcover)
ISBN 0-12-583982-0 (paperback)

ly, many applied branches of biology have contributed information to this subject, particularly plant breeding with a concern for purity of strains and the proper manipulation of plants to achieve the desired amount of crossing (Allard, 1960; Free, 1970) and allergists interested in the mass movement of pollen that may cause public health problems (Edmonds, 1979). Together, these fields have created a backdrop of information on pollen flow dynamics that is increasingly being used in evolutionary biology and population ecology. Many questions remain, however, particularly regarding the precise roles that the interaction of population ecology and pollination dynamics play in plant microevolution; field methods designed to answer these questions are still being developed.

Many components of a plant's life history are involved in controlling gene flow, microevolution, and population differentiation; these components have been reviewed by several workers (Darlington, 1958; Grant, 1958; Stebbins, 1958; Levin, 1978a, 1981; Antonovics and Levin, 1980). They include generation length, cytogenetic factors such as chromosome number and crossing-over frequency, and fertilization controls such as breeding and pollination systems, population size, and external isolation mechanisms (Grant, 1949, 1958; Bradshaw, 1972). My purpose here is to focus attention on the complexities that operate during the pollination stage of a life history, particularly the close relationships between what is traditionally called "population ecology" and the movement of pollen onto a plant.

II. The Measurement of Pollen Flow

A. Indirect Methods

1. Dyes and Powders

Movements of pollen grains from specific sources are often difficult to determine directly, because pollen reaching a target may come from any direction. Consequently, systems of pollen "surrogates" have been used to simulate movements of the grains themselves when directional patterns need to be understood. In these studies, dyes or micronized powders are applied to anthers by the investigator. Then, after a period of pollinator activity, surrounding plants are searched for the marker. Usually this is done in the field, but flowers can also be collected and examined under the microscope for a more precise analysis of movement. Reports using this technique vary widely in their methodology, some investigators counting individual dye particles and others recording simply the presence or absence of the marker. Sometimes only particles on the stigma are recorded; at other times, particles anywhere on the flower are recorded as evidence that the pollinator has visited the flower. Any of these approaches yields some useful information, but it is very unclear whether the more precise counting methods truly represent more accurate information, because of the possible low correlation between these methods and actual gene flow.

Methylene blue dye has been used in agronomic studies of pollen flow in cotton fields and has produced results compatible with data from other methodologies (Stevens and Finkner, 1953; Thies, 1953; Simpson, 1954; Sindu and Singh, 1961). More recent studies have used fluorescent powders which are more easily seen than dyes when present in small amounts (Stockhouse, 1976). In these studies, both hummingbird and bee pollinators have been shown to move the powders, and pollen transport over both short and long distances has been inferred (Linhart, 1973; Price and Waser, 1979; Linhart and Feinsinger, 1980; Thomson, 1981; Waser and Price, 1983).

A high correlation between movement of these powders and of pollen has been reported by Waser and Price (1982) for *Ipomopsis aggregata* (Polemoniaceae), which is visited by hummingbirds. In that study, dye particle deposition on stigmas was examined under a dissecting microscope after flowers had been visited sequentially by hummingbirds. Also, the number of pollen grains on the stigmas was recorded (Fig. 1). There was a very good correlation of deposition patterns between dye and pollen, suggesting, for the pollen size and morphology of this plant species, that dyes may be good analogues of actual pollen movements. The authors caution that their experiments used emasculated flowers and that natural populations of *I. aggregata* may have a different pollen deposition pattern, as competitive interactions between endogenous and introduced pollen on a flower may be complex (i.e., layering or dilution effects).

Another comparison between fluorescent dyes and actual gene flow pattern has been done in our laboratory using test gardens of the weedy mustard, *Brassica campestris*. In this work, a central patch of 16 plants with a dominant marker allele for golden yellow petals was planted in the middle of eight clusters of plants homozygous for a recessive pale yellow petal color. Also, fluorescent dye was applied to anthers of the central plants. A few days after the plants were exposed to natural pollinators, each plant's stigmas were examined under ultraviolet light for the presence of dye. Then the plants were allowed to mature seed in a greenhouse shielded from pollinators, and the distribution of the dominant marker on the progeny of the 87 recessive plants was recorded. The results (Fig. 2) show that on 54% of the plants the fluorescent dye technique "worked": Both dye particles and heterozygote progeny were present *or* both were missing. However, 29% of the plants had the dye but no heterozygote progeny, and 11% of the plants had no dye but did have heterozygote progeny. These results suggest that fluorescent dyes for plants of this floral morphology are useful in suggesting the distance and direction of pollen flight, but offer little help as exact predictors of the pollination dynamics in the population. More calibration experiments are needed to understand the limitations of dyes as pollen analogs; the ease

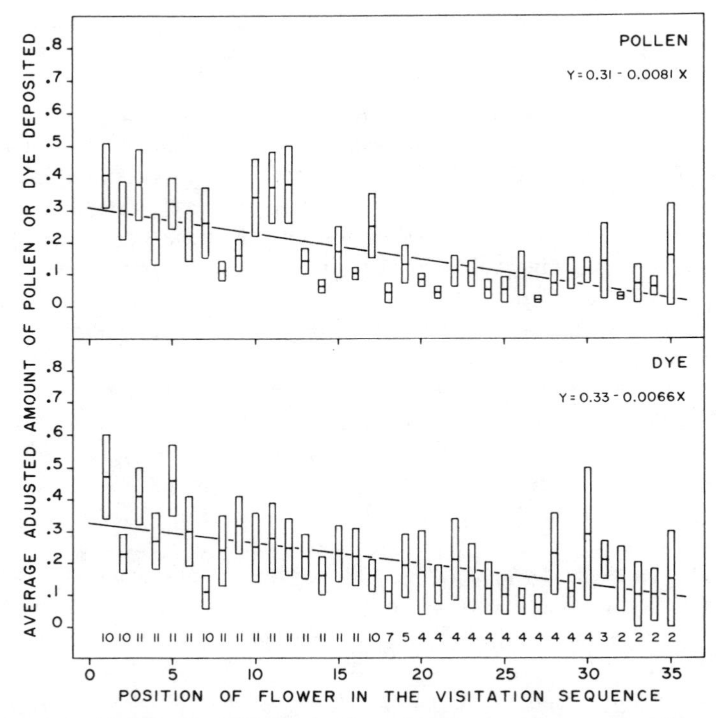

Fig. 1. Deposition of pollen (top) and dye (bottom) as functions of the position of an emasculated *Ipomopsis aggregata* flower in a visitation sequence. Horizontal bars are mean adjusted values for number of dye particles and pollen grains reaching flowers at a given position. Vertical rectangles show 1 SE around the mean. Small numbers along the bottom show sample sizes. Least-squares linear regression equations and lines are also shown. From Waser and Price (1982). Copyright © 1982, the Ecological Society of America.

and inexpensive nature of the technique remain alluring, and many species may still prove quite amenable for this technique.

2. *Chemical Labeling of Pollen Grains*

A potential improvement in accuracy over the use of pollen analogues can be made by recording the movement of chemicals that are introduced directly onto or within the pollen grains. With this procedure, the pollen itself is the vehicle of movement and assumptions about parallel movement dynamics of dye particles and pollen grains do not have to be made. Radiochemicals have been introduced onto pollen grains of both wind-pollinated species (Colwell, 1951) and hummingbird-visited plants (Schlising and Turpin, 1971). After the pollen disperses, plants or pollen traps surrounding the treated flowers are surveyed for the pres-

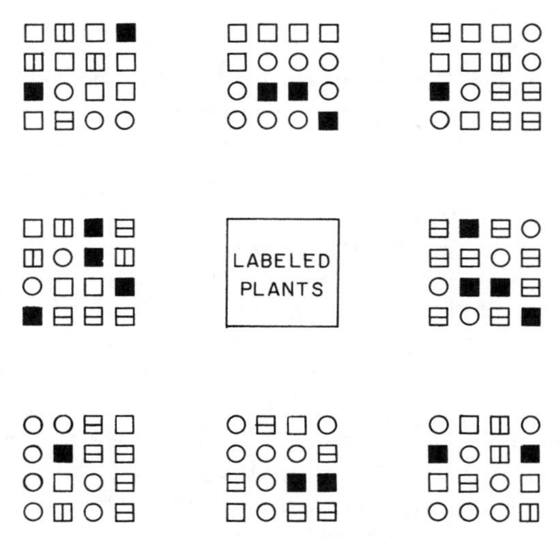

Fig. 2. Movement of fluorescent dye powders and genetic markers in a test garden of *Brassica campestris*. Blocks of 16 plants were arrayed in an 8-m^2 field, with the central block homozygous for the dominant petal color (golden yellow). All other plants were homozygous recessive (lemon yellow) for petal color. Seeds were collected from each plant in the outer clusters and planted, and the progeny were scored for the presence of the dominant allele. Also, each flower in the central patch had fluorescent dye applied to the anthers, and the presence of any dye on the surrounding plants were determined with an ultraviolet light after 3 days of natural pollination. Open square: neither dye nor dominant marker present. Solid square: both dye and dominant marker present. Vertically lined square: only dominant marker present. Horizontally lined square: only fluorescent dye present. Circle: plant had no progeny.

ence of labeled grains. This technique has not been widely used, in part because of local ordinances limiting the use of radionuclides in the environment, and also because of the technical problems involved. The half-life of the radionuclides, for example, must be considered when planning the collection times.

Autoradiography has been suggested as an accurate tool for measuring pollen movements because individual grains can potentially be scored on the developed film (Reinke and Bloom, 1979). With this technique, radiochemicals are injected into the flower bud and must be incorporated into the developing pollen grains before anthesis. After pollen transport, stigmas are collected at set distances from the inoculated plants and are placed on film for developing. If the label has been regularly and evenly incorporated into the grains, the distribution of radioactive grains on the stigmas will represent a very accurate picture of pollen flow. The number of radioactive grains and the percentage of all grains that are from the inoculated plant can be computed. At large distances from the labeled plant, the number of stigmas that must be scanned for labeled grains under a microscope must be very large, but a detailed diagram of the pollen cloud around the original plants could be developed. Details of the developmental pattern and potential uptake of the label by the incipient pollen grains must be known for a particular species before the technique can be used with confidence.

Surface labeling of the pollen grains may also be useful when followed by neutron-activation analysis (Gaudreau and Harden, 1974; Handel, 1976). In this technique, a chemical not usually present in the pollen, such as a rare earth, is topically applied in solution to the dehiscing anther. Grains become coated with the chemical, and surrounding stigmas are then searched for the presence of the chemical marker after an appropriate period of pollen flow. Since the amount of marker chemical on the stigma may be very small, the precision of neutron-activation analysis might yield the best data. Stigmas are individually loaded into containers, irradiated with a neutron source, and then placed in a scintillation counter to assay for the gamma ray energy levels particular to the decay products of the introduced chemical. If the appropriate gamma levels are present, then one can assume that the labeled pollen reached the stigma being analyzed. The advantages of neutron-activation analysis are its precision (very small amounts of pollen can be found) and its convenience for the field researcher (specimens are collected in the field immediately after pollination events, but the actual analysis of the samples can be postponed indefinitely, since the samples do not become radioactive until they are put into a reactor). The researcher does not have to analyze the samples until the end of the field or growing season. Also, the lack of introduction of radioactive material into the environment is appealing and necessary to many investigators. Difficulties with the neutron-activation technique include the need for correlation between estimates of pollen flow based on the activation analysis and actual pollen flow (does the topical application change the flight dynamics of the grains?) and the need for close collaboration with a reactor and scintillation counter facility, which is often busy and requires financial compensation that may be beyond the budget of many pollination biology investigations. Regardless, in situations in which no direct method for tracking pollen is practical, neutron-activation analysis remains a viable technique to answer questions about pollen transport.

3. Pollinator Movements

Watching pollinators move among flowers on a plant and among plants in a population has yielded a tremendous amount of information about insect, bat, and hummingbird foraging behavior, and about the possible role of foraging movements in the evolution of floral morphology and display, as well as in pollen flow dynamics among plants (Heinrich and Raven, 1972; Proctor and Yeo, 1972; Heithaus *et al.*, 1974; Macior, 1974; Levin and Kerster, 1974; Feinsinger, 1978; Levin, 1978b; Brantjes, 1979; Howell, 1979; Pyke, 1981; Waddington, 1981; Waddington and Heinrich, 1981; Linhart and Feinsinger, 1980; Kevan and Baker, 1983). Although most movements by pollinators are to very near neighbors, occasional longer flights also occur. Flight patterns are very sensitive to many ecological factors and to behavioral differences among the pollinators.

Pollen vectors have different efficiencies in transporting pollen. Movement is controlled by their behavior (Waser, Chapter 10, this volume), the size and pelage of their bodies (Primack and Silander, 1975; Motten *et al.*, 1981; Tepedino, 1981; Waser and Price, 1981; Parker, 1982), and the location of pollen grains or pollinia on the insect body (Faegri and van der Pijl, 1966; Proctor and Yeo, 1972). In plant species producing pollinia, such as orchids and milkweeds, the placement of the pollen masses on the pollinators is very precise, and is critical to understanding the eventual deposition pattern of the pollinia and the efficiency of insects as successful pollinators (van der Pijl and Dodson, 1966; Morse, 1982).

Additional problems with extrapolating pollen movement patterns from the behavior of one pollinator to another are caused by the sharply different foraging behavior among taxa. Some tropical hummingbirds, for example, are trapliners and sequentially visit plants over a wide geographic area (Linhart, 1973; Feinsinger, 1976; Linhart and Feinsinger, 1980). Other species are generally territorial and tend to move pollen only among flowers on one or a few adjacent trees (Grant and Grant, 1968; Feinsinger, 1978). These specific areas change with the size of the plants and the size and interactions within the hummingbird guild. Some individuals in territorial species, for example, are subordinates and must move from tree to tree for nectar. These animals cause a low but potentially important degree of xenogamy. Analogous traplining is known to occur for some large euglossine bees of the tropics (Dressler, 1968, 1982; Janzen, 1971a; Williams and Dodson, 1972; Ackerman *et al.*, 1982) and for some bumblebee species in the temperate zone (Heinrich, 1976; Thomson *et al.*, 1982).

Honeybees tend to forage within rather sharply delineated areas at any one time, as do some bumblebees (Free, 1970; Thomson *et al.*, 1982). For example, Butler *et al.* (1943) showed that honeybees would return to visit a few Petri dishes within a much larger homogeneous array once the bees had a positive reward from the few dishes. In large homogeneous test gardens of crop plants, honeybees will tend to remain in small subdivisions of the garden rather than spread out over the field (Gary *et al.*, 1972, 1977). Singh (1950) showed that honeybees remained in 18-m^2 foraging areas. Similarly, a study of honeybee foraging ranges from a hive in a deciduous forest showed that small areas of the forest were visited each day, presumably where high nectar rewards were present, but the hive switched its foraging ground to other nectar sources as they became more attractive (Visscher and Seeley, 1982). The total geographic range visited by the hive during the summer was a circle about 8 km in radius, but only small sections received visitors, and shared pollen, at one time (Visscher and Seeley, 1982).

Little is known about the movements of solitary bees, although attention to this topic is growing with interest in the coevolution of plants with the native fauna (Eickwort and Ginsberg, 1980). Flights of the halictids *Agapostemon texanus, Augochlorella striata, and Lasioglossum* sp. foraging on bindweed, *Convolvulus arvensis,* were all quite short (averaging 7.6–15.8 cm at flower patches of different densities), but the three species of bees had subtly different foraging ranges (Waddington, 1979). Generally, few long flights were seen, and the leptokurtic pattern seen in honeybee and bumblebees was followed (see also Waser, 1982). Two polyphagous bees, *Ceratina calcarata* (Anthophoridae) and *Halictus ligatus* (Halictidae), visiting composite flowers also showed leptokurtic flight distances, most flights being shorter than 50 cm (Ginsberg, 1983). In this study *H. ligatus* showed directionality in its flight pattern, whereas *C. calcarata* did not.

When flowers on one plant species are visited by more than one insect species, the different sorts of insects may have contrasting effects on pollen movements, in part because of differing pollen load capacity but also because of different flight and foraging behaviors. Schmitt (1980) showed that butterflies and bumblebees behaved differently on three *Senecio* species in California. Bumblebees tended to visit near neighbors and visit many heads per plant, whereas butterflies had greater interplant flight distances. Pollen transport would be more restricted in populations of these plants that are visited only by bumblebees, and a few lepidoptera visits may strongly increase the pollen flow and consequently the neighborhood size. When two bee species visit the same crop species, different pollination effects have also been seen. Pedersen (1967) showed that honeybee and leaf-cutter bees visiting test gardens of alfalfa caused different amounts of cross-pollination and different total seed yields, the leaf-cutter bees having lower crossing rates (45.4% vs. 41.2%) but 19% higher yields. Similarly, honeybees carried alfalfa pollen farther than did leaf-cutter bees (Bradner *et al.*, 1965). Squash bees, *Peponapis pruinosa* (Anthophoridae), have different foraging behavior on the summer squash, *Cucurbita pepo,* than do honeybees on the same plants (Tepedino, 1981). Both preference for pistillate flowers and foraging time differed between the bee species. On onion flowers in a commercial field, *Halictus farinosus* pollinators apparently are more efficient at setting seeds than are honeybees (Parker, 1982). In a Costa Rican dry forest, more than 70 bee species were found on *Andira* flowers, but intertree movement was predominantly by six *Centris* (Anthophoridae) species (Frankie *et al.*, 1976). These few studies all suggest that the relative importance of different bee species must be known before pollen movement can be gauged. For example, Waser (1982) has shown that different insects have very similar flight characteristics on three Rocky Mountain wildflowers; this result has been seen in other cases (Motten *et al.*, 1981; Waser and Price, 1981).

There must be plant responses to the diversity of bee species in the community. The possibilities of character displacement in flowering phenology and morphology have been well discussed in the literature, but other responses may also be found. Dafni and Werker (1982) have shown that the stamens of *Sternbergia clusiana* (Amaryllidaceae) are dimorphic, and honeybees collect pollen from the inner, short anthers, whereas hover flies (Syrphidae) collect pollen from the outer, longer stamens. This is interpreted as an adaptive response to maximize pollination efficiency in a harsh environment where pollinators may often be limiting. Similarly, the evolution of distyly in *Jepsonia heterandra* may have been influenced by the contrasting pollinating behaviors of the syrphids and halictids that visit this species (Ornduff, 1975). In these ways, the evolutionary response of the plant illustrates the selective pressure that the contrasting behaviors of two pollinators can exert.

The correlation of pollinator movement patterns with actual pollen transport can be difficult to show, and in some cases the correlation may be weak. Pollen carryover, in which only a fraction of the pollen present on an animal is deposited on a flower and subsequent flower visits get some pollen picked up from several flowers, not just from the last flower visited, must be common. When this happens, pollinator movement and deposition patterns are more complexly related (Lertzman and Gass, 1983).

This general problem has been reviewed by Levin (1981), and has been investigated in studies in which both insect movements and actual gene flow have been followed in the same test garden. In one study (Schaal, 1980), a garden of *Lupinus texensis* was planted with a clump of plants in the center having an isozyme marker not present in the other plants. Insect movements among the flowers were followed, and seeds were collected, planted, and scored for the presence of the marker allele. Schaal found that the insect movements were much more restricted than was the movement of the marker, and concluded that pollen carryover (many flowers receiving some pollen from a donor flower) had to be an important phenomenon in this plant species when served by this pollinator (Fig. 3). Similar evidence of high pollen carryover has now been shown in a variety of plant species with very different floral morphologies: e.g., *Cucumis melo* (Handel, 1982), *Ipomopsis aggregata* (Price and Waser, 1982), *Ipomoea purpurea* (Ennos and Clegg, 1982), and *Diervilla lonicera* (Thomson and Plowright, 1980). Certainly, in cases in which the pollen is attached more firmly than merely by being brushed onto the pollinator's body, carryover would be expected to be high. For example, the well-studied and specialized *Hyptis pauliana* and *Eriope crassiopes* (Lamiaceae) anthers have explosive mechanisms for shooting pollen onto the bodies of pollinators (Harley, 1971; Brantjes and de Vos, 1981), and dense concentrations of grains from individual plants should occur. In an even more complex situation, Vogel (1981) showed that bicellular hairs (*Klebstoffhaare*) on stamens of *Cyclanthera pedata* (Cucurbitaceae) shoot glue onto pollinators when they visit the anthers, and pollen grains become glued

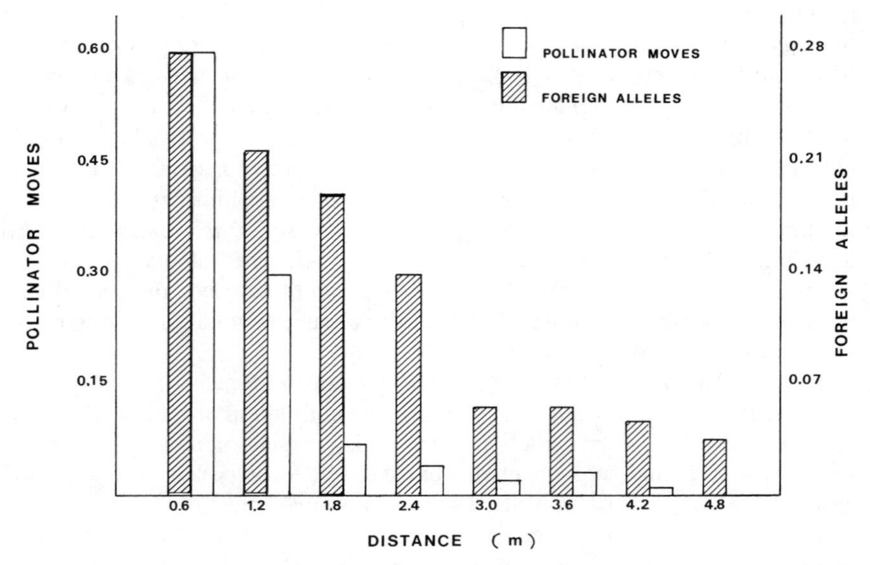

Fig. 3. Gene flow and pollinator foraging distances in *Lupinus texensis*. Gene flow (solid bars) is plotted as the frequency of foreign alleles detected in the F_1 progeny of plants at a given distance. Pollinator moves (open bars) are plotted as the frequency of the total pollinator moves to a given distance. From Schaal (1980). Copyright © 1980, Macmillan Journals Limited.

to the insects' bodies. High carryover must accompany this remarkable natural history phenomenon.

B. Direct Methods

1. Artificial Samplers

Work on air pollution control and a need to understand the dispersal of atmospheric particulate matter that causes public health problems led to the development of sophisticated air samplers and the research field of aerobiology (Gregory and Monteith, 1967; Dimmick and Akers, 1969; Edmonds, 1979). Many of these artificial samplers (vs. natural samplers, i.e., stigmas) have been used to understand pollen flight dynamics for forest trees and for wind-pollinated crop and weed stands. The efficiency and accuracy of these samplers are very sensitive to their mechanical design, particularly the speed of air as it passes through the collector and the size and shape of the collecting surface (Ogden *et al.*, 1974; Raynor, 1979). These same factors that alter the efficiency of mechanical samplers in capturing pollen grains are implicated in the evolution of morphological structures on real plants that must garner pollen from the air (Whitehead, 1969; Niklas, 1981). Differences between the size and shape of stigmas and those of the collection surface on samplers may cause differences in the deposition rate between the natural and mechanical collectors. Another consideration in using machines to sample pollen is that different particle types collect at different rates on any one sampler (Raynor, 1979). For particles above 15 μm in diameter, Raynor (1979) recommends using rotating impact samplers for estimating concentrations in the air. For estimating surface deposition rates, a special deposition sampler such as the Tauber trap is needed.

Studies of movements by wind-transported pollen have been done with this equipment for a variety of herbs and trees. These studies have given detailed information about the change in pollen concentrations with changes in ambient wind speed, pollen type, distance and height of the collection surface from the pollen source, and nature of the source population (see Wright, 1952; Raynor *et al.*, 1970, 1971, 1976; Stern and Roche, 1974).

2. *Censusing Grains on Stigmas*

The use of stigmas, rather than artificial samplers, when counting pollen grains eliminates the possible sources of error that the mechanical devices create. However, very special situations are needed if the pollen recorded on a stigma is to be traced to a specific paternal parent from a known distance and direction from the stigma. For example, Thomson and Plowright (1980) were able to study movement of *Erythronium americanum* pollen directly because this species is dimorphic for grain color (yellow and red) in some populations. They introduced a flower with the red pollen to a bumblebee pollinator and then searched later flowers visited for the presence of the red pollen marker (finding that pollen carryover can be extensive, red appearing sporatically through the first 50 flowers examined). Such dimorphisms are tremendously valuable and work with the *Erythronium* stocks are continuing, but dimorphisms are rare, except in heterostylous species in which different size classes are often found. Some work in natural populations of heterostylous species in which pin and thrum colonies are spatially isolated has been done, addressing questions of rate of pollen collection and movement between pin and thrum flowers (Ganders, 1974; Ornduff, 1978, 1980), population structure of the morphs, and the rate of fruit set (Wyatt and Hellwig, 1979).

Similarly, populations of species that have different pollen morphologies have been used to study interspecific pollen transfer. For example, Levin and Kerster (1967) showed that interspecific pollen transfer between populations of *Phlox pilosa* and *P. glaberrima* was high, so that the pollinating system itself was not a barrier to hybridization between these taxa. Interestingly, in the context of this review, the amount of pollen transfer between taxa was strongly influenced by the distance between the populations. That is, 25% of the stigmas of *P. glaberrima* and 16% of the stigmas of *P. pilosa* had alien pollen when the populations were confluent, but only 16% and 13% of stigmas, respectively, had alien pollen when the populations were contiguous. A test garden experiment with a patch of *P. glaberrima* within a field of *P. pilosa* confirmed that pollen transfer fell abruptly with distance when the large pollen grains from the central species were counted on the stigmas of the surrounding plants (Levin and Kerster, 1974). Increased distance causing less pollen transfer is known from test populations of *P. divaricata* and *P. bifida* in which alien pollen was censused on stigmas (Levin, 1972a). In that study, both the number of contaminated stigmas and the number of alien grains decreased from confluent populations to ones that were 30–35 m apart. In these ways, censusing of pollen on stigmas can address a variety of evolutionary questions.

Natural situations, however, in which dimorphic grains are present in appropriate arrangements, are very uncommon, and some manipulative approaches have been tried to enable the investigator to count grains that emanated from a known source. Waser and Price (1982, 1983) have emasculated flowers of *I. aggregata* and followed the deposition pattern of pollen onto these flowers when the autochthonous pollen is absent. This is a tedious procedure and may be operationally impossible with certain floral morphologies, but in the Price and Waser studies, clear deposition patterns could be measured. Results from emasculated flowers were also compared with results from controls that were left intact (Price and Waser, 1982). Control flowers that still had pollen-carrying anthers had a steeper decline in the amount of dye from a known source present on the stigma, showing that the use of manipulated flowers alone would not give a completely accurate survey of deposition studies for *I. aggregata* (Price and Waser, 1982; and Fig. 1 cited therein).

An alternative approach to tracking individual grains, autoradiography, has been suggested by Reinke and Bloom (1979) for situations in which there is no natural marker on pollen grains of a source plant to distinguish its grains from those of the surrounding population (see Section II,A,2).

3. Progeny Analysis

An additional direct approach used to determine where pollen has been carried is to search the progeny for a marker gene that is nonrandomly distributed in the parental population. The position of seedlings with the marker gene can be compared with the location of parentals with the appropriate genotype, and dispersal characteristics of pollen in the population can be computed. This basic technique has a long history of use by plant breeders and agronomists, who have been interested in the distance necessary to keep cultivars reproductively isolated so that pure stocks of seeds could be maintained (Fryxell, 1957; Allard, 1960; Gorz and Haskins, 1971). Alternatively, there is interest in achieving the most efficient planting design to enable maximum hybridization among strains.

A detailed account of these procedures is supplied in the work of Griffiths (1950) and Bateman (1947a,b), who interplanted genotypes of several crops and described the distribution of the dominant marker in the progeny at increasing distances from the patch of parental dominants. Analogous work has been done with test garden plantings of *Zea mays* (Jones and Brooks, 1950; Paterniani and Short, 1974), *Gossypium hirsutum* (Simpson and Duncan, 1956), *Brassica oleracea* (Nieuwhof, 1963), *Allium cepa* (van der Meer and van Bennekom, 1968), *Raphanus sativus* (Crane and Mather, 1943), and *Trifolium pratense* (Williams and Evans, 1935). This work has produced the general background to our understanding of pollen dynamics in populations. One important consideration, however, is that work with cultivars tends to minimize the effects of the overall genetic heterogeneity in most natural plant populations, factors that must be important in determining progeny gene frequencies in nature. Differences among plants such as phenotype, competitive ability among pollen tubes in the style, and size differences among plants (as this influences pollen capture rates and assortative mating) are not present in most test garden cases. The appearance of any of these biotic differences among plants in a natural population could make it difficult to extrapolate from the crop experiments to a natural situation. For example, in a study of test gardens of *Cucumis sativus* (Handel, 1983a), adjacent populations were shown to have significantly different directional components of pollen flow, as determined by progeny analysis. This was thought to be the result of different distances between the populations and a nearby apiary; in natural populations, the geometry of pollinator nests vis-à-vis the plants may have similar effects and limit the applicability of test garden results.

In natural populations, where monomorphic patches of a marker gene are rare, progeny analysis is useful for describing very local subdivision of gene frequency patterns (Hamrick and Allard, 1972; Hamrick and Holden, 1979; Schaal, 1975; Allard, 1975; Jain, 1976b; Bosbach and Hurka, 1981; Rai and Jain, 1982), and for dealing with the relative importance of selection and gene migration through pollen for creating and maintaining the observed genotypic patterns. Surveys of isozyme frequencies among populations, in particular, are providing needed information on genetic variation, actual breeding systems, and paternity patterns in plant populations (Moran and Brown, 1980; Mitton *et al.*, 1981; Shen *et al.*, 1981; Hamrick, 1982). When the various ecological factors that intrude on the generally even distribution of pollen seen in crop experiments can be identified and measured, then progeny analysis in natural populations can supply needed information about the actual movement of pollen in diverse habitats and among species of widely different floral and structural morphologies.

III. The Effects of Population Geometry

A. Population Size

As the number of plants in a population increases, the targets for a fixed amount of pollen introduced from outside the population increase, and the average rate of fertilization by that pollen should decrease. This general effect has been seen in several agronomic studies that have established the role of "varietal mass" in limiting contamination of cultivars. For example, Bateman (1947a) set up a cross-shaped population of radish (*Raphanus sativus*), with the central plants carrying a dominant marker allele and the four arms of the test garden carrying the recessive. The arms each had a different width, with 1, 3, 6, and 12 rows of plants, respectively. When the progeny of the plants in each arm were scored, Bateman found that there was a significant decrease in contamination of an arm as the mass of the arm increased. This result was compounded by position effects within each arm, the border plants of the wider arms having more contamination than the internal plants. A similar effect was seen in radish by Crane and Mather (1943), and in test plantings of cotton (Fryxell, 1956), beans (Bond and Pope, 1974), and red clover (Williams and Evans, 1935). In the studies of red clover (*Trifolium pratense*), test plots 48 and 270 yd^2 in area were planted adjacent to sources of marker alleles; the contamination level in the smaller plot was 21.7%, and in the larger plot it was 11.1%. Additional effects of flower densities within these fields were also noted (see below).

Another observed effect of increasing the size of a population in these experiments was a substructuring of gene frequencies between the outer and the internal plants. Fryxell (1956) recorded greater panmixia at the borders of cotton fields than in the center, and Haskell (1954) found 64% more outcrossing in the end plants of a field of blackberries than in the internal plants in a row. The position effects become more pronounced as the size of the entire population increases, and the extinction of the introduced pollen occurs farther from the center of the population. Other ecological factors may counter this presumed general effect, as was found in test plantings of *Vicia faba* by Bond and Pope (1974). In some cases, there was more cross-breeding in the center of the fields than in the borders. Using hilum color as a genetic marker, the authors reported that there were fewer flowers at the center of the plots during a time of relatively low bee visitation, and crossing there was enhanced. The lower flower density at the center of the originally homogeneous planting was attributed to disease and local competitive effects. These stochastic factors did not appear the next year of testing, and the level of crossing between center and borders was then the same. The intrusion of such factors as disease and unequal pod development in agronomic cases suggests that natural populations may also have very localized problems that can override the effect of varietal mass alone.

In addition, large populations may occasionally become substructured because some members are far from the major activities of important pollinators. In particular, if selection of foraging and nesting microsites limits the local distribution of the pollinators, plants far from these sites may be effectively isolated from the plants adjacent to the pollinators' area of major activity. This phenomenon is seen, for example, in certain milkweed species that grow in fields; plants near the border of the field are visited at night by geometrid moths that prefer the cover of adjacent woods, whereas plants in the center of the field are visited by noctuid moths that forage more commonly in the open (D. Morse, personal communication). Invasion of the edges of the field by the milkweed population causes a shift in the total pollinator pool and creates a potential for differing pollen movement patterns on the edge and at the center of the population.

In an analogous way, an increase in the size of a population may cause the borders of the population to extend to a microsite that is unattractive to pollinators and foraging to be restricted to one part of the population. Watson (1969) studied a large population of *Potentilla erecta* in England that spread over pasture and grassland habitats. There were many more flowers in the pasture because the species was denser there and flowering is more common on plants growing among low grasses. In the rougher grassland, *Potentilla* flowering was reduced. The sharp ecotypic differences Watson discovered between plants in the two areas were hypothesized to be caused, in part, by the restriction of the honeybee pollinators to the more attractive plants in the pasture.

These empirical results have facilitated the construction of several computer simulations exploring the effect of population size on contamination. Pedersen *et al.* (1969) has developed a model based on a series of test plantings of white alfalfa (*Medicago sativa*) over 3 years, showing that as field size increases, the predicted contamination of the field decreases. The same result occurs by increasing the distance of the field from a source of exogenous pollen (Fig. 4). These results are supported by work by Antonovics (1968) showing, in a computer simulation, that extraneous pollen is more likely to increase in a population that is small; selection alone does not determine what happens to genes from introduced pollen. In the work by Pedersen *et al.* (1969), the authors caution that the actual amount of crossing registered in real fields during three consecutive years was 16.9, 10.7, and 4.7%, and that this huge variation was due to environmental factors; varietal and experimental design differences must be produced in one growing season to get useful comparisons. In an elegant study of a natural population of *Eucalyptus delegatensis,* Moran and Brown (1980) also showed that crossing rates may vary from year to year on the same plants. This insect-pollinated species carries its fruits for several years; the fruits collected from 30 trees showed that outcrossing rates changed from 85% to 60% over 3 consecutive years.

The equilibrium gene frequencies in a plant population are dependent on several geometric and ecological factors that control pollen flow. Levin (1978a; see also Levin and Wilson, 1978) has shown that gene frequency is influenced by both the patch size and the dispersal schedule of the pollen (Fig. 5). Within the simulated population, clusters of homozygous dominant and recessive plants at a fixed density were altered in size from 2×2 through 32×32 plants per cluster. The total population contained 4096 plants. With increasing patch size, equilibrium gene frequency is depressed for all pollen flight schedules except random

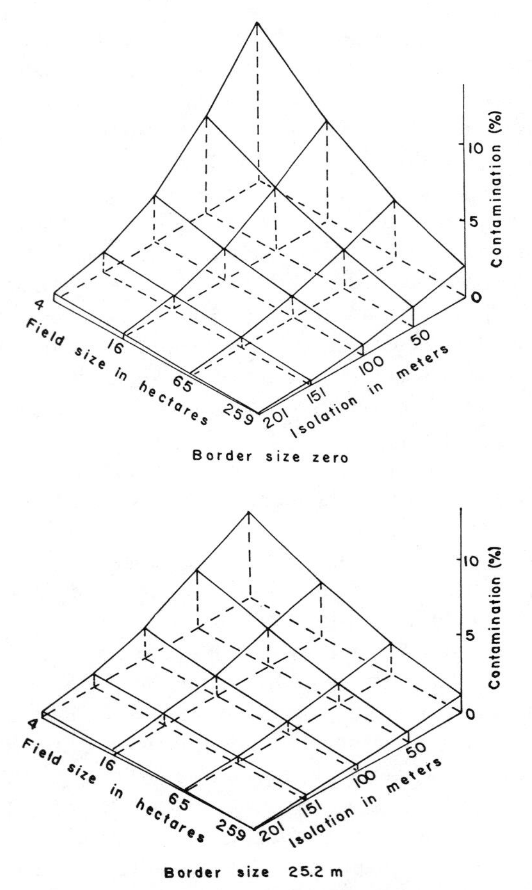

Fig. 4. Predicted contamination in alfalfa fields in relation to isolation distance and field size in square fields. This computer simulation is based on the predicted contamination (%) of the field using the formula:

$$C = \frac{45^{(1 - .02X)}}{X^{.5} + e^{.1I}}$$

where 45 = the predicted maximum percentage of contamination between adjacent fields; X = distance into the field from the edge; e = base of the natural logarithms; and I = isolation in units of 5.03 m (rods) from the edge of the field to the contaminant (from Pedersen *et al.*, 1969. Published by permission of the Crop Science Society of America). An additional graph was created by Pedersen *et al.* showing that the contamination level at all points was reduced when the field was surrounded by a 25.2-m border, the maximum contamination level at 4 ha and 0 m isolation falling to about 12%.

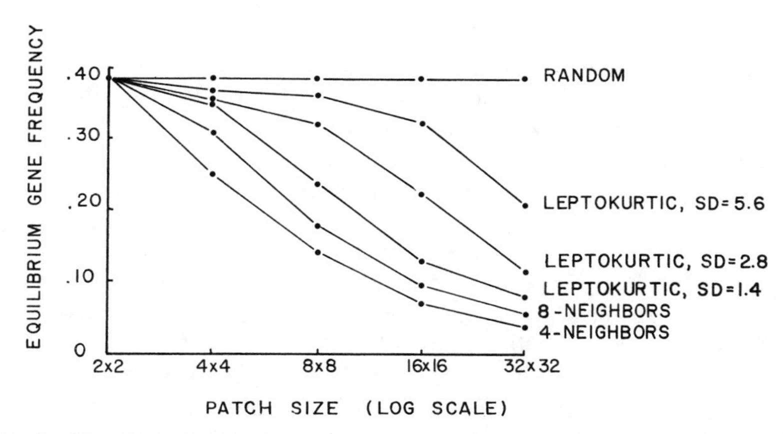

Fig. 5. The effect of patch size on the mean gene frequency of a maladapted gene, from a computer simulation. Equilibrium gene frequency declines sharply for all dispersal patterns tested except random dispersal. Details of the model's assumptions and parameters are in Levin (1978a, pp. 202 ff.).

dispersal, which is not applicable for actual plant populations. The leptokurtic dispersal pattern (SD = 2.8 units), more reasonable from what we know about pollen dispersal, causes the equilibrium frequency to fall from 0.38 to 0.11 for the largest clusters. Increasing patch size in this simulation also causes a reduction in equilibrium gene frequency, whether the population is outcrossing or 50% selfing (Levin and Wilson, 1978; and Fig. 14 cited therein) and whether the patches are contiguous or have gaps between them. Discontinuous patches have a faster decline in equilibrium gene frequency than do continuous ones at all sizes (Levin and Wilson, 1978; and Fig. 16 cited therein). The genetic consequences of different geometries complement the other evolutionary effects of differing plant population size, such as the rate of extinction of local populations and the effects of seed pools on genetic change and composition. Similarly, Turner *et al.* (1982) showed that patch formation, persistence, and spread in a simulated population could be largely explained as a consequence of nearest-neighbor pollination schedules, without the imposition of outside selective pressures. The arrangement of monomorphic genotype patches themselves in the population effected the long-term gene frequency distributions.

B. Population Density

1. Correlation between Plant Density and Pollinator Flight Patterns

The movements of pollinators are determined in large part by a balance between the energetic costs and rewards of foraging on flowers (Heinrich, 1975; Pyke, 1981). Several studies have shown that there is a strong correlation between the flight dynamics of insect and vertebrate pollinators and the density of the flowers being visited. Usually, pollinator flights are shorter where plant spacing means are smaller. Levin and Kerster (1969a,b, 1974) summarized the existence of this relationship in natural populations of varying densities of several species (*Echinops sphaerocephalus, Vernonia fasiculata, Eupatorium maculatum, Liatris aspera, Lythrum alatum, L. salicaria, Monarda fistulosa, Pychnanthemum virginianum, and Veronicastrum virginicum*). This work showed that the neighborhood sizes of these populations and the movement of pollen through them were very dependent on plant density (Fig. 6). Similar correlations between plant spacing and pollinator flight parameters have been recorded for several *Viola* species (Beattie, 1976, 1978) and *Pyrrhopappus carolinianus* (Estes and Thorp, 1975), which were serviced by several bee species.

A direct test of this relationship in natural populations of *Helianthus annuus* used analysis of allozyme frequencies to determine whether different densities were correlated with differences in the apparent rate of crossing (Ellstrand *et al.*, 1978). The outcrossing rate among populations varied significantly from 0.54 to 0.91 ($\bar{x}$ = 0.76) and was negatively correlated with population density. The authors also noted that sparse populations were more common in dry marginal habitats, and such populations must be expected to have more outcrossing for precisely this reason. For outcrossing to vary with population density, in addition to density dependence of pollinator flight distances, the relatedness of plants must decline with distance (Ellstrand *et al.*, 1978). This is certainly true in many long-lived plant populations (Schaal, 1975; Levin, 1981), but may be less com-

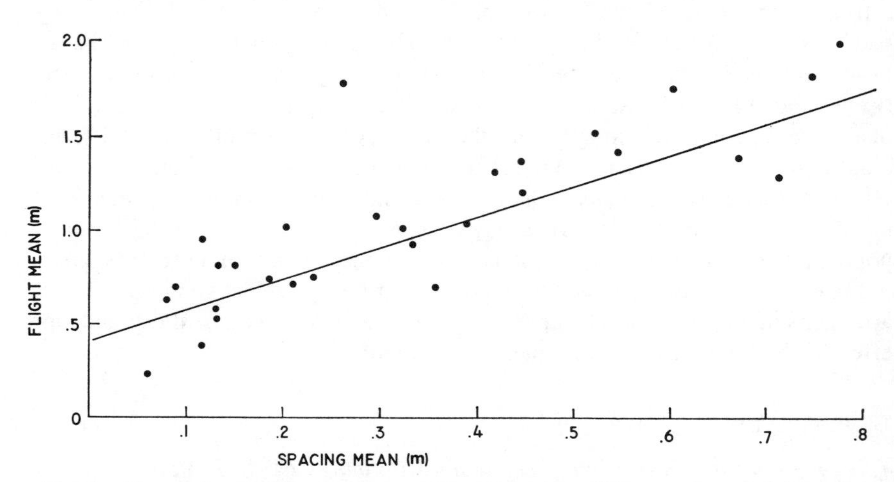

Fig. 6. The relationship between plant spacing mean and bee flight mean for populations of nine plant species occurring at various densities. The correlation coefficient for this relationship was 0.90, and for plant spacing with flight the SD was 0.83. The authors concluded that bees respond similarly to plant spacing parameters of many species. From Levin and Kerster (1969a).

mon in ephemeral or weedy situations in which most recruitment is from outside the local population, not from progeny of persisting plants.

In wind-pollinated systems, plant density and pollen dispersal distance are also related. Bannister (1965) suggested that plants in dense populations of *Pinus radiata* have potentially fewer mates than do individuals in sparse populations. Raynor's group has also shown that the dynamics of wind-carried pollen are sensitive to the density and structure of the source population (Raynor *et al.*, 1970, 1971). Using Griffiths' (1950) data on actual gene flow in a test garden of *Lolium perenne* as the basis for a computer model, Gleaves (1973) related density and clustering of plants to interception of pollen. The empirical model, not including the effects of selection on the frequencies of the alleles, showed that pollen movement, $f(x)$, could be estimated by

$$f(x) = 1/x^k$$

where x is the distance and k is an empirically derived constant that varies with the plant species and the pollinator pool of any one population.

Test garden work on the effects of density changes alone on pollen transport patterns confirms these conclusions. Haskell (1944), for example, reported on two varieties of tomatoes. When the varieties were in rows that were 6 ft apart, there was 1.4% crossing; rows that were 8 ft apart had a "negligible amount" of crossing (tomatoes have a low crossing potential because of the structure of the staminal cone around the pistil).

Bateman (1947a) constructed a large series of plantings of radish (*Raphanus sativus*) at four different densities in which the ratio of the central patch of plants with a dominant marker to the surrounding plants with a recessive trait was kept constant while the density of the test gardens was changed. The spread of contamination in the progeny (Fig. 7) in the different gardens occurred at a similar rate (the slopes in Fig. 7 are not significantly different), but dominant genes traveled less distance in dense than in sparse populations. A detailed plant-by-plant analysis in the gardens uncovered variations in contamination of different plants at the same isolation distance from the central patch. Bateman attributed this result to subtle differences in the degree of self-incompatibility within the gardens, variation in seed set on plants with time (if the contamination rate varied slightly with time), and interplant differences in floral mass, causing the influence of varietal mass to be expressed at the level of single inflorescences. Bateman's study has been influential for its detailed support of the concept that density and pollen flow are closely related and for showing, through the discussion of variation among plants in the garden, that very local factors intrinsic to individual plants can modify the population mean in contamination rate.

The modification of the density effect by number of flowers on a plant, rather than number of plants in a population, has been illustrated by Williams and

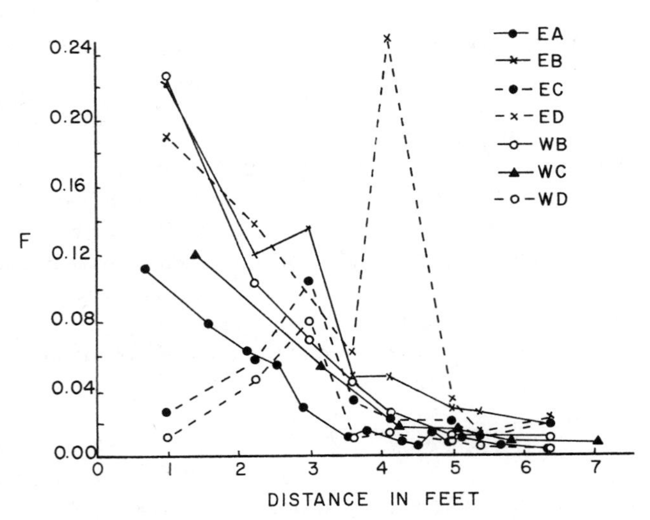

Fig. 7. Contamination (F) of pollen from a central patch of plants with a dominant marker introduced into seven test plots of different densities. Solid lines represent squares with highly significant distance variances; broken lines represent squares with less significant variances. Four densities in equal-area plots were tested: A = 15 × 15; B = 7 × 7 (staggered arrangement); C = 8 × 7 (grid arrangement); D = 4 × 4. Two replicates of each density were planted (''E'' and ''W''), but one field (WA) failed and was eliminated from the analysis. From Bateman (1947a). Published by Cambridge University Press.

Evans (1935) in test gardens of red clover (*Trifolium pratense*) that were studied for 2 consecutive years. In a 400 yd^2 plot of clover surrounded by fields with contrasting genotypes, only 5% of the plants had any inflorescences, and then usually only one to four. The seeds collected in this population yielded progeny with 44.8% contamination. The second year, however, the population had an ''abundance of blooms,'' and only 1.52% of the progeny bore the leaf-mark trait from the surrounding fields. Contamination was greater in progeny from the edge plants (4.01%) than from those in the central plants (1.95%). Repeating this observation in another field, Williams and Evans (1935) found contamination falling from 5.88% to 1.30% over 2 years as flower density rose from ''fairly numerous'' to ''numerous.'' If aging of a population, as it controls the increased ability of the population to flower, can produce effects on the pollen transport patterns and gene migration events identical to those caused by adding additional plants or increasing the density of plants, then the lability of pollen flow parameters is increased significantly. Another temporal influence on pollen flow is the effect of changing plant density in a population from year to year. Levin (1979) has shown that the pattern of shifts in plant density over time will have direct consequences for the neighborhood size and area and the potential for the population to undergo selective differentiation (Fig. 8). Populations with large, frequent

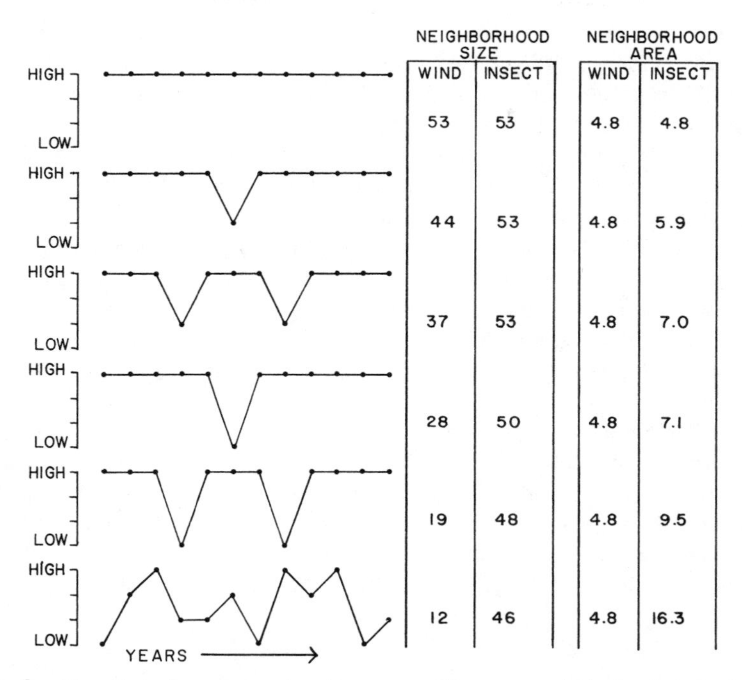

Fig. 8. The effect of population size changes over 12 years on neighborhood size and area with density-dependent and density-independent pollination in hypothetical situations. Mean neighborhood area increases with declining density in bee-pollinated plants but remains constant in wind-pollinated plants. The densities used in the model were 1, 3.25, 5, and 11 plants per m^2; the variances of pollinator flight distances were 2.6, 1.4, 0.8, and 0.04 m^2, respectively, for increasing plant densities. Density is represented by the "High–Low" scale. From Levin (1979).

fluctuations in density (lower part of Fig. 8) will have much larger neighborhood areas if the flowers are insect pollinated.

2. *Internal Arrangements in Populations*

The position of plants relative to others within its population will contribute to its fate as a pollen recipient or donor. Few natural plant populations have regular or random distributions of individuals (Greig-Smith, 1964; Harper, 1977a, Chapters 8–10), and the clustered distributions that are so common should have very different pollen flow patterns from those in the evenly arrayed agronomic studies. Studies with natural crossing rates in populations of thyme (*Thymus vulgaris*) showed that "spatial structure," the distribution and size of plants in the immediate vicinity of an individual, were the main factors that determined its breeding system (Brabant *et al.*, 1980). The selfing rate of individual thyme plants, measured by the distribution of alleles that control the production of thymol and carvacrol, was also examined in a series of gardens in different localities by Dommée (1981). Analysis of the outcrossing rates of identical plants in different sites showed that there was a significant locality effect, *le valeur pollinisatrice,* that complemented the genotypic effect, *le valeur allogamie,* in determining the outcrossing index. This elegant study also isolated the effect of the size of the plants in the populations as important in determining the outcrossing rate, with larger plants having increased selfing. In a direct study of spatial arrangement per se, Ennos and Clegg (1982) planted replicate fields of *Ipomoea purpurea* in two patterns: on opposite sides of some test gardens and in a random arrangement in others. The outcrossing rate measured by contrasting genotypes (at an esterase locus) was significantly higher in the random arrangement, showing that breeding system estimates made from marker loci in natural or test populations would change depending on the spatial arrangement.

Gleaves (1973) has modeled the effect of aggregation in a simulated population of 225 plants in a 280-cm^2 grid. Three arrangements (no aggregation, 20-cm spacing; 10-cm spacing within aggregations; 5-cm spacing within aggregations) were compared to determine the amount of endogenous pollen received. The within-population fertilization rate was less for regularly spaced individuals than for aggregated individuals (Fig. 9). Clumping of plants imposed lower gene flow

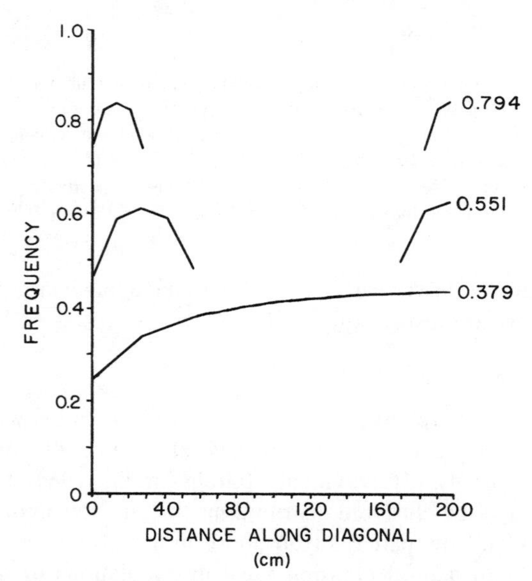

Fig. 9. Comparison of intrapopulation fertilization frequencies with different patterns of aggregation in a computer simulation. The numbers of the right-hand side of the figure are the averages of the intrapopulation fertilization frequencies for the 225 plants in the model. The gaps in the top two lines occur because there were no plants at the intermediate distances in the aggregated populations; the lowest line represents frequencies in the population that were continuous, not aggregated. From Gleaves (1973). Published by Longman Group Ltd.

TABLE I

COMPARISON OF GENE FLOW PATTERNS IN TWO EXPERIMENTAL POPULATIONS OF *BRASSICA CAMPESTRIS* THAT HAVE THE SAME NUMBER OF PLANTS AND THE SAME OVERALL DENSITY, BUT DIFFERENT SPATIAL PATTERNS WITHIN THEIR BOUNDARIES[a]

	Experiment 82-1	Experiment 82-5
Arrangement of plants	Dense clusters	Dense ranks
Number of seedlings scored	5,287	4,032
Percentage of plants with any dominant seedlings	45.0	65.8
Percentage of progeny seedlings with dominant allele	5.8	10.1
Appearance of dominant allele among clusters of parent plants		
Mean percentage	6.7	10.1
S.D. of mean	8.0	5.3
Coefficient of variation (S.D./$\bar{x}$)	119%	52%

[a]From S. N. Handel and J. L. Mishkin, in preparation.

from outside the population and acted the same as a general increase in population density. We have tested Gleaves' model in test garden populations of mustard (*Brassica campestris*) and found that gene flow from the center of a population to aggregations at the border of a garden is significantly less than in a garden with the same overall density, but with plants regularly arranged. Both the total contamination and the number of plants with any contamination are less in the aggregated population. The variation in gene flow among sections of the field was also much higher in tight clusters than in an even array (Table I).

3. *Pollinator Activity and Plant Density*

Most models and experiments on the effects of plant density on pollination systems make the assumption that the number and activities of the pollinators are ecological constants, once the threshold of attractiveness of the plants to the pollinators has been crossed. Further, pollinator limitation of seed set is not a significant factor that interferes with an understanding of pollen transport patterns. There is some information on the role of pollinator limitation on seed set in natural populations; the few detailed studies that have been done show that this is only sometimes present (Schemske *et al.*, 1978; Waser, 1979; Zimmerman, 1980; Lee and Bazzaz, 1982a). The activities of pollinators probably are quite sensitive to plant density and population size, however, and may be an additional variable acting on gene flow patterns as the plant population structure changes. Bateman (1956) stated that low plant densities caused decreased attractiveness to bees and fewer visits per flower in his crop experiments. Consequently, at low densities, a self-incompatible species should have a decline in seed set and a self-compatible species should have a rise in the percentage of self-fertilization (if crossing usually occurs in the population). This opinion was supported by Richards and Ibrahim (1978), who cautioned that the lower frequency of insect visits in sparse plant populations would make neighborhood size calculations themselves highly density dependent, and low pollinator abundance would not be completely compensated for by wider foraging forays by the pollinators that did service the sparse populations.

Field evidence for these concepts is available from so-called opportunistic pollinators which visit populations that have relatively sudden, large increases in flower density; these pollinators do not visit populations that have low but steady floral availability. Populations of the tropical shrub *Hybanthus prunifolius* (Violaceae) have episodes of synchronous flowering that occur after heavy rains (Augspurger, 1980, 1981). Pollination by bees (*Melipona interrupta*) is enhanced by this synchrony at both the population and individual levels. In asynchronous populations with low flowering density, plants with few flowers were not visited, and on some days of observation no plants were visited; in synchronous populations, all plants were visited and the proportion of flowers that formed fruit was higher (Augspurger, 1980). Plants with visitors averaged 200 flowers each, whereas plants with no visitors averaged about 50 flowers each. Flowers in the synchronous populations yielded a 86% fruit set and a 78% ovule set, whereas flowers in the asynchronous, low-density plots yielded a 58% fruit set and a 40% ovule set (Augspurger, 1981). The asynchronous plants also had a higher rate of fruit infestation by bruchid seed predators, which contributed to the difference in reproductive success between the two situations. In these ways, high flowering density had effects on the pollen flow dynamics not explained by simple changes in the flight parameters of the pollinators. A similar phenomenon is known from tropical species in the Bignoniaceae, in which flowering is contracted either into repeated short bursts of high-density blooming or the ''big bang'' strategy similar to *Hybanthus* (Gentry, 1974). In situations in which there is competition for bee visitation, there is apparently selection for high-reward clusters of flowers. A high density of flowers attracting an abundant pollinator guild can have a lingering effect on pollen flow. In the temperate tree *Catalpa speciosa,* mass flowering in the central part of the flowering season attracts a large pool of pollinators (Stephenson, 1982). During the final days of flowering, the pollinators continue to service the trees, and the percentage of flowers that set seed reaches its maximum: 36% of flowers over 2 years as compared to 17% for the initial and peak flowering times combined. In this case, more flowers cause an increase in visitation, but the pollination dynamics in the later period of low density can be understood only by reference to the plant–animal interactions that occurred earlier in the year.

The change in pollinator number during periods of high flowering density is sometimes accompanied by a change in pollinator behavior (Shapiro, 1970; Levin and Kerster, 1974; Levin, 1981). Frankie (1975, 1976) noted an increase in aggressive interactions among territorial or group-foraging bee species in the tropics when mass flowering resulted in the interaction of many species on single trees. As a consequence, the number of interplant movements by these bees increased, and presumably the amount of cross-pollination as well. Gentry (1978) noticed that high bee density could result in high interplant movements because of an interaction at a higher trophic level: In cases in which large numbers of pollinators were on tropical trees, bee-eating birds congregated and drove bees away from the tree. These birds (flycatchers in his example) preferentially forage at trees which are mass flowering, as these have the greatest concentration of pollinators.

Floral preference is also increased by high-density arrays of flowers. Consequently, the number of legitimate pollinations should increase, and the amount of hybrid seeds produced in the community should be reduced (Levin, 1979; Antonovics and Levin, 1980). In an experiment with populations of artificial flowers that had the same mean reward of nectar but different variances, bumblebees tolerated a high variance when flowers were clumped rather than in an even array and increased their visitation rate (Real *et al.*, 1982). Similarly, honeybees have shorter flight distances on dense arrays of nectar-rich flowers than on nonrewarding clusters (Waddington, 1980). These results show that flower density may have manifold effects on the energetics of the plant–pollinator interaction, and consequently on the movement of bees as it is determined by the need for high energetic reward (see also Heinrich and Raven, 1972; Heinrich, 1975).

4. *Other Ecological and Evolutionary Consequences of Changes in Floral Density*

An increase in flowering density, like any change in an important component of an organism's life history, influences many aspects of the biology of plants. Simply focusing the flowering season into a tight synchrony may have effects beyond a mere shift in floral density. For example, there may be changes in the potential for outcrossing, isolating mechanisms, energetics of pollinators and the ability to attract them, the ability to avoid serious damage by seed and fruit predators, and the demography of seedlings that are produced in a short period of time by a synchronous maturation of fruits (Augspurger, 1981). Also, density changes in populations affect the morphology of a plant and its ability to initiate flowers; there is a nonlinear relationship between density and the number of flowers produced per plant (Harper, 1977a, Chapter 7).

Changes in plant density have also been investigated from the perspective of damage and destruction to seeds. This has immediate effects on the success of the density–pollination interaction. Many seed predators (insects and vertebrates) are attracted to and oviposit or forage on dense populations, although the predators would not attack seeds or fruits in sparse populations (Janzen, 1971b; Vandermeer, 1975; Harper, 1977a, Chapters 15–17; Antonovics and Levin, 1980).

The complex interaction between plant density, seed predation, and pollination can be illustrated by seed set patterns in *Cassia biflora* (Silander, 1978). Seed set was higher in dense populations, and pollinator activity (by carpenter bees) was positively correlated with high density. However, pod infestation by bruchids was also higher. Seed infestation was density independent in Silander's study, allowing for the overall increase in the number of seeds despite the activities of the beetles. Silander cautioned that the lower seed set of isolated plants may have been caused by less favorable microsites and higher levels of interspecific competition, rather than directly by pollinators or seed predators; the interaction of these several factors must be understood to predict the effect of density per se on seed set in a plant. Density dependence of pollination success, seed predation, and of seed pod herbivory was observed for the tropical shrub *Bauhinia ungulata* (Heithaus *et al.*, 1982). Eventual seed production by the shrub was jointly determined by the three processes, not by the change in pollination events alone with shifts in density. Manipulation experiments by Platt *et al.* (1974) on populations of *Astragalus canadensis* showed that pollinator success did increase with plant density, but attacks on seeds were higher at low densities by curculionid beetles. Apparently, weevil populations occur at higher densities on more isolated plants and destroy a larger proportion of seeds on these plants. In contrast, Ralph (1977) found that seed damage by *Oncopeltus fasciatus* on milkweed, *Asclepias syriaca*, increased with increased plant density. The pattern of seed production vis-à-vis pollination intensity and seed predation is clearly quite dependent on the population ecologies of both the animal and the plant, and the relative success of plant defenses from animal search and herbivory rates. Pollination biology can never be isolated from the effects of density on subsequent components of reproductive success. There is also growing evidence that plant density shifts cause changes in the attractiveness of the population to many types of insect herbivores (Root, 1973; Kareiva, 1982).

Internal controls over seed production after pollination and fertilization are also important for many species, although most work in this area has been done with crop plants that have been artificially selected for high and relatively synchronous seed set (Bollard, 1970; Yoshida, 1972; Stephenson, 1981; Lee and Bazzaz, 1982b). Physiological inhibition of later-developing seeds by earlier ones is known for many species. Regardless of the pollination intensity at later stages of plant growth, when flower density may be at a maximum, the pollen flow events early in the growing season may be the only ones that actually result in mature seeds. This is the case with cultivated cucumbers, for example, in which the maturation of early fruits will stop development of late fruits, even if flower density and pollinator visitations remain very high (Golínska, 1925; Handel and Mishkin, 1983).

There are also effects of density shifts on the mating systems of plants, both in the direct proportion of flowers that become selfed and in the eventual evolutionary effect of density on the development and persistence of self-incompatibility in the population (Baker, 1955; Jain, 1976a; Antonovics and Levin, 1980). With increased flower density, there is commonly an increase in geitonogamous fertilization, particularly in plants that are clonal (Bateman, 1956; Kalin de Arroyo, 1976). This, in turn, may cause a decline in seed set in self-incompatible species or an increase in inbreeding (and possibly in the negative effects of inbreeding depression) in self-compatible species. The percentage of flowers that self may change from one period of the flowering season to the next as population density changes (Carpenter, 1976; Stephenson, 1982). In some species, such as *Lithospermum caroliniense,* there is a relatively larger percentage of cleistogamous flowers as the plant density increases, changing the inbreeding proportion in the population by a developmental pathway rather than by a manipulation of the pollinators (Levin, 1972b).

Evolutionary responses to regularly high floral densities may occur, such as a decrease in nectar reward per flower (Heinrich and Raven, 1972; Heinrich, 1975; Waser and Price, 1983). Plants that regularly occur at low densities may experience selection for a shift to genotypically controlled self-fertility. This has been seen in populations of *Armeria maritima* that are found on poisoned soils (Lefèbvre, 1970). These populations are in colonizing situations in which population densities are usually very low and plants with self-fertility would be at an advantage, just as in other examples of selfing and colonizing associated under "Baker's law" (Baker, 1955). The *Armeria* example is somewhat different from that seen in populations on mine dumps, in which selfing probably lowers the percentage of pollen from surrounding, nontolerant populations that reaches the plants on poisoned soils (Antonovics, 1968). No nontolerant populations were recorded near the *Armeria* populations; density shifts alone were implicated in the change in breeding system.

High densities and restricted gene flow in plants can cause the establishment of very local subdivisions in populations, which can enhance the ability of these plants to respond to local selection pressure and to undergo genetic drift (Levin and Kerster, 1975; Levin, 1981). Local pollen dispersal patterns by bees have been frequently mentioned as components of local genetic subdivision in plants (cf. Free, 1966; Lefèbvre, 1976; Levin, 1979; Handel, 1983a). The leptokurtic pollen dispersal pattern suggested for many plants produces a much more restricted gene flow pattern than stepping-stone or random patterns, and also affects the average number of generations until near-fixation of a gene occurs (Levin and Kerster, 1975; and Table 4 cited therein). This interplay has been modeled by Turner *et al.* (1982), showing that the nearest-neighbor pollination

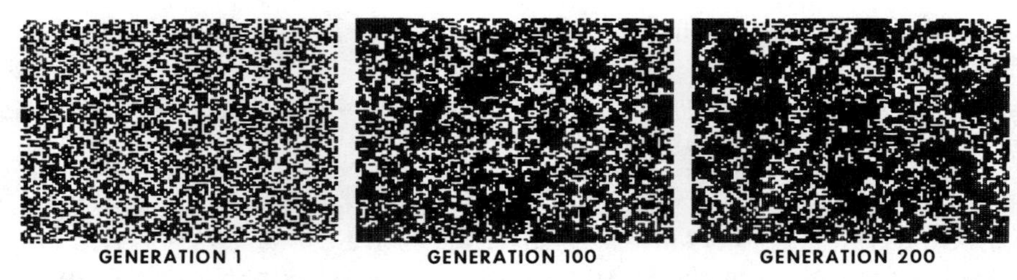

Fig. 10. Spatial distribution of genotypes in a simulated plant population at intervals of 100 generations, assuming nearest-neighbor pollinations. Development of homozygous patches occurs relatively quickly and is well established by generation 100, even without the imposition of selection pressure. From Turner *et al.* (1982).

patterns can itself cause inbreeding, an increase in homozygosity, and local differentiation (''patchiness'') within the population (Fig. 10). No selection was needed to explain the production of large areas of homozygous individuals, even after only 100 generations. The interplay of pollinator behavior and local pollen transport alone, both ultimately dependent on the density parameters of the plant population, created marked subdivisions in the population.

C. Population Shape

Natural plant populations are rarely square or rectangular; more often they follow some ecotone boundary or spread at different rates in various directions to produce an irregular outline. Many populations are quite linear, following specific habitat types such as beach fronts, roadsides, or limestone or serpentine outcrops. Models of gene flow have shown that a linear arrangement of organisms has different consequences for gene frequencies and movement than does a two-dimensional population (Maruyama, 1971, 1972; Nagylaki, 1976). Certainly, there is field evidence that in linear populations, gene flow and selection pressure are best expressed in terms that reflect the particular geometry of the population. Roadside populations of *Plantago lanceolata* and *Cynodon dactylon,* for example, are lead tolerant, whereas populations of these species that are parallel to the road axis, but 4 m or more away from the road, have very low lead tolerance (Wu and Antonovics, 1976). The interesting pollen flow story for that situation would be the mass and tempo of movement perpendicular to the road axis, not simply at a radial distance from the population center.

The influence of patch shape on pollen flow has been explored in a computer simulation by Levin and Wilson (1978; see also Levin, 1978a) based on a model of some 4000 plants placed in a 64 × 64 grid and containing a single locus of alleles with differing relative fitnesses. The plants were assumed to be self-incompatible, and after pollen flow of varying patterns, selection, and scaling of gene frequencies, the pattern and amount of alleles present were computed. The population shape contributed directly to the eventual genotypic pattern. As the shape of the population changed from 1 × 64 to 8 × 8 patches, the amount of contact between the contrasting patch types shifted and the amount of alien pollen that entered each patch increased from 55 to 26%. Equilibrium gene frequency of the maladapted gene also shifted from 0.38 in the 1 × 64 arrangement to 0.16 in the 8 × 8 arrangement.

Another simulation by Pedersen *et al.* (1969) compared two fields of similar density and area but different shape (0.4 × 0.4 vs. 0.2 × 0.8 km). The rectangular field was more susceptible to contamination from exogenous pollen than the square field, except at high isolation distances at which very little pollen reached the field (Table II). In addition, changing the isolation distances between the sides of the fields (a more realistic situation) caused complex patterns and differences between the two plots (Pedersen *et al.*, 1969; and Table 3 cited therein, for all combinations tested). For example, if the rectangular field had isolation distances of 50.3 m on three sides and 402.4 m on one side, there will be 4.6% contamination if the 402.4-m isolation distance is on the short side of the field but 2.9% contamination if it is on the long side of the field.

Some field experiments have confirmed that population shape alone may directly influence pollen transport patterns. Rai and Jain (1982) planted test gardens of barley, *Avena barbata,* in both cross and spiral patterns, with plants bearing dominant morphological markers in the center. The mean pollen flow, measured by scoring the progeny, was significantly higher in the cross-shaped population (2.94 vs. 0.98 m). There was much directionality of pollen movement in their experiments, and many of the pollen flow distributions were platykurtic (Fig. 11). A complex pattern of gene frequency arrangements was also seen in natural populations of *A. barbata,* and the authors concluded that the "real joyful diversity of natural patterns (or lack of them)" (p. 405) cautioned against a too willing acceptance of theoretical models of gene flow. Our own work with

TABLE II

Comparison of the Amount of Contamination of Exogenous Pollen into Fields of Similar Size and Density, But Different Shape, from a Computer Simulation[a]

Isolation distance (m)	Shape of field (km) 0.4 × 0.4	0.2 × 0.8	Difference in contamination (%)
0	6.58	8.16	24
50.3	4.1	5.03	23
402.4	0.01	0.01	0

[a]Pedersen *et al.* (1969).

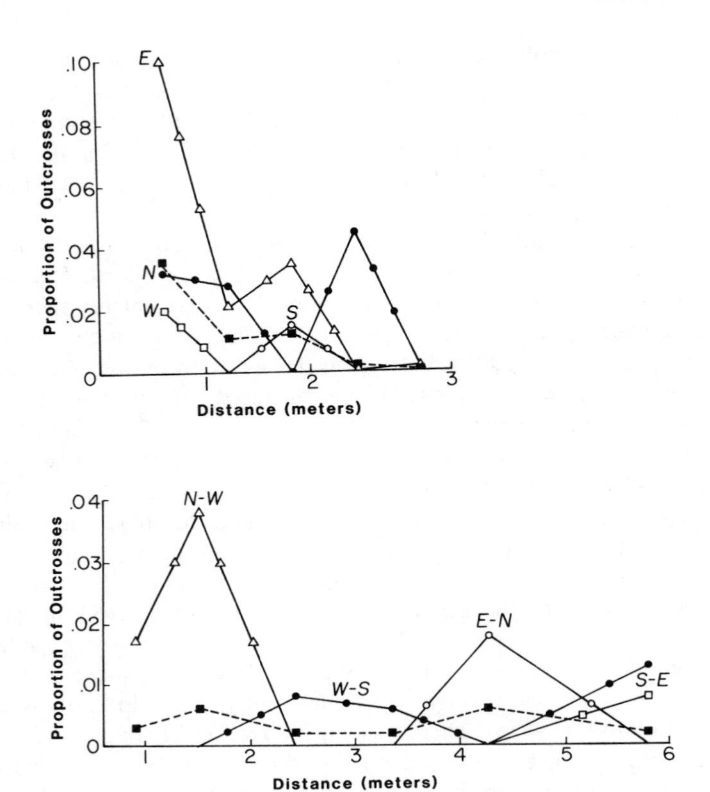

Fig. 11. Outcrossing rates at a locus controlling leaf sheath pubescence in test gardens of *Avena barbata* set up in different arrangements. The *x* axis shows the distance from the center of the population. (Top) Results from gardens with plants in a cross-shaped array. (Bottom) Results from gardens with plants arranged in a spiral pattern emanating from the central patch of plants. From Rai and Jain (1982).

test gardens of *B. campestris* has shown that the shape of the internal patch of dominant-allele-bearing plants can also have a big effect on gene flow in the population (S. N. Handel and J. L. Mishkin, in preparation). For example, in one experiment we arranged the dominant plants in a single straight line across the center of the population rather than in a central block, with all the plants 80 cm apart. After scoring 3098 progeny seedlings, we found that 87.8% of plants surrounding the dominants had some dominant seedlings and 34.4% of the progeny carried the dominant allele. These percentages are significantly higher than the results from populations in which the dominant plants were in a block, as noted in Table I.

A comprehensive series of test plots (interplanted kale and cabbage) by Nieuwhof (1963) showed that the patch shape of varieties within large populations directly influences pollen flow, but only after a threshold is reached. Interplanting the two cultivars in different patch shapes caused the following contamination patterns: single rows, 43–47%; double rows, 39–42%; four-row groups, 19–21%. The effect of patch shape here is obviously influenced by a varietal mass factor, but the influx of pollen seems to be clearly related to the proportion of plants in each patch that borders on the opposite variety. In an additional experiment, Nieuwhof (1963) planted square and rectangular arrays (Fig. 12), with the former array having one central patch of recessive plants and the latter two. The progeny of the recessives in the rectangular plot had 56% and 51% contamination on either side; the square plot had 52% contamination along the edge rows of the recessives, but only 35% in the center. Here, changing the dimensions of the recessive patch and splitting it caused elimination of plants that had very low contamination rates. Similar results were obtained by changing the dimensions of radish (*Raphanus sativus*) (Crane and Mather, 1943) and of turnip (*Brassica rapa*) (Bateman, 1947a) plots.

The impact of patch shape and edge effects on pollen flow is due not only to the extinction of pollen flow with distance, as discussed previously, but also to the influence of patch shape on other peculiarities of pollinator behavior. Ginsberg (1983), for example, showed that the solitary bee *Halictus ligatus* tended to forage at the edge of flower patches, and that little movement into the patch centers was observed. Therefore, increasing the proportion of border plants by altering patch shape could cause increased contamination.

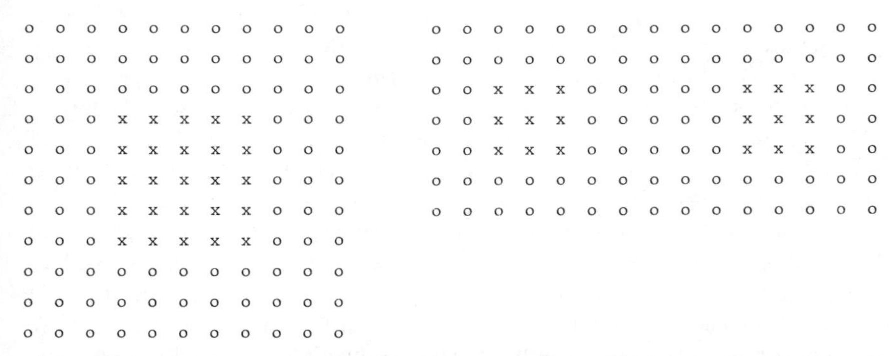

Fig. 12. The arrangement of plants in Nieuwhof's (1963) experiment showing the effect of different patch shapes on the contamination of seed stocks by alien pollen. 0 = red cabbage (dominant); X = white cabbage (recessive). Interplant distances were 65 × 65 cm for the red cabbage and 60 × 60 cm for the white cabbage.

IV. Internal Structure of the Population

A. Clonal Growth

When growth is by stolons or rhizomes, flowers become increasingly surrounded by inflorescences of the same plant. If pollen flow is restricted because pollinators tend to visit nearest-neighbor flowers, large clones will tend to have more geitonogamous pollinations than do small clones. The occurrence of plants with a clonal growth form is widespread, with many clones reaching enormous sizes (Harberd, 1967; Oinonen, 1967). In situations in which clonal growth is an important part of population structure, an understanding of pollen flow dynamics must be modified to incorporate the distribution and frequency of clones (Levin and Kerster, 1971). The inaccuracy of simply recording flower densities when many of the flowers are parts of the same "inflorescence," but spread in two rather than three dimensions, can destroy an understanding of actual gene flow patterns.

An example of this interplay between clonal growth and potential change in the breeding system has been seen in the wind-pollinated, self-compatible sedge, *Carex platyphylla* (Handel, 1983b). This species has very restricted pollen flow, with most pollen falling within 20 cm of a staminate spikelet (Handel, 1976). When clones become large, an increasing percentage of the pollen that can fall on a stigma comes from anthers that are on different parts of the same plant. Consequently, large plants should have a greater percentage of self-pollination than small clones which are more closely surrounded by conspecific plants of different genotypes (Fig. 13).

This shift in the breeding system should also occur in self-incompatible species that are clonal if pollinator movements from flower to flower are typically very short. This has been seen in regard to *Trifolium repens,* where bee flights are between neighboring inflorescences and clonal growth is present (Heinrich, 1979; Handel, 1983b). There is great variation in the number and placement of inflorescences on a clone and in the growth rate of plants even within single natural populations (Burdon, 1980), but in most situations large clone size should result in fewer fertilizations, as more pollen becomes loaded on a bee, and deposited on other flowers of the same clone, where the pollen is incompatible. This particular interaction changes when large size results in the intergrowth of clones. If this occurs, many nearest-neighbor inflorescences would be from different parents, and consequently probably compatible, and fertilization could occur.

There are other possible evolutionary escapes from the trap of increased selfing with large clone size, as has been seen in *Aralia hispida,* in which parts of a

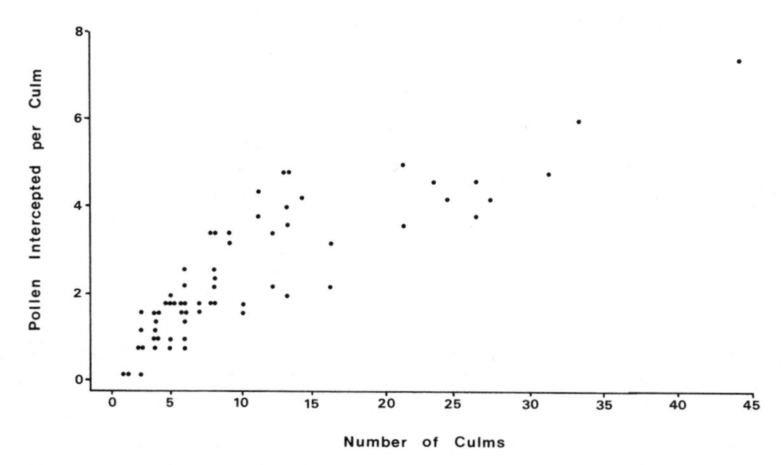

Fig. 13. Pollen deposition patterns within clones of *Carex platyphylla*. The average amount of endogenous pollen on each culm increases sharply until a clone has about 10 culms; then the increase in endogenous pollen as the clone continues to grow is small. Smaller clones have a relatively lower rate of self-pollination, given a fixed deposition rate of exogenous pollen. From Handel (1983b).

clone have synchronized protandry, enforcing outbreeding (Thomson and Barrett, 1981; Barrett and Thomson, 1982). Also, large plants with long flowering periods may effectively set seed only during the early part of the season; early fruits of many species inhibit the development of later-pollinated flowers (Yoshida, 1972; Stephenson, 1982). Pollinations occurring when the clones are large on these species would have a minor effect on the genotypic character of the seeds, because few seeds would be set from late-appearing flowers.

Large clone size is analogous to some differences in pollination biology that occur in any plant when the inflorescence size increases (Wyatt, 1982; Waser, Chapter 10, this volume). Pollinators often follow set behaviors within an inflorescence, and the behavior may change with inflorescences of different geometries. In a study of *Oenothera fruticosa*, for example, bees visited 5.2 flowers on large plants (defined as having more than 40 flowers) but only 2.1 flowers on small plants (fewer than 20 flowers) (Silander and Primack, 1978). On the larger plants, more than 35% of the pollen transferred was from within the inflorescence and self-incompatible. The advantage of having more flowers and a relatively higher visitation rate was somewhat offset by the higher frequency of incompatible pollen that was deposited. The joint effects of pollinator behavior, deposition rates of compatible pollen, clonal growth rate, and clone size require populations with clonal growth to be mapped and the geometric structure of the population interpreted with reference to the internal distribution of the various genotypes.

B. Genotypic and Physiological Variation

Plants within a population can vary tremendously in both microhabitat and genotypic differences; the extent of this variation is now being surveyed at the molecular level (Brown, 1979; Hamrick, 1982). Components of this variation can directly affect the pollination biology of plants in the population. There is direct evidence that the amount of crossing varies among plants in any one population. This was perhaps most clearly shown in test garden studies such as those by Jones and Brooks (1950) on maize and by Rick (1947) on tomatoes (*Lycopersicon esculentum*). Changes in gene frequency were recorded in different positions of the field, with certain zones having much more crossing than others, a patchwork of hybridization. The same sort of variation in crossing has been seen in test plots of *Cucumis melo* and *C. sativus,* in which the position and percentage of outcrossing have been recorded for individual fruits (Handel, 1982, 1983a). Given the variety of outcrossing rates in homogeneous fields, natural populations might be expected to show even more complex patterns. The variation in crossing might be caused by small differences in the distance to a source of compatible pollen, as in Rick's (1947) study of tomatoes. There, interplanting sterile and fertile plants in different arrangements caused immediate differences in the ability to set fruit and seeds. Contrasting plants were put in four-plant clusters, a checkerboard arrangement, and then in completely mixed plots where each grid position had both a sterile and a fertile plant. The number of seeds per fruit in these arrangements was 25, 22.8, and 35.5, respectively. The number of fruits produced per 100 flowers was 1.5, 2.7, and 4.7, respectively. Changing the location of pollen sources within the population over a matter of centimeters caused a three-fold increase in the frequency of fruit set.

In addition to variation in the local pollen environment, there can be differences in selfing rates that are genotypically controlled by the maternal parent. This is well documented by interpopulation differences in crossing rates for several species (Vasek, 1965, 1967; Allard *et al.*, 1968; Levin, 1978a; and Table 2 cited therein). The differences can sometimes be quite high, as in the 13–97% range in outcrossing in ruderal populations of *Lupinus succulentus* in California (Harding and Barnes, 1977). The percentage of crossing, however, can also vary significantly among plants within a population and among flowers of the same plant (Humphreys and Gale, 1974; Hamrick, 1982).

Additional intrapopulation life history variations that modify the breeding system have recently been reviewed by Levin (1981). Realizing that pollen flow parameters alone were not the sole determinants of progeny genotypes, Levin listed other factors that complement pollen dispersal patterns. These include differences in the seed set and seed abortion rates that occur after close or far neighbors cross (Waser and Price, 1983); the known variation in seed and seedling mortality rates; and the variation in the fitness of offspring produced after crossing occurs between near or distant parent plants. These factors, in addition to the lowered pollen–pistil compatibility that often occurs among near neighbors, form a more comprehensive picture of the dynamic controls over progeny genotypes within a population.

There are also temporal shifts in flowering time among plants which could limit pollen transfers. Even plants that differ in flowering time by a single day may be effectively isolated if the period of pollen viability in natural populations is as short as has been seen in many cultivated species. Kraai (1962), for example, showed that pollen viability in six species ended after 12 hours. However, these species were all highly bred for relatively simultaneous flowering; there was no selective pressure for long pollen life. In natural populations a rather different pattern may eventually be found, but little information along these lines is currently available. Pollen longevity is known to be strongly influenced by temperature and relative humidity (Stanley and Linskens, 1974, Chapter 5). Although certain families of plants (Rosaceae, Primulaceae, Amaryllidaceae) have species whose pollen lasts a relatively long time (31–38 days) under natural conditions, other families, notably the Graminaceae, have pollen that typically is viable for only 1–2 days (Johri and Vasil, 1961). The latter species may have nearby populations that are regularly noninterbreeding for solely this reason. Additionally, several species are known in which flowering initiation and duration vary widely among individuals (see Gentry, 1974; Stiles, 1975; Wilken, 1977; Leverich and Levin, 1979; Motten, 1982). Also, different populations of the same species may have different flowering times, causing different interactions with the available pollinator pool (Schmitt, 1983). These are the sorts of species in which flower density at any one time would give only a vague representation of the actual pollen transfer patterns that exist in the population.

C. Effect of Microclimate

In natural populations there are often many subtle changes in the local habitats of plants, sometimes over scales that should be measured in meters or centimeters, not kilometers. These changes have manifold effects on many aspects of population biology, including recruitment, competitive interactions, and mortality schedules. These effects are voluminously cited by Harper (1977a) and are vividly summarized by his statement (Harper, 1977b, p. 156): "I suspect that the most important developments in terrestrial plant ecology may well come from exploring intra-specific variation within populations and the intra-habitat variation in space and time. . . . Gerald Manley Hopkins wrote 'Glory be to God for dappled things.' No part of . . . ecology has yet faced the fact that to an organism and its progeny the environment is a richly dappled, minutely speckled, thing."

Small-scale changes in the physical environment must have an impact on the pollination systems of plants, but this interplay is not well known. Some secondary interactions, such as the proximity of plant populations or their parts to pollinator nests (which themselves are placed after consideration of the microhabitat), and a limitation of flower density by plant competitors have already been mentioned, but there are other possible interactions. Insolation, for example, can have a strong influence on flowering phenology and genetic subdivisions in populations. Jackson (1966) studied the flowering phenology of several spring herb populations that spread across a ravine whose slopes were 46 m apart and faced north and south. The difference in flowering of these species on the two slopes varied from 3 to 11 days (Table III). Many of these species usually have only a short period of flowering, and plants on one slope would be temporally isolated from ones on the other slope despite their spatial proximity (Table III). Jackson (1966) showed that the slope-to-slope phenological difference was as great as would be expected from 110 mil of latitude.

Even along a single slope, relatively small elevational differences can cause genetic isolation because of temporal differences in flower appearance and pollinator service. Mahall and Bormann (1978) studied the appearance of woodland herbs along a slope in New Hampshire. Several of the herbs showed significant differences in leaf maturation time (which is correlated with flowering time) with elevation, although one herb, *Uvularia sessilifolia,* did not, probably because its life cycle starts somewhat later in the growing season (Table IV). The possibility for microdifferentiation among the segments of these populations consequently exists, but the species each have different phenologies up and down the slope.

TABLE III

MICROCLIMATIC INFLUENCES ON FLOWERING TIME AND TEMPORAL ISOLATION AMONG PLANTS[a,b]

Flower life (days)	Species	Difference in flowering time (days) between north- and south-facing slopes
3	*Claytonia virginica*	4
2	*Sanguinaria canadensis*	3
4	*Dentaria laciniata*	6
11	*Dicentra cucullaria*	6
	Arisaema triphyllum	7
	Arabis laevigata	6
	Phacelia bipinnatifida	11
	Smilacina racemosa	4

[a]Many spring herbs have short flowering times and are extremely sensitive to the effect of slope aspect.

[b]Data on flower life are from Schemske *et al.* (1978). Data on flowering phenology are from Jackson (1966).

TABLE IV

VARIATION IN TIMING OF LEAF MATURATION IN SPRING HERBS ALONG AN ELEVATIONAL GRADIENT IN THE WHITE MOUNTAINS OF NEW HAMPSHIRE[a,b]

	Elevation (m)			
	554	664	747	782
Erythronium americanum	1	4	6	15
	MHT	LT	LT	LMH
Claytonia caroliniana	1	6	9	16
	MHT	LHT	LMT	LMH
Uvularia sessilifolia	1	1	1	1
Aster acuminatus	1	11	16	17
	MHT	L	L	L
Oxalis montana	1	6	32	5
	HT	H	LMT	LH

[a]Data are the number of days from the onset of flowering in each species at the lowest elevation. The letters indicate which elevations are significantly different in flowering phenology from the particular entry (L = low elevation; M = middle elevation, 664 m; H = high elevation, 747 m; T = top elevation).

[b]Data are from Mahall and Bormann (1978). Published by the University of Chicago Press.

Generalizations about the role of the microhabitat on temporal isolation may be difficult to make. Also, the differences that have been seen in these studies suggest that the irregular topography in forests may permit more opportunity for genetic isolation through pollen flow restriction than in pastures or grasslands, where the microhabitat is generally more even.

The sex ratio in plant populations is often not equal, and microhabitat segregation by the different sexes can contribute to this pattern (Levin, 1978a). Several species of plants have female and male individuals that sort out according to short-range environmental variables such as soil moisture and nutrients. This is true for many species in the western United States (e.g., *Thalictrum fendleri, Ephedra viridis, Distichlis spicata,* and *Acer negundo*) and in other areas (Freeman *et al.*, 1976; Bawa, 1980; Cox, 1981). Since an unequal sex ratio will immediately affect pollen distribution amounts and patterns, microhabitat differences play an important role for these species. The ratio of male and female flowers on monoecious plants is also affected by microhabitat differences and can have analogous effects on pollen movement (Freeman *et al.*, 1981).

There are additional ways in which small-scale ecotones may restrict pollen movement. In a population of *Armeria maritima,* Eisikowitch and Woodell (1975) observed that bumblebees visited plants on the dunes and shingle beach, but not those in the adjacent marsh. There are several hypothesized reasons for this, such as the bees' distaste for getting wet as the inflorescences on the marsh bent into the water. The authors concluded that at their study site the possibility for disruptive selection and local race differentiation because of the nonrandom movements of the bees was quite strong. An even more small-scale example of local environmental restriction on pollen movement was noticed by Beattie (1971) in the coastal forests of San Mateo County, California. Here pollinator movements were strongly correlated with the appearance of ephemeral sunflecks of the forest floor. As the sunflecks moved along the ground, the temperature of the illuminated flowers rose abruptly, and the number of flower visitors jumped from none to about 20 during the time of illumination. Consequently, the pollinators tracked the sunflecks through the forest, rather than density shifts or population boundaries of the plants themselves. Beattie (1971) commented that the opportunity for outbreeding appeared most frequently on those plants that were located in the path of the sunbeam. These paths may change from year to year with the changing geometry of the forest canopy, facilitating pollination on different plants during different years. Conversely, other pollinators such as fungus-gnats (Sciaridae and Mycetophilidae) prefer the shade and are most active as flower visitors away from sunflecks (Mesler *et al.*, 1980). These studies show that the physical environment around individual plants or small clusters may be critical in mediating the pattern of pollen flow, and may override or complement the effects of population geometry by itself. Together with their strong influences on seed dispersal dynamics (Howe and Smallwood, 1982; Moore and Wilson, 1982), small-scale topographic differences are basic to an understanding of plant microevolutionary events.

V. Summary: "The Pollination Milieu"

Although the early and extensive work in pollination biology concentrated on the adaptive nature of the morphological structures of flowers (Waser, Chapter 10, this volume) and the types and movements of pollinators that visited different species, the growth of evolutionary ecology has demanded that the interplay between plant and pollinator be understood in much greater detail. In consequence, new work has shown that the pollination biology of any one species is not constant among populations, but may change with the nature of the pollinators that are available, and with the demography, community structure, and population ecology that is best defined for each population separately. Detailed work, particularly by agronomists dealing with genetically homogeneous crops growing in simple environments, has shown that pollination biology, particularly the amount of cross-pollination, is very sensitive to the geometry of the population: the number of plants that are massed together, the density of flowers within the population, and the shape of the population. Different sizes, densities, and shapes of populations of the same species can elicit different responses from pollinators and differences in the amount of pollen that is transferred. In some cases, the identity of the pollinators, as well as the abundance of individuals and of flower visits, change as the population geometry is altered.

Natural stands probably involve even more complex interactions among population geometry, pollinator behavior, and gene flow because of community interactions and microhabitat differences that can change the breeding structure of the plants. This may be particularly true in areas where habitat heterogeneity is frequent, such as in forests, and plant response (flowering time and amount) may be quite sensitive to these differences. Also, natural populations have great genotypic and physiological diversity, and not all plants may be able to share pollen equally. Plants that have clonal growth have even more complex interactions between pollinator behavior and pollen movement because many pollinator visits are between flowers of the same genotype. Pollinator movement and flower density measurements for these plants must be modified to express the relatedness of flowers on the different clones, the interplay of pollination events, and the physiological control over fruit and seed development.

Natural populations rarely have monomorphic patches of plants that can supply unambiguous genetic markers with which to track pollen flow (as are available in crop plants). Work in these populations usually requires an indirect method for tracking pollen movement. None of the methods that have been tried, including dyes, fluorescent powders, observation of pollinator movements, neutron-activation analysis, and autoradiography by labeled pollen grains, are useful in all situations. An investigator must choose an indirect method carefully, and calibrate the method for the particular plant species and the common pollinators that visit the test populations. Different pollinators can have very different efficiencies as pollen transportors, and each species that visits a population must be characterized separately.

Haskell (1954) referred to the ''pollination milieu'' of any population, meaning that breeding systems and pollen movements must be defined by reference to the immediate and idiosyncratic nature of any one population, rather than by some generalization based on cursory observations of one stand that may be considered typical of the species. The typological viewpoint, now discredited as inadequate to describe species or community characteristics, must be removed from the study of pollination biology if we are to understand the actual diversity of processes that occur in populations that may vary in only the most subtle ways. The remarkable responsiveness of plant breeding systems to small changes in population ecology suggests that many populations may be unique in their relationships to pollinators, and microevolutionary processes such as population differentiation may follow very different paths and tempos in very similar, but not identical, stands.

Acknowledgments

I thank Judith Mishkin for her help throughout my work in pollination biology, and my other co-workers, Jan Roueche, Catherine S. McFadden, Deborah Levine, Maureen Stanton, Blair Simpson, and Hester Parker, for their many contributions. This review has been improved by the comments of Les Real, Beverly Rathcke, and Nick Waser, to whom I am grateful. Support for my work that is mentioned here was provided by NSF Grants DEB-79-27022 and DEB-81-06934.

References

Ackerman, J. D., Mesler, M. R., Lu, K. L., and Montalvo, A. M. (1982). Food-foraging behavior of male Euglossini (Hymenoptera: Apidae): Vagabonds or trapliners? *Biotropica* **14,** 241–248.

Allard, R. W. (1960). "Principles of Plant Breeding." Wiley, New York.

Allard, R. W. (1975). The mating system and microevolution. *Genetics* **79,** 115–126.

Allard, R. W., Jain, S. K., and Workman, P. L. (1968). The genetics of inbreeding populations. *Adv. Genet.* **14,** 55–131.

Antonovics, J. (1968). Evolution in closely adjacent plant populations. VI. Manifold effects of gene flow. *Heredity* **23,** 507–524.

Antonovics, J., and Levin, D. A. (1980). The ecological and genetic consequences of density-dependent regulation in plants. *Annu. Rev. Ecol. Syst.* **11,** 411–452.

Augspurger, C. K. (1980). Mass-flowering of a tropical shrub (*Hybanthus prunifolius*): Influence of pollinator attraction and movement. *Evolution (Lawrence, Kans.)* **34,** 475–488.

Augspurger, C. K. (1981). Reproductive synchrony of a tropical shrub: Experimental studies on effects of pollinators and seed predators on *Hybanthus prunifolius* (Violaceae). *Ecology* **63,** 775–788.

Baker, H. G. (1955). Self-compatibility and establishment after long-distance dispersal. *Evolution (Lawrence, Kans.)* **9,** 347–349.

Baker, H. G. (1979). Anthecology: Old testament, new testament, apocrypha. *N. Z. J. Bot.* **17,** 431–440.

Bannister, M. H. (1965). Variation in the breeding system of *Pinus radiata*. *In* "The Genetics of Colonizing Species" (H. G. Baker and G. L. Stebbins, eds.), pp. 353–372. Academic Press, New York.

Barrett, S. C. H., and Thomson, J. D. (1982). Spatial pattern, floral sex ratios, and fecundity in dioecious *Aralia nudicaulis* (Araliaceae). *Can. J. Bot.* **60,** 1662–1670.

Bateman, A. J. (1947a). Contamination of seed crops. I. Insect pollination. *J. Genet.* **48,** 257–275.

Bateman, A. J. (1947b). Contamination of seed crops. III. Relation with isolation distance. *Heredity* **1,** 303–336.

Bateman, A. J. (1956). Cryptic self-incompatibility in the Wallflower: *Cheiranthus cheiri* L. *Heredity* **10,** 257–261.

Bawa, K. S. (1980). Evolution of dioecy in flowering plants. *Annu. Rev. Ecol. Syst.* **11,** 15–40.

Beattie, A. J. (1971). Itinerant pollinators in a forest. *Mandroño* **21,** 120–124.

Beattie, A. J. (1976). Plant dispersion, pollination and gene flow in *Viola*. *Oecologia* **25,** 291–300.

Beattie, A. J. (1978). Plant-animal interactions affecting gene flow in *Viola*. *In* "The Pollination of Flowers by Insects" (A. J. Richards, ed.), pp. 151–164. Academic Press, New York.

Bollard, E. G. (1970). The physiology and nutrition of developing fruits. *In* "The Biochemistry of Fruits and their Products" (A. C. Hulme, ed.), Vol. 1, pp. 387–425. Academic Press, New York.

Bond, D. A., and Pope, M. (1974). Factors affecting the proportions of cross-bred and self-bred seed obtained from field bean (*Vicia faba* L.) crops. *J. Agric. Sci.* **83,** 343–351.

Bosbach, K., and Hurka, H. (1981). Biosystematic studies on *Capsella bursa-pastoris* (Brassicaceae): Enzyme polymorphisms in natural populations. *Plant Syst. Evol.* **49,** 73–94.

Brabant, Ph., Gouyon, P. H., Lefort, G., Valdeyron, G., and Vernet, Ph. (1980). Pollination studies in *Thymus vulgaris L.* (Labiatae). *Oecol. Plant.* **1,** 37–45.

Bradner, N. R., Frakes, R. V., and Stephen, W. P. (1965). Effects of bee species and isolation distance on possible varietal contamination in alfalfa. *Agron. J.* **57,** 247–248.

Bradshaw, A. D. (1972). Some of the evolutionary consequences of being a plant. *Evol. Biol.* **5,** 25–47.

Brantjes, N. B. M. (1979). Sensory responses to flowers in night-flying moths. *In* ''The Pollination of Flowers by Insects'' (A. J. Richards, ed.), pp. 13–20. Academic Press, New York.

Brantjes, N. B. M., and de Vos, O. C. (1981). The explosive release of pollen in flowers of *Hyptis* (Lamiaceae). *New Phytol.* **87,** 425–430.

Brown, A. H. D. (1979). Enzyme polymorphisms in plant populations. *Theor. Pop. Biol.* **15,** 1–42.

Burdon, J. J. (1980). Intra-specific diversity in a natural population of *Trifolium repens. J. Ecol.* **68,** 717–735.

Butler, C. G., Jeffree, E. P., and Kalmus, H. (1943). The behaviour of a population of honeybees on an artificial and on a natural crop. *J. Exp. Biol.* **20,** 65–73.

Carpenter, F. L. (1976). Plant-pollinator interactions in Hawaii: Pollination energetics of *Metrosideros collina* (Myrtaceae). *Ecology* **57,** 1125–1144.

Colwell, R. N. (1951). The use of radioactive isotopes in determining spore distribution patterns. *Am. J. Bot.* **38,** 511–523.

Cox, P. A. (1981). Niche partitioning between sexes of dioecious plants. *Am. Nat.* **117,** 295–307.

Crane, M. B., and Mather, K. (1943). The natural cross-pollination of crop plants with particular reference to the radish. *Ann. Appl. Biol.* **30,** 301–308.

Dafni, A., and Werker, E. (1982). Pollination ecology of *Sternbergia clusiana* (Ker-Gawler) Spreng. (Amaryllidaceae). *New Phytol.* **91,** 571–577.

Darlington, C. D. (1958). ''The Evolution of Genetic Systems,'' 2nd ed. Oliver and Boyd, Edinburgh.

Dimmick, R. L., and Akers, A. B., eds. (1969). ''An Introduction to Experimental Aerobiology.'' Wiley (Interscience), New York.

Dommée, B. (1981). Rôles du milieu et du génotype dans le régime de la reproduction de *Thymus vulgaris* L. *Oecol. Plant.* **2,** 137–147.

Dressler, R. L. (1968). Pollination by euglossine bees. *Evolution* (*Lawrence, Kans.*) **22,** 202–210.

Dressler, R. L. (1982). Biology of the orchid bees (Euglossini). *Annu. Rev. Ecol. Syst.* **13,** 373–394.

Edmonds, R. L., ed. (1979). ''Aerobiology: The Ecological Systems Approach.'' Dowden, Hutchinson, and Ross, Stroudsburg, Pennsylvania.

Eickwort, G. C., and Ginsberg, H. S. (1980). Foraging and mating behavior in Apoidea. *Annu. Rev. Entomol.* **25,** 421–446.

Eisikowitch, D., and Woodell, S. R. J. (1975). Some aspects of pollination ecology of *Armeria maritima* (Mill.) Willd. in Britain. *New Phytol.* **74,** 307–322.

Ellstrand, N. C., Torres, A. M., and Levin, D. A. (1978). Density and the rate of apparent outcrossing in *Helianthus annuus* (Asteraceae). *Syst. Bot.* **3,** 403–407.

Ennos, R. A., and Clegg, M. T. (1982). Effect of population substructuring on estimates of outcrossing rate in plant populations. *Heredity* **48,** 283–292.

Estes, J. R., and Thorp, R. W. (1975). Pollination ecology of *Pyrrhopappus carolinianus* (Compositae). *Am. J. Bot.* **62,** 148–159.

Faegri, K., and van der Pijl, L. (1971). ''Principles of Pollination Ecology,'' 2nd rev. ed. Pergamon, Oxford.

Feinsinger, P. (1976). Organization of a tropical guild of nectarivorous birds. *Ecol. Monogr.* **46,** 257–291.

Feinsinger, P. (1978). Ecological interactions between plants and hummingbirds in a successional tropical community. *Ecol. Monogr.* **48,** 269–287.

Frankie, G. W. (1975). Tropical forest phenology and pollinator-plant coevolution. *In* "Coevolution of Animals and Plants" (L. Gilbert and P. H. Raven, eds.), pp. 192–209. Univ. of Texas Press, Austin.

Frankie, G. W. (1976). Pollination of widely dispersed trees by animals in Central America, with an emphasis on bee pollination systems. *In* "Tropical Trees: Variation, Breeding and Conservation" (J. Burley and B. T. Styles, eds.), pp. 151–159. Academic Press, New York.

Frankie, G. W., Opler, P. A., and Bawa, K. S. (1976). Foraging behavior of solitary bees: Implications for outcrossing of a neotropical forest tree species. *J. Ecol.* **64,** 1049–1058.

Free, J. B. (1966). The foraging behavior of bees and its effect on the isolation and speciation of plants. *In* "Reproductive Biology and Taxonomy of Vascular Plants" (J. G. Hawkes, ed.), pp. 76–92. Pergamon, Oxford.

Free, J. B. (1970). "Insect Pollination of Crops." Academic Press, New York.

Freeman, D. C., Klikoff, L., and Harper, K. (1976). Differential resource utilization by the sexes of dioecious plants. *Science* **193,** 597–599.

Freeman, D. C., McArthur, E. D., Harper, K. T., and Blauer, A. C. (1981). Influence of environment on the floral sex ratio of monoecious plants. *Evolution (Lawrence, Kans.)* **35,** 194–196.

Fryxell, P. A. (1956). Effect of varietal mass on the percentage of outcrossing in *Gossypium hirsutum* in New Mexico. *J. Hered.* **47,** 299–301.

Fryxell, P. A. (1957). Mode of reproduction of higher plants. *Bot. Rev.* **23,** 135–233.

Ganders, F. R. (1974). Disassortative pollination in the distylous plant *Jepsonia heterandra. Can. J. Bot.* **52,** 2401–2406.

Gary, N. E., Witherell, P. C., and Marston, J. (1972). Foraging range and distribution of honey bees used for carrot and onion pollination. *Environ. Entom.* **1,** 71–78.

Gary, N. E., Witherell, P. C., Lorenzen, K., and Marston, J. M. (1977). The interfield distribution of honey bees on carrots, onions, and safflower. *Environ. Entomol.* **6,** 637–640.

Gaudreau, M., and Hardin, J. (1974). The use of neutron activation analysis in pollination ecology. *Brittonia* **26,** 316–320.

Gentry, A. H. (1974). Flowering phenology and diversity in tropical Bignoniaceae. *Biotropica* **6,** 64–68.

Gentry, A. H. (1978). Anti-pollinators for mass-flowering plants? *Biotropica* **10,** 68–69.

Ginsberg, H. S. (1983). Foraging movements of *Halictus ligatus* (Halictidae) and *Ceratina calcarata* (Anthophoridae). *Psyche* (in press).

Gleaves, J. T. (1973). Gene flow mediated by wind-borne pollen. *Heredity* **31,** 355–366.

Golínska, J. (1925). Recherches sur la croissance des fruits et la fructification des concombres (Cucumis sativus). *Acta Soc. Bot. Pol.* **3,** 97–114.

Gorz, H. J., and Haskins, F. A. (1971). Evaluation of cross-fertilization in forage crops. *Crop Sci.* **111,** 731–734.

Grant, V. (1949). Pollination systems as isolating mechanisms in angiosperms. *Evolution (Lawrence, Kans.)* **3,** 82–97.

Grant, V. (1958). The regulation of recombination in plants. *Cold Spring Harbor Symp. Quant. Biol.* **23,** 337–363.

Grant, K. A., and Grant, V. (1968). "Hummingbirds and their Flowers." Columbia Univ. Press, New York.

Gregory, P. H., and Monteith, J. L., eds. (1967). "Airborne Microbes." Cambridge Univ. Press, London and New York.

Greig-Smith, P. (1964). "Quantitative Plant Ecology," 2nd ed. Butterworth, London.

Griffiths, D. J. (1950). The liability of seed crops of perennial ryegrass (*Lolium perenne*) to contamination by wind borne pollen. *J. Agric. Sci.* **40,** 19–38.

Hamrick, J. L. (1982). Plant population genetics and evolution. *Am. J. Bot.* **69,** 1685–1693.

Hamrick, J. L., and Allard, R. W. (1972). Microgeographical variation in allozyme frequencies in *Avena barbata. Proc. Natl. Acad. Sci. U.S.A.* **69,** 2100–2104.

Hamrick, J. L., and Holden, L. R. (1979). Influence of microhabitat heterogeneity on gene frequency distribution and gametic phase disequilibrium in *Avena barbata. Evolution (Lawrence, Kans.)* **33,** 521–533.

Handel, S. N. (1976). Restricted pollen flow of two woodland herbs determined by neutron-activation analysis. *Nature (London)* **260,** 422–423.

Handel, S. N. (1982). Dynamics of gene flow in an experimental garden of *Cucumis melo* (Cucurbitaceae). *Am. J. Bot.* **69,** 1538–1546.

Handel, S. N. (1983a). Contrasting gene flow patterns and genetic subdivision in adjacent populations of *Cucumis sativus* (Cucurbitaceae). *Evolution (Lawrence, Kans.)* **37** (in press).

Handel, S. N. (1983b). The intrusion of clonal growth patterns on plant breeding systems. *Am. Nat.* (in press).

Handel, S. N., and Mishkin, J. L. (1983). Temporal shifts in gene flow: Evidence from an experimental population of *Cucumis sativus. Evolution (Lawrence, Kans.).* (Submitted).

Harberd, D. J. (1967). Observations on natural clones of *Holcus mollis. New Phytol.* **66,** 401–408.

Harding, J., and Barnes, K. (1977). Genetics of *Lupinus.* X. Genetic variability, heterozygosity and outcrossing in colonial populations of *Lupinus succulentus. Evolution (Lawrence, Kans.)* **31,** 247–255.

Harley, R. M. (1971). An explosive mechanism in *Eriope crassipes,* a Brazilian labiate. *Biol. J. Linn. Soc.* **3,** 159–164.

Harper, J. L. (1977a). "Population Biology of Plants." Academic Press, New York.

Harper, J. L. (1977b). The contributions of terrestrial plant studies to the development of the theory of ecology. *In* "Changing Scenes in Natural Sciences, 1776–1976" (C. E. Goulden, ed.), pp. 139–157. Academy of Natl. Sci., Philadelphia, Pennsylvania.

Haskell, G. (1944). The pollination and spatial isolation of vegetable seed crops. *Northwest Nat.* **19,** 34–44.

Haskell, G. (1954). The genetic detection of natural crossing in blackberry. *Genetica* **27,** 162–172.

Heinrich, B. (1975). Energetics of pollination. *Annu. Rev. Ecol. Syst.* **6,** 139–170.

Heinrich, B. (1976). The foraging specializations of individual bumblebees. *Ecol. Monogr.* **46,** 105–128.

Heinrich, B. (1979). Resource heterogeneity and patterns of movement in foraging bumblebees. *Oecologia* **40,** 235–245.

Heinrich, B., and Raven, P. H. (1972). Energetics and pollination ecology. *Science* **176,** 597–602.

Heithaus, E. R., Opler, P. A., and Baker, H. G. (1974). Bat activity and pollination of *Bauhinia pauletia:* plant-pollinator coevolution. *Ecology* **55,** 412–419.

Heithaus, E. R., Stashko, E., and Anderson, P. K. (1982). Cumulative effects of plant-animal interactions on seed production by *Bauhinia ungulata,* a neotropical legume. *Ecology* **63,** 1294–1302.

Howe, H. F., and Smallwood, J. (1982). Ecology of seed dispersal. *Annu. Rev. Ecol. Syst.* **13,** 201–228.

Howell, D. J. (1979). Flock foraging in nectar-eating bats: Advantages to the bats and the host plants. *Am. Nat.* **114,** 23–49.

Humphreys, M. O., and Gale, J. S. (1974). Variation in wild populations of Papaver dubium. VIII. The mating system. *Heredity* **33,** 33–42.

Jackson, M. T. (1966). Effects of microclimate on spring flowering phenology. *Ecology* **47,** 407–415.

Jain, S. K. (1976a). The evolution of inbreeding in plants. *Annu. Rev. Ecol. Syst.* **7,** 469–495.

Jain, S. K. (1976b). Patterns of survival and microevolution in plant populations. *In* "Population

Genetics and Ecology'' (S. Karlin and E. Nevo, eds.), pp. 49–90. Academic Press, New York.

Janzen, D. H. (1971a). Euglossine bees as long-distance pollinators of tropical plants. *Science* **171,** 203–205.

Janzen, D. H. (1971b). Seed predation by animals. *Annu. Rev. Ecol. Syst.* **2,** 465–492.

Johri, M., and Vasil, I. K. (1961). Physiology of pollen. *Bot. Rev.* **27,** 325–381.

Jones, M. D., and Brooks, J. S. (1950). Effectiveness of distance and border rows in preventing outcrossing in corn. *Okla. Agric. Exp. Stn. Tech. Bull.* No. T-38.

Kalin de Arroyo, M. T. (1976). Geitonogamy in animal pollinated tropical angiosperms. A stimulus for the evolution of self-incompatibility. *Taxon* **25,** 543–548.

Kareiva, P. (1982). Experimental and mathematical analyses of herbivore movement: Quantifying the influence of plant spacing and quality on foraging discrimination. *Ecol. Monogr.* **52,** 261–282.

Kevan, P. G., and Baker, H. G. (1983). Insects as flower visitors and pollinators. *Annu. Rev. Entomol.* **28,** 407–453.

Kraai, A. (1962). How long do honey-bees carry germinable pollen on them? *Euphytica* **11,** 53–56.

Lee, T. D., and Bazzaz, F. A. (1982a). Regulation of fruit and seed production in an annual legume, *Cassia fasciculata. Ecology* **63,** 1363–1373.

Lee, T. D. and Bazzaz, F. A. (1982b). Regulation of fruit maturation pattern in an annual legume, *Cassia fasciculata. Ecology* **63,** 1374–1388.

Lefèbvre, C. (1970). Self fertility in maritime and zinc mine populations of *Armeria maritima* (Mill.) Willd. *Evolution* (*Lawrence, Kans.*) **24,** 571–577.

Lefèbvre, C. (1976). Breeding system and population structure of *Armeria maritima* (Mill.) Willd. on a zinc-lead mine. *New Phytol.* **77,** 187–192.

Lertzman, K. P., and Gass, C. L. (1983). Alternative models of pollen transfer. *In* ''Handbook of Experimental Pollination Biology'' (C. E. Jones and R. J. Little, eds.), pp. 474–489. Van Nostrand-Reinhold, New York.

Leverich, W. J., and Levin, D. A. (1979). Age-specific survivorship and reproduction in *Phlox drummondii. Am. Nat.* **113,** 881–903.

Levin, D. A. (1972a). Pollen exchanges as a function of species proximity in *Phlox. Am. Nat.* **101,** 387–400.

Levin, D. A. (1972b). Plant density, cleistogamy, and self-fertilization in natural populations of *Lithospermum caroliniense. Am. J. Bot.* **59,** 71–77.

Levin, D. A. (1978a). Some genetic consequences of being a plant. *In* ''Ecological Genetics: The Interface'' (P. F. Brussard, ed.) pp. 189–214. Springer-Verlag, Berlin and New York.

Levin, D. A. (1978b). Pollinator behavior and the breeding system of plant populations. *In* ''The Pollination of Flowers by Insects'' (A. J. Richards, ed.), pp. 133–150. Academic Press, New York.

Levin, D. A. (1979). Pollinator foraging behavior: Genetic implications for plants. *In* ''Topics in Plant Population Biology'' (O. T. Solbrig, S. Jain, G. B. Johnson, and P. H. Raven, eds.), pp. 131–153. Columbia Univ. Press, New York.

Levin, D. A. (1981). Dispersal versus gene flow in plants. *Ann. Mo. Bot. Gard.* **68,** 233–253.

Levin, D. A., and Kerster, H. W. (1967). An analysis of interspecific pollen exchange in *Phlox. Am. Nat.* **101,** 387–400.

Levin, D. A., and Kerster, H. W. (1969a). The dependence of bee mediated pollen and gene dispersal on plant density. *Evolution* (*Lawrence, Kans.*) **23,** 560–572.

Levin, D. A., and Kerster, H. W. (1969b). Density-dependent gene dispersal in *Liatris. Am. Nat.* **103,** 61–74.

Levin, D. A., and Kerster, H. W. (1971). Neighborhood structure in plants under diverse reproductive methods. *Am. Nat.* **105,** 345–354.

Levin, D. A., and Kerster, H. W. (1974). Gene flow in seed plants. *Evol. Biol.* **7,** 139–220.

Levin, D. A., and Kerster, H. W. (1975). The effect of gene dispersal on the dynamics and statics of gene substitution in plants. *Heredity* **35,** 317–336.

Levin, D. A., and Wilson, J. B. (1978). The genetic implications of ecological adaptations in plants. *In* "Structure and Functioning of Plant Populations" (A. H. J. Freysen and J. W. Woldendorp, eds.), pp. 75–98. North-Holland Publ., Amsterdam.

Linhart, Y. B. (1973). Ecological and behavioral determinants of pollen dispersal in hummingbird-pollinated *Heliconia*. *Am. Nat.* **107,** 511–523.

Linhart, Y. B., and Feinsinger, P. (1980). Plant-hummingbird interactions: Effects of island size and degree of specialization on pollination. *J. Ecol.* **68,** 745–760.

Macior, L. W. (1974). Behavioral aspects of coadaptations between flowers and insect pollinators. *Ann. Mo. Bot. Gard.* **61,** 760–769.

Mahall, B. E., and Bormann, F. H. (1978). A quantitative description of the vegetative phenology of herbs in a northern hardwood forest. *Bot. Gaz. (Chicago)* **139,** 467–481.

Maruyama, T. (1971). The rate of decrease of heterozygosity in a population occupying a circular or linear habitat. *Genetics* **67,** 437–454.

Maruyama, T. (1972). The rate of decrease of genetic variability in a two-dimensional continuous population of finite size. *Genetics* **70,** 639–651.

Mesler, M. R., Ackerman, J. D., and Lu, K. L. (1980). The effectiveness of fungus gnats as pollinators. *Am. J. Bot.* **67,** 564–567.

Mitton, J. B., Linhart, Y. B., Davis, M. L., and Sturgeon, K. B. (1981). Estimation of outcrossing in ponderosa pine, *Pinus ponderosa* Laws, from patterns of segregation of protein polymorphisms and from frequencies of albino seedlings. *Silvae Genet.* **30,** 117–121.

Moore, L. A., and Willson, M. F. (1982). The effect of microhabitat, spatial distribution, and display size on dispersal of *Lindera benzoin* by avian frugivores. *Can. J. Bot.* **60,** 557–560.

Moran, G. F., and Brown, A. H. D. (1980). Temporal heterogeneity of outcrossing rates in alpine ash (*Eucalyptus delegatensis* R. T. Bak.) *Theor. Appl. Genet.* **57,** 101–105.

Morse, D. H. (1982). The turnover of milkweed pollinia on bumble bees, and implications for outcrossing. *Oecologia* **53,** 187–196.

Motten, A. F. (1982). Autogamy and competition for pollinators in *Hepatica americana* (Ranunculaceae). *Am. J. Bot.* **69,** 1296–1305.

Motten, A. F., Campbell, D. R., Alexander, D. E., and Miller, H. L. (1981). Pollination effectiveness of specialist and generalist visitors to a North Carolina population of *Claytonia virginica*. *Ecology* **62,** 1278–1287.

Nagylaki, T. (1976). The decay of genetic variability in geographically structured populations. *Theor. Pop. Biol.* **10,** 70–82.

Nieuwhof, M. (1963). Pollination and contamination of *Brassica oleracea*. *Euphytica* **12,** 17–26.

Niklas, K. J. (1981). Airflow patterns around some early seed plant ovules and cupules: Implications concerning efficiency in wind pollination. *Am. J. Bot.* **68,** 635–650.

Ogden, E. C., Raynor, G. S., Hayes, J. V., Lewis, D. M., and Haines, J. H. (1974). "Manual for Sampling Airborne Pollen." Hafner, New York.

Oinonen, E. (1967). The correlation between the size of Finnish bracken (*Pteridium aquilinum* (L.) Kuhn) clones and certain periods of site history. *Acta For. Fenn.* **83,** 1–51.

Ornduff, R. (1975). Complementary roles of halictids and syrphids in the pollination of *Jepsonia heterandra* (Saxifragaceae). *Evolution (Lawrence, Kans.)* **29,** 370–373.

Ornduff, R. (1978). Features of pollen flow in dimorphic species of *Lythrum* section *Euhyssopifolia*. *Am. J. Bot.* **65,** 1077–1083.

Ornduff, R. (1980). Heterostyly, population composition, and pollen flow in *Hedyotis caerulea*. *Am. J. Bot.* **67,** 95–103.

Parker, F. D. (1982). Efficiency of bees in pollinating onion flowers. *J. Kans. Entomol. Soc.* **55,** 171–176.

Paterniani, E., and Short, A. C. (1974). Effective maize pollen dispersal in the field. *Euphytica* **23,** 129–134.

Pedersen, N. W. (1967). Alfalfa cross-pollination studies involving three varieties and two pollinator species. *Crop Sci.* **7,** 59–62.

Pedersen, N. W., Hurst, R. L., Levin, M. D., and Stoker, G. L. (1969). Computer analysis of the genetic contamination of alfalfa seed. *Crop Sci.* **9,** 1–4.

Platt, W. J., Hill, G. R., and Clark, S. (1974). Seed production in a prairie legume (*Astragalus canadensis* L.). *Oecologia* **17,** 55–63.

Price, M. V., and Waser, N. M. (1979). Pollen dispersal and optimal outcrossing in *Delphinium nelsoni*. *Nature* (*London*) **277,** 294–297.

Price, M. V., and Waser, N. M. (1982). Experimental studies of pollen carryover: hummingbirds and *Ipomopsis aggregata*. *Oecologia* **54,** 353–358.

Primack, R. B., and Silander, J. A. (1975). Measuring the relative importance of different pollinators to plants. *Nature* (*London*) **255,** 143–144.

Proctor, M., and Yeo, P. (1972). ''The Pollination of Flowers.'' Taplinger, New York.

Pyke, G. H. (1981). Optimal foraging in nectar-feeding animals and coevolution with their plants. *In* ''Foraging Behavior: Ecological, Ethological, and Psychological Approaches '' (A. C. Kamil and T. D. Sargent, eds.), pp. 19–38. Garland STPM Press, New York.

Rai, K. N., and Jain, S. K. (1982). Population biology of *Avena* IX. Gene flow and neighborhood size in relation to microgeographic variation in *Avena barbata*. *Oecologia* **53,** 399–405.

Ralph, C. P. (1977). Effect of host plant density on populations of a specialized, seed-sucking bug, *Oncopeltus fasciatus*. *Ecology* **58,** 799–809.

Raynor, G. S. (1979). Sampling techniques. *In* ''Aerobiology: The Ecological Systems Approach'' (R. L. Edmonds, ed.), pp. 151–172. Dowden, Hutchinson and Ross, Stroudsburg, Pennsylvania.

Raynor, G. S., Ogden, E. C., and Hayes, J. V. (1970). Dispersion and deposition of ragweed pollen from experimental sources. *J. Appl. Meteorol.* **9,** 885–895.

Raynor, G. S., Ogden, E. C., and Hayes, J. V. (1971). Dispersion and deposition of timothy pollen from experimental sources. *Agric. Meteorol.* **9,** 347–366.

Raynor, G. S., Ogden, E. C., and Hayes, J. V. (1976). Dispersion of fern spores into and within a forest. *Rhodora* **78,** 473–487.

Real, L., Ott, J., and Silverfine, E. (1982). On the trade-off between the mean and the variance in foraging: Effect of spatial distribution and color preference. *Ecology* **63,** 1617–1623.

Reinke, D. C., and Bloom, W. L. (1979). Pollen dispersal in natural populations: A method for tracking individual pollen grains. *Syst. Bot.* **4,** 223–229.

Richards, A. J., and Ibrahim, H. (1978). Estimation of neighborhood size in two populations of *Primula veris*. *In* ''The Pollination of Flowers by Insects'' (A. J. Richards, ed.), pp. 165–174. Academic Press, New York.

Rick, C. M. (1947). The effect of planting design upon the amount of seed produced by male-sterile tomato plants as a result of natural cross-pollination. *Proc. Am. Soc. Hortic. Sci.* **50,** 273–284.

Root, R. B. (1973). Organization of a plant-arthropod association in simple and diverse habitats: The fauna of collards (*Brassica oleracea*). *Ecol. Monogr.* **43,** 95–124.

Schaal, B. A. (1975). Population structure and local differentiation in *Liatris cylindracea*. *Am. Nat.* **109,** 511–528.

Schaal, B. A. (1980). Measurement of gene flow in *Lupinus texensis*. *Nature* (*London*) **284,** 450–451.

Schemske, D. W., Wilson, M. F., Melampy, M. N., Miller, L. J., Verner, L., Schemske, K. M., and Best, L. B. (1978). Flowering ecology of some spring woodland herbs. *Ecology* **59,** 351–366.

Schlissing, R. A., and Turpin, R. A. (1971). Hummingbird dispersal of *Delphinium cardinale* pollen treated with radioactive iodine. *Am. J. Bot.* **58,** 401–406.

Schmid, R. (1975). Two hundred years of pollination biology: An overview. *Biologist* **57,** 26–35.

Schmitt, J. (1980). Pollinator foraging behavior and gene dispersal in *Senecio* (Compositae). *Evolution (Lawrence, Kans.)* **34,** 934–943.

Schmitt, J. (1983). Density-dependent pollinator foraging, flowering phenology, and temporal pollen dispersal patterns in *Linanthus bicolor. Evolution (Lawrence, Kans.)* (in press).

Shapiro, A. M. (1970). The role of sexual behavior in density-related dispersal of pierid butterflies. *Am. Nat.* **104,** 367–372.

Shen, H. H., Rudin, D., and Lindgren, D. (1981). Study of the pollination pattern in a Scots pine seed orchard by means of isozyme analysis. *Silvae Genet.* **30,** 7–14.

Silander, J. A., Jr. (1978). Density-dependent control of reproductive success in *Cassia biflora. Biotropica* **10,** 292–296.

Silander, J. A., Jr., and Primack, R. B. (1978). Pollination intensity and seed set in the evening primrose (*Oenothera fruticosa*). *Am. Midl. Nat.* **100,** 213–216.

Simpson, D. M. (1954). Natural cross-pollination in cotton. *U.S. Dep. Agric. Tech. Bull.* No. 1094, pp. 1–17.

Simpson, D. M., and Duncan, E. N. (1956). Cotton pollen dispersal by insects. *Agron. J.* **48,** 305–308.

Sindu, A. S., and Singh, S. (1961). Studies on the agents of cross pollination of cotton. *Indian Cotton Grow. Rev.* **15,** 341–353.

Singh, S. (1950). Behavior studies on honeybees in gathering nectar and pollen. *Mem. N.Y. Agric. Exp. Stn. (Ithaca)* No. 288.

Stanley, R. G., and Linskens, H. F. (1974). "Pollen—Biology, Biochemistry, Management." Springer-Verlag, Berlin and New York.

Stebbins, G. L. (1958). Longevity, habitat, and release of genetic variability in the higher plants. *Cold Spring Harbor Symp. Quant. Biol.* **23,** 365–378.

Stephenson, A. G. (1981). Flower and fruit abortion: Proximate causes and ultimate functions. *Annu. Rev. Ecol. Syst.* **12,** 253–279.

Stephenson, A. G. (1982). When does outcrossing occur in a mass-flowering plant? *Evolution (Lawrence, Kans.)* **36,** 762–767.

Stern, K., and Roche, L. (1974). "Genetics of Forest Ecosystems." Springer-Verlag, Berlin and New York.

Stevens, S. G., and Finkner, M. D. (1953). Natural crossing in cotton. *Econ. Bot.* **7,** 257–269.

Stiles, F. G. (1975). Ecology, flowering phenology, and hummingbird pollination of some Costa Rican *Heliconia* species. *Ecology* **56,** 285–301.

Stockhouse, R. (1976). A new method for studying pollen dispersal using micronized fluorescent dusts. *Am. Nat.* **96,** 241–254.

Tepedino, V. J. (1981). The pollination efficiency of the squash bee (*Peponapis pruinosa*) and the honey bee (*Apis mellifera*) on summer squash (*Cucurbita pepo*). *J. Kan. Entomol. Soc.* **54,** 359–377.

Thies, S. A. (1953). Agents concerned with natural crossing of cotton. *Agron. J.* **45,** 481–484.

Thomson, J. D. (1981). Spatial and temporal components of resource assessment by flower-feeding insects. *J. Anim. Ecol.* **50,** 49–59.

Thomson, J. D., and Barrett, S. C. H. (1981). Temporal variation of gender in *Aralia hispida* Vent. (Araliaceae). *Evolution (Lawrence, Kans.)* **35,** 1094–1107.

Thomson, J. D., and Plowright, R. C. (1980). Pollen carryover, nectar rewards, and pollinator behavior with special reference to *Diervilla lonicera. Oecologia* **46,** 68–74.

Thomson, J. D., Maddison, W. P., and Plowright, R. C. (1982). Behavior of bumble bee pollinators of *Aralia hispida* Vent. (Araliaceae). *Oecologia* **54,** 326–336.

Turner, M. E., Stephens, J. C., and Anderson, W. W. (1982). Homozygosity and patch structure in plant populations as a result of nearest-neighbor pollination. *Proc. Natl. Acad. Sci. U.S.A.* **79,** 203–207.

Vandermeer, J. (1975). A graphical model of insect seed predation. *Am. Nat.* **109,** 147–160.
van der Meer, Q. P., and van Bennekom, J. L. (1968). Research on pollen distribution in onion seed fields. *Euphytica* **17,** 216–219.
van der Pijl, L., and Dodson, C. (1966). "Orchid Flowers: Their Pollination and Evolution." Univ. of Miami Press, Coral Gables, Florida.
Vasek, F. C. (1965). Outcrossing in natural populations. II. *Clarkia unguiculata. Evolution (Lawrence, Kans.)* **19,** 152–156.
Vasek, F. C. (1967). Outcrossing in natural populations. III. The Deer Creek population of *Clarkia exilis. Evolution (Lawrence, Kans.)* **21,** 241–248.
Visscher, P. K., and Seeley, T. D. (1982). Foraging strategy of honeybee colonies in a temperate deciduous forest. *Ecology* **63,** 1790–1801.
Vogel, S. (1981). Die Klebstoffhaare an den Antheren von *Cyclanthera pedata (Cucurbitaceae). Plant Syst. Evol.* **137,** 291–316.
Waddington, K. D. (1979). Flight patterns of three species of sweat bees (Halictidae) foraging at *Convolvulus arvensis. J. Kan. Entomol. Soc.* **52,** 751–758.
Waddington, K. D. (1980). Flight patterns of foraging bees relative to density of artifical flowers and distribution of nectar. *Oecologia* **44,** 199–204.
Waddington, K. D. (1981). Factors influencing pollen flow in bumblebee-pollinated *Delphinium virescens. Oikos* **37,** 153–159.
Waddington, K. D., and Heinrich, B. (1981). Patterns of movement and floral choice by foraging bees. *In* "Foraging Behavior: Ecological, Ethological, and Psychological Approaches" (A. C. Kamil and T. D. Sargent, eds.), pp. 215–230. Garland STPM Press, New York.
Waser, N. M. (1979). Pollinator availability as a determinant of flowering time in ocotillo (*Fouquieria splendens*). *Oecologia* **39,** 107–121.
Waser, N. M. (1982). A comparison of distances flown by different visitors to flowers of the same species. *Oecologia* **55,** 251–257.
Waser, N. M., and Price, M. V. (1981). Pollinator choice and stabilizing selection for flower color in *Delphinium nelsonii. Evolution (Lawrence, Kans.)* **35,** 376–390.
Waser, N. M., and Price, M. V. (1982). A comparison of pollen and fluorescent dye carry-over by natural pollinators of *Ipomopsis aggregata* (Polemoniaceae). *Ecology* **63,** 1168–1172.
Waser, N. M., and Price, M. V. (1983). Optimal and actual outcrossing in plants, and the nature of plant-pollinator interaction. *In* "Handbook of Experimental Pollination Biology" (C. E. Jones and R. J. Little, eds.), pp. 341–359. Van Nostrand-Reinhold, New York.
Watson, P. S. (1969). Evolution in closely adjacent plant populations. VI. An entomophilous species, *Potentilla erecta,* in two contrasting habitats. *Heredity* **24,** 407–422.
Whitehead, D. R. (1969). Wind pollination in the angiosperms: evolutionary and environmental considerations. *Evolution (Lawrence, Kans.)* **23,** 28–35.
Wilken, D. H. (1977). Local differentiation for phenotypic plasticity in the annual *Collomia linearis* (Polemoniaceae). *Syst. Bot.* **2,** 99–108.
Williams, N. H., and Dodson, C. H. (1972). Selective attraction of male euglossine bees to orchid floral fragrances and its importance in long distance pollen flow. *Evolution (Lawrence, Kons.)* **26,** 84–95.
Williams, R. D., and Evans, G. (1935). The efficiency of spatial isolation in maintaining the purity of red clover. *Welsh J. Agric.* **11,** 164–171.
Wright, J. W. (1952). Pollen dispersion of some forest trees. United States Department of Agriculture, Northeast Forest Experiment Station, Paper 46.
Wu, L. and Antonovics, J. (1976). Experimental ecological genetics in *Plantago* II. Lead tolerance in *Plantago lanceolata* and *Cynodon dactylon* from a roadside. *Ecology* **57,** 205–208.
Wyatt, R. (1982). Inflorescence architecture: How flower number, arrangement, and phenology affect pollination and fruit set. *Am. J. Bot.* **69,** 585–594.

Wyatt, R., and Hellwig, R. L. (1979). Factors determining fruit set in heterostylous bluets, *Houstonia caerulea* (Rubiaceae). *Syst. Bot.* **4,** 103–114.

Yoshida, S. (1972). Physiological aspects of grain yield. *Annu. Rev. Plant Physiol.* **23,** 437–464.

Zimmerman, M. (1980). Reproduction in *Polemonium*: Competition for pollinators. *Ecology* **61,** 497–501.

CHAPTER 9

Foraging Behavior of Pollinators

KEITH D. WADDINGTON

Department of Biology
University of Miami
Coral Gables, Florida

I. Introduction

The importance of the foraging behavior of pollinators as a force shaping plant populations and communities has long been speculated upon and scientifically investigated. Classic theories on speciation in flowering plants likewise place great emphasis on pollinator foraging behavior as a mechanism (Grant, 1963; Grant and Grant, 1968). This book is largely an essay on the patterns which exist in plant populations and communities which are believed to be explained in part by pollinator foraging behavior. Thus, floral morphology and plant architecture, phenological patterns of flowering, the diversity of sexual systems, and gene

POLLINATION BIOLOGY

ISBN 0-12-583980-4 (hardcover)
ISBN 0-12-583982-0 (paperback)

flow and population structure all depend in part on the foraging behavior of pollinators.

The purpose of this chapter is to present a summary of the information on pollinator foraging behavior. Animals, including many species of birds and insects and some mammals, feed on the nectar or pollen in flowers and thereby act as pollinators; however, the research on their foraging behavior is not evenly distributed among the various taxa. This chapter reflects that taxonomic bias by focusing on research on bees and hummingbirds. The aim is to join isolated bits of research in order to discover general patterns and concepts. This, however, is not always possible since many studies stand alone, without verification or without analogous studies on various systems. Such gaps are the stimuli for additional research.

II. A Model for Studying Foraging Behavior

A number of approaches have been used to investigate the foraging behavior of pollinators. From these studies have arisen the general concepts and current theories addressed in this chapter. The earliest studies of pollinator behavior, before the 1930s, were observational and based primarily on understanding the pollination biology of certain plants. Examples of behavior described include how pollinators manipulate flowers to obtain food and which pollinators choose which flowers. No critical quantitative assessment of pollinator behavior was made and few general concepts of behavior were formulated.

The era of experimental investigations of pollinator behavior was ushered in by Plateau (1899), von Frisch (1967, for review), and others with studies of sensory physiology, perception, and orientation behavior of bees. The scientific investigation of pollinator foraging behavior has matured in parallel with and profited from general developments in ethology.

The observational approach to studying foraging behavior placed considerable limitations on what could be learned and on the extent to which results could be generalized to various situations. The reason for this is that any aspect of foraging behavior observed in the field is a function of both intrinsic and extrinsic factors (Fig. 1). Intrinsic factors include the sensory information that can be received and processed before and during foraging, decisions based on information, memory, learning, and stereotypic behavior. These mechanisms, or "rules" (Waddington and Heinrich, 1981), may either be built into the genome or may be the result of previous experience. Extrinsic factors are the aspects of the environment to which the pollinator responds, including floral color, shape, and density, number and distribution of plant species, wind velocity, and the distribution of nectar rewards. Thus, the foraging behavior of a pollinator in a particular floral array is a result of a set of behavioral and physiological mecha-

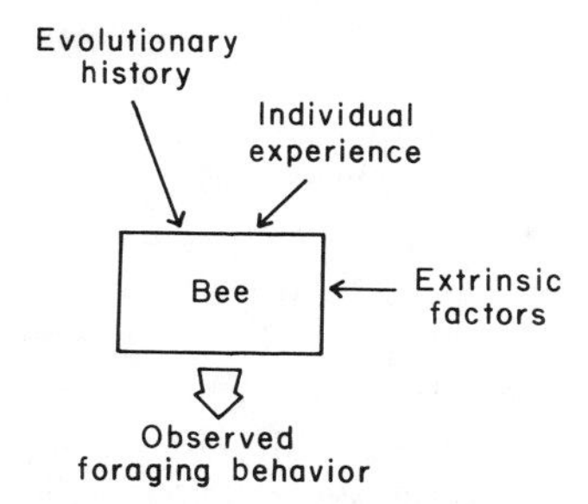

Fig. 1. Foraging behavior is a result of intrinsic factors, including the animal's evolutionary history and its own experiences, and extrinsic factors. The intrinsic factors must be known to predict an animal's behavior in a novel floral array.

nisms which are superimposed on an environmental background. An observed foraging behavior confounds intrinsic and extrinsic factors and is not sufficient information for predicting behavior in, say, a different floral array. If the goal is to predict and understand pollinator behavior in a variety of field conditions, then efforts should be directed toward elucidating further the mechanisms of behavior. Once a mechanism or set of mechanisms is well understood, a behavioral pattern can presumably be predicted against a number of environmental backgrounds.

To illustrate this point, consider the following simple example. Assume that a pollinator has one "hard-wired" rule: Go to the closest unvisited flower. Its foraging behavior would be very different depending on the local floral patch. Flight distance between flowers would vary greatly depending on the arrangement of flowers on plants (singly, in racemes, etc.) and on the plant density. The quantitative sequence of visits to different species depends on the number and distribution of species present. Clumping would result in a high frequency of transitions between flowers of the same species, but other spatial patterns would yield more interspecific flights.

This approach, of investigating mechanisms for predicting behavior, may be a terribly complex task since the number of mechanisms underlying a particular behavioral pattern may be very large and may even vary with changes in the environment. However, all the complexities may not need to be understood; it may be possible to understand just a few mechanisms in order to predict and explain, with acceptable accuracy, many patterns of foraging behavior. This is the approach adopted throughout this chapter for understanding foraging behavior; its usefulness of course, will have to await further studies. The hope is that once we understand some basic principles of pollinator foraging behavior, we may better understand the vast array of floral colors, patterns, displays and arrangements, population structure, and community patterns that are found in nature.

III. Sensory Information

A full understanding of any aspect of foraging behavior requires answering two questions: (1) What information is received and processed? (2) How is this information used, that is, what decisions are made based on this information? With the answers to these questions, an animal's foraging behavior in a particular floral array should be predictable. In this section, I will provide an overview of what is known about the sensory capabilities of various pollinators, the processing of the acquired sensory information, and the recollection and use of the information for making foraging decisions.

Pollinators find their way about and make foraging decisions based primarily on vision, olfaction, taste, and time. These are raw inputs from which useful information must be extracted. Not all pollinators use these senses to the same extent, nor do they perceive the same range of stimuli with the same acuity. Floral biologists have long understood the importance of learning the sensory worlds of pollinators in order to understand the evolution of the diversity of colors and odors of flowers (Lovell, 1918). However, we have advanced knowledge on just one pollinator, the honeybee, and even for the honeybee the story is incomplete. This account of the pollinators' sensory world is brief. This is due in part to a limitation on space, but also reflects our ignorance of these matters for most pollinators.

A. Vision

Of the three sensory modalities most used by pollinators during foraging, we understand vision best. Most studies of vision have been performed on animals that are not pollinators. However, we can support our knowledge of pollinators by extrapolating from other animals to pollinators that are close relatives.

Light is reflected from objects, and the eye serves to project an image onto a group of sensitive receptors. The structure of eyes varies among taxa and imposes different contraints for viewing the outside world. Hummingbirds, like other vertebrates, have a lens for focusing an image on the retina. Bees and other insect pollinators have compound eyes composed of numerous closely packed photoreceptive units; the ommatidium has its own lens and a stack of receptors. Each ommatidium points in a unique direction and receives light from a small spot in the total visual field, perhaps creating a mosaic image.

There is considerable information on the vision of chickens and pigeons, but little work has been done on vision in avian pollinators. The best we can do is link some uniform findings on bird vision with some behavioral observations on hummingbirds. The avian eye contains cones that are wavelength-specific receptors; the cones of most avian retinas contain brightly colored yellow, orange, or red oil droplets. It has been suggested that light passing through these colors intensifies them but reduces discrimination of other colors (e.g., blue and violet). A number of variously designed behavioral experiments have been conducted to examine the perceived spectral range and color preferences of birds. McCabe (1961) found that diurnal birds are more stimulated by red-yellow than by other colors. Bene (1941), Wagner (1946), and Lyerly *et al.* (1950) did not detect any innate color preferences. One thing is clear: Solid experimental work is needed to understand the visual constraints on avian pollinators.

Honeybees, and probably other bees and some butterflies, have trichromatic vision. The bee's ommatidia are divided into eight cells; some of the cells have photopigments sensitive to green light, and the others are blue or ultraviolet (UV) receptors. The range of the honeybee's sensitivity is between yellow-orange (650 nm in wavelength) and UV (300 nm), and the bees can discriminate several colors within this range. Bees are blind to red, but some butterflies are not.

The structure of the insect eye also permits the perception of the plane of polarization of light. Birds do not have this capability. The pattern of polarized UV light is symmetrical around the sun and exists even when the sun is hidden by clouds. Thus, by being able to "read" the pattern of polarization, insects such as bees can orient to the sun's position whether or not the sun is in view. Bees are known to use the sun as an orientation cue for finding home and food patches, and may also use the sun to orient them while flying between flowers.

B. Olfaction

The biology of olfaction is not well understood, but certainly the olfactory worlds of different pollinators are very different. Hummingbirds apparently have very limited olfactory capabilities, and odors are frequently lacking in "hummingbird flowers" (Grant and Grant, 1968). By contrast, the olfactory world of most insect pollinators is richer. Von Frisch and colleagues (1967, for review) used various essential oils (e.g., lemon, orange, peppermint) and extracts prepared from many wildflowers to detect, via behavioral studies, the olfactory abilities of honey bees. They found that the bees can discriminate many fragrances, that their olfactory acuity is similar to that of humans (although, not surprisingly, the bees detect more dilute flowery scents), and that they can discriminate various odor intensities above the detection threshold.

Kugler's (1950, 1951, 1956) experiments on several genera of flies were not designed to examine the range of the flies' olfactory capabilities, but they do demonstrate the flies' use of olfactory cues for making foraging decisions. Butterflies and moths likewise use olfactory cues for orientation to flowers and for discriminating flowers, but again the appropriate experimental work has not yet been done to discern their range of capabilities (for review, see Proctor and Yeo, 1972).

C. Taste

The function of taste is to encourage the ingestion of nutrients and to discriminate among available foods. Most studies of pollinators have dealt with the taste of sweet substances and thus apply to questions on foraging for nectar; little is known about the taste of constituents of pollen.

Honey bees discriminate through taste among the three primary sugars of nectar: sucrose, glucose, and fructose, and mixes of these sugars. They detect solutions as low as 2–4% sugar (von Frisch, 1967) and can discriminate approximately 5% differences in sucrose concentration within the normal range of nectar concentration (i.e., 10–45%) (Jamieson and Austin, 1958; Moffett *et al.*, 1976; Waller, 1972). Taste is probably the cue used in these discriminations; however, some measurement of the solution's viscosity may also be important. When viscosity is increased (by adding tasteless raffinose) independently of sweetness, the probability of dancing after foraging increases (as sugar concentration is increased) (von Frisch, 1967).

Using preference experiments, Hainsworth and Wolf (1976) examined some of the gustatory capabilities of five species of hummingbirds. The birds could discriminate 2% differences in sucrose solutions at low concentrations; their ability to discriminate (as indicated by preference) decreased as the absolute concentration increased. The birds did discriminate different sugars and mixes of sugars on some occasions [sucrose (S), glucose (G), and fructose (F) were used]. Hainsworth and Wolf ranked sugars on the basis of an acceptance/rejection ratio; the ranking was SFG = SF > S > FG > SG > F > G. The mixture of all three sugars was equally preferred to sucrose + fructose, sucrose was the most preferred single sugar, and glucose was least preferred compared to all other solutions. The preference rankings of SFG, F, and G are similar to the relative compositions found among hummingbird flowers. They also found that the birds showed no preferences for amino acids occurring at the low concentrations found in nectar; however, higher concentrations were detected and repulsed the birds.

D. Time

It is critical that pollinators have a sense of time. Plant species generally produce nectar (and pollen) during only a part of the day, and each species produces forage at different times. The trick, then, is for the pollinator to learn when to forage each day at particular species. Studies of the honeybee indicate the presence of a 24-hour internal "clock." Bees trained to feed for several days at a particular resource came on time to the location in the absence of the resource. Similarly, bees can be trained to come to the same location several times per day and to different locations at different times. The work of Beling (1929), Wahl (1932), and Bennett and Renner (1963) on the honeybee clock is reviewed by von Frisch (1967). Other pollinators probably possess a sense of time which can be integrated with information on food value and site location for making foraging decisions.

IV. Sensory Integration, Learning, and Memory

In order for an animal to find flowers in the field, to make profitable choices among the flowers, and to return to the nest, it must receive and integrate the sensory information described above, make associations between different bits of information, store the information, and recall it for use. This is a complex process, and we are largely ignorant of how it works for any animal. However, some progress has been made using the honey bee; a summary of one line of work is presented.

Karl von Frisch (1967, for review) recognized the importance of associative learning to the foraging bee and exploited it in order to elucidate some of the bee's sensory capabilities. For example, he discovered color vision by first training bees to associate a reward with one color; then, in a later test, the bees chose between variously colored empty dishes. The bees discriminated among the colors and preferentially visited the "training" color. Associations can also be made between reward and many odors and between solutions of various combinations (and concentrations) of sugars and odors and colors.

Menzel and Erber (1978) and other neurophysiologists have looked more closely at the neural mechanisms of learning and memory. The details of this story are exciting.

To investigate learning, bees are first trained to visit a feeding platform. A spontaneous choice test using two spectral colors is run (with no rewards), and the radiation density of the colors is adjusted so that neither color is preferred over the other. The bee is then rewarded at one flower and shortly after is tested by being given a choice between the two colors. This process is repeated after each subsequent reward from the same flower. An example of a learning curve is shown in Fig. 2. The bees learn to visit the rewarding flower in about 90% of the visits after just five or six rewards (odor learning is quicker).

Not all learning curves are alike, and these differences can tell us something about the bees' built-in learning programs and the neural basis of memory. For example, learning is affected by color. The bees learn most quickly those colors which are most likely to be food signals (e.g., violet, blue, and bee-purple) (Daumer, 1958; Menzel *et al.*, 1974). Background colors (e.g., green) are not learned as quickly. Menzel calls this learning program "phylogenetic pre-learning" and notes that the color learning system is indeed a complex integration process. (Surprisingly, rate of learning is independent of sugar concentration and quantity of the reward.)

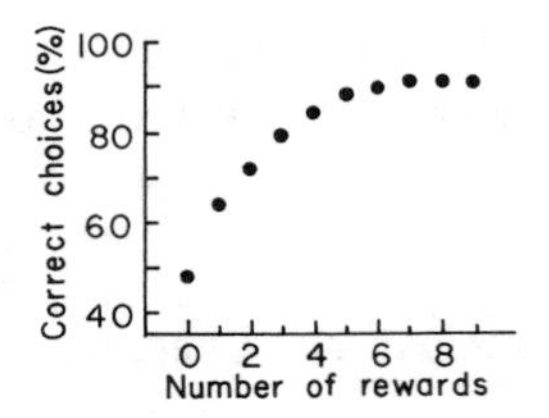

Fig. 2. Example of a honeybee learning curve. Learning improves quickly after successive reward–signal associations. After Menzel *et al.* (1974).

Menzel and Erber (1978) reviewed their own work on memory. First, memory is very stable if just three associations are made between a color signal and a sugar reward. After just one association, a different result hints at the actual neuronal processes of memory. Bees had a high rate of correct choices between two colors immediately after one association between one of the colors and a reward. However, over the next 2 minutes their performance decreased, only to rise again to the initial level of correct choices (Fig. 3). These results indicate that "The physiological mechanisms underlying the association between a signal and a reward . . . need time . . . and most probably go through phases in which the neural substrate for memory is different" (Menzel and Erber, 1978, p. 104). The following mechanisms, which are similar to those in vertebrates, are suggested.

First, information is stored in a sensory memory. This information is lost almost immediately if it is not reinforced—in this case, by stimulation of the sugar receptors on the proboscis. The bees will learn a color signal only if the signal is presented within 2–3 seconds before a reward, and it is learned only during the approach to a flower (Opfinger, 1931). (Thus, information remains in sensory storage for only 2 seconds.) If the color is first turned on after the approach, while the reward is ingested, learning does not occur. After the reward, the information passes into a short-term memory, shown in Fig. 3 by the

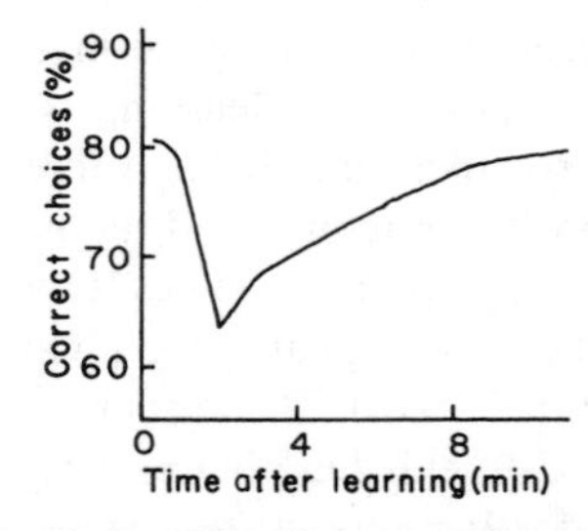

Fig. 3. Percentage of correct choices following a one-trial training period. Accuracy was high immediately after learning. After a decline in correct responses, accuracy again improved as the information passed into long-term memory. After Menzel and Erber (1978).

high level of correct choices immediately after the association. The information then passes into a long-term memory which appears to be spatially separated from the site of short-term memory in the brain. The passage is indicated by the rise in performance after the decline.

Some results are counterintuitive, but they make sense in light of this learning mechanism. For example, a bee learns an association between color and reward more quickly if the reward is made available at a rate lower than the bee's sucking rate than if the reward is amply available for the same duration; a signal associated with a poorer reward is learned more quickly. This is explained by the occurrence of interruptions in sucking when the reward is slowly made available. A signal is sent to short-term memory each time sucking is initiated (just once when the reward is amply available), resulting in a higher reinforcement of the signal associated with the poorer reward and faster learning of that signal–reward association. This mechanism has also been used to explain bumble bee floral preferences (Waddington *et al.*, 1981).

Thus, animals such as honeybees have special programs for learning, including what to learn and when to learn it (Gould, 1982). Studies such as those by Menzel and his colleagues are extremely important in understanding foraging behavior. Since some of the mechanisms of the honeybee so far discovered are similar to those found in vertebrates, some very general rules and concepts of associative learning and memory may result from future studies.

V. The Cost–Benefit Approach

Recently, studies of pollinator foraging behavior have shifted from descriptions of behavior to attempts to understand and predict behavioral patterns. The cost–benefit approach has been used extensively to this end, and the framework for investigation has been the mathematical model. Since this approach will be covered in the remainder of this chapter, some background is provided here.

Studies using the cost-benefit approach applied to foraging formally began with theoretical models by MacArthur and Pianka (1966) and Emlen (1966); however, much discussion of forager efficiency occurred before 1966 (e.g., Darwin, 1883, speculated on the efficiency of bees). The recent body of literature concerning a variety of animals is collectively known as "optimal foraging theory." Current thinking on optimal foraging theory is presented in several papers in Kamil and Sargeant (1981). The key assumption of all these studies is that the animal maximizes some efficiency expression (usually stated as net caloric gain per time) because fitness is positively correlated with foraging efficiency. Since an optimization approach is assumed, mathematical models employing optimization techniques are used (see Oster and Wilson, 1978, for uses and a critical discussion of optimization models).

Heinrich and Raven (1972) first advocated and supported with data an energetic approach to studying the foraging behavior of pollinators. Since then, some of the most productive and enlightening studies of optimal foraging theory have used pollinators as model animals. Reasons for this success include the following: (1) food (nectar) can be easily quantified and manipulated (e.g., Waddington, 1979b); (2) the movements of many pollinators are slow enough so that the foraging path, time budget, and choice of flowers can be quantified; and (3) predation and mate choice are not always important factors influencing observed behavioral patterns (Pyke, 1978a). Furthermore, it has been possible to quantify the energetic budgets of many pollinators (e.g., Heinrich, 1975).

The optimal foraging approach has proved fruitful and is a reasonable first approach to understanding and predicting foraging behavior; however, note the following caveat. Many of the major assumptions of the models have not been verified. For example, we do not know the functional relationship between fitness and foraging efficiency, and there are no data to show how profitability, (i.e., a resource's net energetic value) is measured by pollinators (Waddington, 1982b). These behavioral and physiological mechanisms must be verified in future studies.

VI. Pollinator Foraging Behavior

At the beginning of their foraging careers, pollinators are naive. They have little or no information about where or when forage exists. The benefits and costs of the various available flowers are, of course, also unknown. Thus, resources must be searched for and "sampled" before economically prudent foraging decisions can be made. Little is known about the initial search behavior of any pollinator; however, it is likely that they employ strategies similar to those of other animals (Jander, 1975). Without specific spatial information on floral resources, the most efficient strategy is to move straight ahead (ranging), presuming that flowers are patchily distributed (Jander, 1975). Once visual or olfactory information is received for detecting floral position, a local search is started, implemented by a convoluted, looping path. Presumably numerous flowers, perhaps in many local patches, are sampled before areas are chosen for repeat visits. The experienced pollinator, foraging within a patch of flowers which may contain one to many flowering species, must make numerous complex decisions.

Here, I will first describe what is known about the foraging range of some pollinators. Second, I will discuss individual foraging behavior once the pollinator is located at a floral patch. What rules of movement do pollinators use when foraging within and between inflorescences? How are these decisions influenced by such factors as the abundance and distribution of food, the floral arrangement, and the plant density? Third, I will discuss the factors that influence the choice of pollinators among various floral types (morphological variants or species).

A. Foraging Range

The foraging ranges of animals, including pollinators, vary in shape and size and are dependent on a number of factors, including resource density and distribution and the density of conspecifics and potential competitors.

The honeybee may fly several kilometers (Eckert, 1933) to forage, but usually forages closer than 1 km from the hive when a suitable resource is available. Vansell (1942) found that bees foraging on a productive resource of *Pyrus* trees decreased markedly in number at 90 m from the hive. As noted by Frankie (1976), authors have found that as floral resources diminish over time, the search areas of bees increase (e.g., Free, 1966).

Numerous studies indicate that individual honeybees exhibit considerable area fidelity (Butler *et al.*, 1943; Singh, 1950; Gary *et al.*, 1977; Weaver, 1957). The bees return repeatedly (even over several days) to the same relatively small area of a field of flowers. This is probably not a strategy which yields the highest rewards, but it may be a low-variance strategy which minimizes the risk of doing very poorly in unknown areas (the low-risk strategy is preferred in the context of floral choice; see below).

Bumble bees exhibit area fidelity under some conditions. I (unpublished study) marked and released bumble bees arriving at a patch containing monkshood (*Aconitum columbianum*) and larkspur (*Delphinium barbeyi*). Most bees never returned to the patch (these bees may have been sampling when captured), but those that returned did so repeatedly, in many cases for several days. Heinrich (1976) found that bumble bees did not return to specific areas in a featureless environment such as a large hayfield; however, in fields containing patches of flowers, individual bees repeatedly visited the same patches. One *Bombus fervidus* worker foraged in an area less than 7 m in length for several days.

Many bumble bees that Heinrich (1976) observed were not only area specific but also repeatedly flew between the same flower clumps. The bees learn the spatial position of the clumps and the flight path between them. This foraging strategy is called "traplining behavior," by analogy with the trapper checking his traps in a set sequence on a routine basis. Euglossine bees (Janzen, 1971) and several anthophorid species (Eickwort and Ginsberg, 1980), all strong fliers, exhibit the trapline strategy. This strategy may be very important to the pollination of widely dispersed, long-blooming tropical shrubs and trees.

Male bees of some species are territorial; a small area of flowers is used as forage and is defended against nonterritorial flower visitors and other bees vying for territories. *Centris* species have been well studied in the Costa Rican dry forest by Frankie and Baker (1974) and Frankie *et al.* (1980). An area 1–4 m in diameter is patrolled regularly for up to 5 hours in a day, and feeding at flowers takes place on the territory. Intruders are chased and usually displaced by the territory holder. The foraging behavior of intruders (their choice of where and on what to forage) may be substantially influenced by territorial bees. The territorial behavior by *Centris* may cause intruders to leave a tree in order to forage elsewhere (Frankie and Baker, 1974).

Some nectivorous birds, under certain conditions, also exhibit territoriality. The recent conceptual framework for studying territoriality is "economic defendability" (Brown, 1964); the idea is that an animal should be territorial only if the gains accrued minus the costs of defense are greater than they would be if the animal were not territorial. Nectivores are excellent animals for investigating this notion because the animal's energy budget can be determined and the energy resources are relatively easy to quantify. Carpenter and MacMillen (1976) constructed a model to predict when a Hawaiian honeycreeper will defend a feeding territory. When a resource is small (below a threshold value), the birds should not defend it since the exclusion of other birds does not increase its worth enough to offset the additional costs of defense. When the resource is very plentiful, birds again should not defend it because energy needs are adequately met even when other birds are feeding in the area. In general, a good match was found between the authors' predictions of honeycreeper foraging behavior in areas with different numbers of flowers. Studies by Gill and Wolf (1975a) on sunbirds in Kenya also seem to support the notion that territorial defense is economically profitable.

Pyke (1979a) developed an optimization model for predicting the territory size (and time budget) of nectivorous birds. Using Gill and Wolf's (1975a) data, Pyke calculated that sunbirds defend territories that are optimal in size in the sense that they result in minimization of the birds' daily energy costs.

Hummingbirds exhibit two foraging strategies which influence foraging range: They too may be territorial, and some use the traplining strategy (Feinsinger and Chaplin, 1975). Linhart (1973), for example, studied hummingbirds utilizing *Heliconia* spp. Nonterritorial birds fed on the dispersed individuals of forest species, and the forest-edge species were fed upon primarily by territorial birds. Feinsinger and Chaplin (1975) hypothesized that selection to maximize foraging efficiency should result in different energy requirements for flight in birds employing the two strategies. Traplining species, which spend more time in flight than territorial species, have longer wings relative to body size and have lower energy demands for flight than do territorial birds. The same pattern exists within single species. Hovering flight is cheaper for *Lampornis calolaema* females than for the more strongly territorial males. Hummingbird flight energetics and foraging strategy are correlated, and studies of other taxa (e.g., insects) may yield additional interesting patterns between morphology and energetics and foraging behavior.

B. Within Floral Patch Foraging Behavior

Once a pollinator arrives at an area with flowers, information must be received and processed, recalled from memory, and decisions made. The nature of the decisions and the resulting behavior will depend on the extrinsic factors mentioned at the beginning of this chapter. Below, two general cases will be treated. In the first case, just one species of plant occurs at the patch. What is the path pollinators take between plants (individual, spaced flowers or inflorescences)? What path is taken between flowers on inflorescences? What extrinsic factors influence the nature of these movement patterns? In the second case, with two or more species of flowers at the patch, what is the pollinator's choice between the flowers and how is the decision made?

1. Foraging in a Single-Species Patch

Plants often grow in single-species patches. If a pollinator chooses to feed in such a patch, its choice of floral species on each flight between plants is determined. However, decisions have to be made regarding which particular flower of that species to visit next. Research has shown that the choice is not random, and we are learning the "rules" that some pollinators use to make choices.

Within-patch foraging behavior is divided into two parts by others (e.g., Pyke, 1978a, 1979b): patterns of movement between and patterns of movement within inflorescences.

Several investigations have been made of the movement patterns of bumble bees and honeybees foraging at what are essentially two-dimensional arrays of inflorescences or single, spaced flowers (some of this literature is reviewed by Waddington and Heinrich, 1981). Two parameters are used to quantify pollinator movements between flowers. The flight distance is the linear distance between two successfully visited flowers. This represents an underestimate of the true flight distance because the flight path is rarely linear. The change in direction is the angular difference between the direction of arrival at an inflorescence and the direction of departure from the inflorescence (Fig. 6, inset).

Distributions of flight distance have a similar shape for a variety of pollinators (bumble bees: Pyke, 1978a; Waddington, 1981; Zimmerman, 1979; honeybees: Levin and Kerster, 1969; Waddington, 1980; butterflies: Levin and Kerster, 1968; several species of wild bees: Levin and Kerster, 1969; Waddington, 1979a; H. S. Ginsberg, unpublished manuscript; and hummingbirds: Pyke, 1981). In each case, the distributions are leptokurtic, with most flights made to a near-neighbor flower (Fig. 4). For bees, at least, this pattern is independent of the plant species involved (Levin and Kerster, 1969). Levin and Kerster (1968)

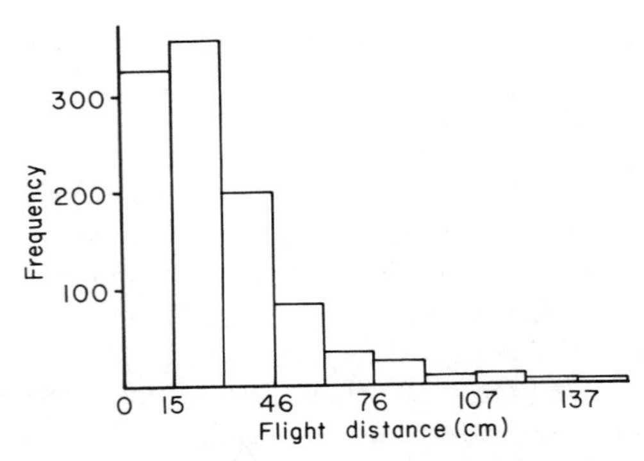

Fig. 4. Flight distances of bumble bees (*Bombus americanorum*) between successively visited inflorescenses of *Delphinium virescens* ($N = 1061$). After Waddington (1981).

attribute the highly leptokurtic distributions of butterflies to the superimposition of two distributions, one composed of feeding flights and the other of exploratory flights. This may be a general reason; however, Zimmerman (1979) found that the distribution of near-neighbor distances of *Polemonium* plants was leptokurtic, which may give rise to a leptokurtic distribution of bumble bee flight distances. Plant density does not seem to alter the shape of distance distributions; however, Levin and Kerster (1969) and Waddington (1980) found a negative relationship between mean flight distance and floral density. This effect is due primarily to visits to near-neighbor flowers.

Thus, pollinators generally make flights to near-neighbor flowers and minimize flight costs, but a closer examination shows that bees adjust their flight distance in a response to the volume of recently received nectar rewards. Pyke (1978a) found a negative relationship between the distance bumble bees fly and the number of flowers visited (used as an indicator of nectar volume) at the departed inflorescence. Waddington (1981) harvested the two bottom flowers of larkspur (*Delphinium virescens*) inflorescences, measured their nectar volumes, and found that these volumes could be used to estimate nectar in the upper flowers. Bumble bees departing these inflorescences with high rewards flew short distances, whereas relatively long flights were made after visits to nectar-poor inflorescences (Fig. 5). Bumble bees (previously excluded by screening) foraging at patches of nectar-rich white clover (*Trifolium repens*) flew about half as far between floral heads as when foraging in nectar-poor depleted patches (Heinrich, 1979a).

What information do bees use to make such foraging decisions? Do they use the magnitude of the last reward, or is information from previous experience used also? Waddington (1980) addressed this question by arraying artificial flowers singly (not in inflorescences) on a horizontal plane. Some flowers contained a reward of sugar solution, whereas other flowers were empty. Each flight from a flower was scored in terms of how many immediately preceding rewarding or nonrewarding visits were made sequentially. The distance was determined for each flight and categorized. The bees appeared to use more information than just that gained from the previous floral visit. Mean flight distance increased as the number of immediately preceding, uninterrupted, nonrewarding visits increased; however, flight distance did not change significantly (although, in five of six experiments, regression lines were negative and one was significantly different from zero) with the number of successive rewarding visits. When rewards are encountered, bees minimize their flight distance, but when nonrewarding flowers are encountered, nearby flowers are passed. Given that nectar may generally be clumped in space (Pleasants and Zimmerman, 1979), these behaviors localize bees in "hot" spots with nectar and minimize the bees' time in nectar-poor areas, thereby enhancing the rate of foraging returns.

The degree of change in direction can also influence the rate of forage returns. An increase in the degree of turning raises the probability of a revisit to a just-emptied flower (Pyke, 1978a), thereby lowering the mean reward per visit. On the other hand, as the angular sector within which a bee is willing to forage is narrowed (degree of turning decreased), the flight distance to a neighboring flower is increased. Although not yet verified, the observations described below may represent a compromise between these considerations.

Change in direction has been observed in several pollinators, and two patterns emerge. The most frequently observed pattern is a tendency to move straight ahead by exhibiting considerable directionality on successive flights. The fre-

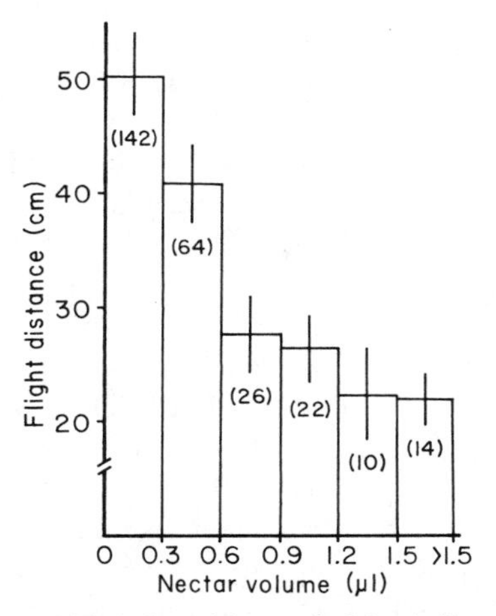

Fig. 5. Mean distances (± SE indicated by vertical bars) flown by bumble bees (*Bombus americanorum*) after visiting inflorescences of *Delphinium virescens* with the indicated mean volume of nectar in the two lowest flowers ($N = 228$). The two flowers are good predictors of rewards in other flowers on the same inflorescence. After Waddington (1981).

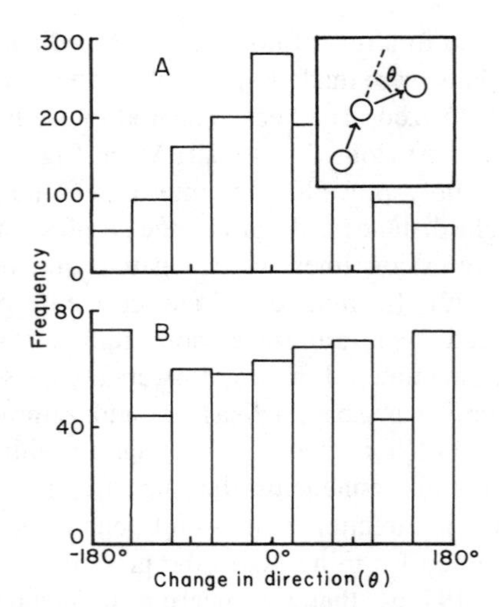

Fig. 6. Frequency distributions of changes in direction (0°) (see inset). (A) Honeybees foraging between artificial flowers. After Waddington (1980). (B) Bumble bees (*Bombus flavifrons*) foraging randomly (with respect to direction) between clusters of flowers of *Polemonium foliosissimum*. After Zimmerman (1979).

quency distribution in Fig. 6A illustrates this behavior for honeybees foraging from an array of artificial flowers; however, bumble bees (Pyke, 1978a), sweat bees (Halictidae) (Waddington, 1979a; H. S. Ginsberg, unpublished manuscript), and butterflies (Levin *et al.*, 1971) behave similarly. The distributions are each unimodal and symmetrical, with the mean angle (and mode) approximately 0°. A second pattern is shown in Fig. 6B. Bumble bees foraging for nectar (Zimmerman, 1979) and pollen (Hodges and Miller, 1981; Zimmerman, 1982), nectivorous sunbirds (Gill and Wolf, 1977), and hummingbirds (Pyke, 1981) all have been observed exhibiting apparently random foraging behavior with respect to change in direction. Zimmerman (1979) argues that this strategy enhances the energy gain when a return to an area still results in low probability of revisitation. Bumble bees, for example, visit a very small proportion of the flowers each time they land on a *Polemonium foliosissimum* plant; thus, a return to a plant means few revisits. Zimmerman (1982) likewise found that bumble bees rarely revisit *Potentilla gracilis* flowers for pollen. The bees are able to detect a recent visit (perhaps by odor) before alighting and pass by these flowers. The bees are able to localize their movement by exhibiting no directionality and flying to near neighbors, but maintain low costs associated with repeat visits.

The amount of nectar in flowers just visited may influence the change in direction. Heinrich (1979a), for example, found much greater change in direction in a nectar-rich area of clover heads than in a neighboring depleted area. Pyke (1978a) also observed bumble bees making greater directional changes after visits to inflorescences with higher rewards than after visits to inflorescences with lower rewards. This behavior would localize the bees' position in nectar-rich areas. Honey bees foraging between single flowers (not inflorescences) did not exhibit this pattern (Waddington, 1980). The bees may not be responding directly to recent rewards. Waddington and Heinrich (1981) suggest other mechanisms, which have little to do with enhancing foraging returns, to explain the same observation. This problem deserves experimental study.

The path of movement is further described by the sequence of directional changes. Pyke (1978a) found that bumble bees tend to alternate left and right turns, which would contribute to a tendency for forward movement. Other studies on hummingbirds (Zimmerman, 1982) demonstrate no such pattern. Again, this is a point for further consideration.

In summary, the combination of flight distance and change in direction, which describe the pollinator's path between plants, seems to enhance foraging returns. Future work is needed to understand more fully the mechanisms of these and related (not yet discovered!) behavioral patterns.

2. *Behavior on the Inflorescence*

Numerous observations have been made of the behavior (i.e., motor patterns) of pollinators on individual flowers (e.g., Knuth, 1906; Macior, 1970); usually these behaviors have been viewed in terms of pollinator effectiveness. We are just beginning to understand the behavior of pollinators on multiple-flower inflorescences. A pollinator approaching an inflorescence must decide which flower to visit first. After the first visit and after each succeeding visit, the next flower must be chosen on the inflorescence, or the animal must decide to leave for another inflorescence. All of the details regarding how these decisions are made are not known, but it is known that the decisions depend on the distribution and abundance of rewards on the inflorescence and on the rewards expected at other inflorescences.

Waddington and Heinrich (1979) constructed simple, artificial inflorescences made up of five vertically arranged flowers (Fig. 7, inset). The distribution of rewards was manipulated in order to learn how bumble bees (*Bombus edwardsii*) decide where to arrive at and depart from an inflorescence, and what they do in between. The bees move upward between the flowers and rarely fly by a flower to visit another one above it. This strong propensity to move upward did not change with the distribution of nectar rewards; however, the start and depart points did (Fig. 7A–C). When rewards are biased toward the bottom of the inflorescence, bees learn to start on a low flower and usually depart before reaching the topmost flowers. When rewards are greatest at the upper flowers, the bees tend to start near the middle of the inflorescence and depart from the top. An even distribution of rewards results in a start near the bottom and a departure near the top. In the field, bumble bees have been observed starting at the bottom

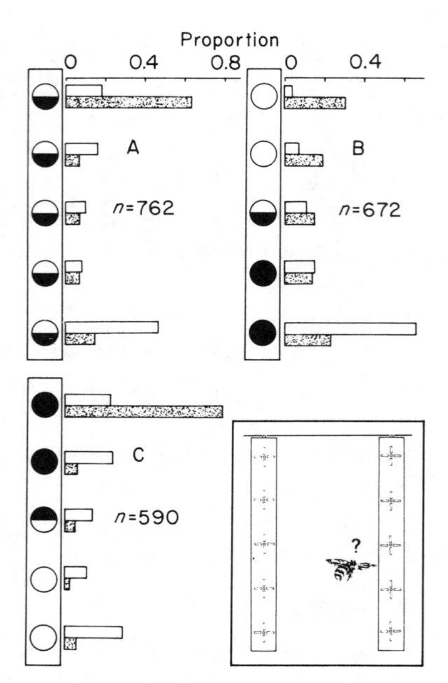

Fig. 7. The proportion of starts (open histogram) by bumble bees (*Bombus edwardsii*) at the different flowers of artificial inflorescences and the proportion of departures (shaded histogram) from the flowers to another inflorescence. (N = number of starts and departures in each of the three experiments.) (A) Reward evenly distributed among flowers, (B) reward in bottom flowers, and (C) reward in top flowers. The following symbols indicate the amount of 30% sucrose solution available at each of the five flower positions: ○—none, water only; ◒—1 μl; ●—2 μl.

on relatively nectar-rich flowers, moving upward, and departing before reaching the top of the inflorescence (e.g., Pyke, 1978b; Waddington, 1981; Best and Bierzychudek, 1982). These behavior patterns are of interest in relation to the distribution of flower sex and are probably important as mechanisms for out-crossing. Bees carrying pollen from another plant first visit the female flowers at the bottom and then move up the inflorescence, picking up pollen from the male flowers before departing to another inflorescence.

Pyke (1978c) viewed inflorescences as patches and asked, how many floral visits should an efficient hummingbird make at an inflorescence before flying to another inflorescence? He found that the decision to leave is a function of the number of flowers already visited, the number of flowers on the inflorescence (more visits at larger inflorescences), and the amount of nectar obtained from the last flower. Although a nice match is observed between the number of flowers visited and the model's predictions, additional work is needed to understand the integration of various cues for making a decision. Best and Bierzychudek (1982) have built a simple model for predicting the departure of bumble bees from *Digitalis* inflorescences.

3. Floral Choice in Mixed-Species Patches

How do pollinators choose from among the myriad flowers seemingly available to them as forage? This question has intrigued biologists for more than a century (Darwin, 1883; Grant, 1950). Most studies have documented the proportion of visits to different floral species by pollinators. In general, pollinators often exhibit considerable fidelity to single species. However, as pointed out by Waddington (1983), few attempts have been made to understand the variation in pollinator behavior (proportions). For example, in most studies before 1960, no information is given on the number of species available and the distribution and degree of intermingling of these species. Recently, studies have more carefully quantified the sequence of visits among different species and important aspects of floral resources. Also, controlled experiments in the field and laboratory are underway to discern how select pollinators make choices among flowers. The cost-benefit framework is the current prevalent approach.

Pollinators sample flowers and are thought to visit more frequently flowers which are perceived to provide the highest net intake of calories per unit time (Gill and Wolf, 1975b; Heinrich, 1979b). Bees, for example, learn which species are best based on how quickly they handle and fly between the flowers and on the magnitude of reward received. Using this as an assumption, two models were designed to predict the floral visitation sequences of pollinators (specifically, honey bees and bumble bees); the models of Waddington and Holden (1979) and Oster and Heinrich (1976) are discussed in detail elsewhere (Waddington, 1983). A third model by Charnov (1976) and others was not derived specifically for pollinators but may also prove useful (Waddington, 1982a). Each model considers important the quantitative costs and rewards for an animal to forage on different food items (i.e., floral species). These models will need modification or replacement as more is learned about pollinator behavior.

The caloric worth of nectar rewards depends on the volume and concentration of rewards. Preferences based on concentration are well known, but the results on volume are perplexing. For example, I found that honey bees preferred a 2-μl reward to a 1-μl reward (of the same sugar concentration) when foraging from a deep corolla tube, but when the corolla tube was shallow, half the bees visited only the 1-μl tubes (K. D. Waddington, unpublished manuscript). Rewards in the field fluctuate quickly on an hourly or daily basis, and bumble bees respond by changing their preferences with some lag time (Heinrich, 1976; Heinrich *et al.*, 1977).

An item not considered important in foraging decisions until recently is the variability in nectar rewards; this variability is a component of the risk associated with foraging. Risk-sensitive foraging has been suggested in theory (Real, 1980a,b). Briefly, it may be beneficial for an animal to minimize the uncertainty (due to reward variance) associated with a foraging decision in order to minimize the risk of doing very poorly in its foraging (Fig. 8). Real (1981) discovered that

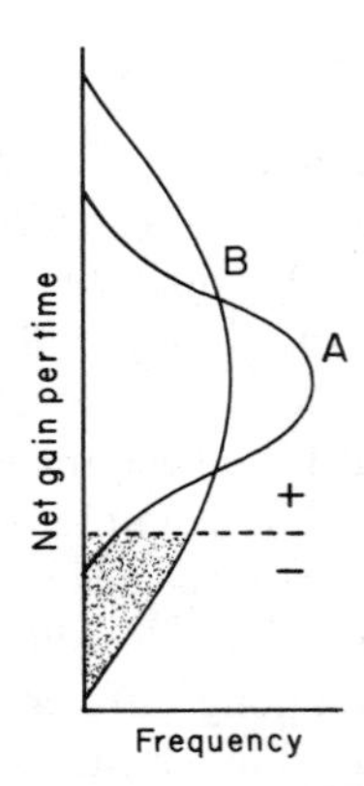

Fig. 8. The expected rewards are the same for a pollinator foraging at plant species A and B; however, the variances differ. The risk of doing very poorly is greater if species B is chosen. The dashed line indicates the hypothetical energetic break-even point for the pollinator; below the line, the pollinator is operating at a negative energy budget (probability indicated by the shaded area).

bumble bees and paper wasps foraging on a patch of two colors of artificial flowers prefer the color with the lower variance in reward even though the expected rewards were equal for both colors. Waddington *et al.* (1981) found that bumble bees prefer continuous to intermittent rewards. Real *et al.* (1983) learned that the preference due to low variance could be offset by raising the mean of the high-variance flowers.

Floral density affects inversely the costs of flying between flowers. Honey bees repeatedly given a choice between a yellow and a blue artificial flower usually chose the flower closer to the previously visited flower, despite the color of the closer flower. In a mixed-species floral array, the probability of a flower of a particular species being the closest flower increases with increased density (Marden and Waddington, 1981). Levin (1972) found that a rare color variant of *Phlox* is visited less frequently by butterflies than expected. Bumble bees foraging on two subalpine species concentrated their foraging in dense patches (i.e., there was a positive correlation between per-flower visitation rate and density), perhaps as a response to lower flight costs (Thomson, 1981).

The relationship between pollinator morphology and floral morphology affects the time needed to land on (or hover in front of) a flower for obtaining the rewards. There is a correlation between proboscis length and time spent by bumble bees on flowers. Bumble bees with short tongues forage more rapidly on goldenrod (with a short corolla) than do long-tongued bees (Morse, 1979). Similarly, Inouye (1980) found that long-tongued bumble bees forage more quickly than short-tongued bees on the long-corolla larkspur. The foraging rate (time per flower) of *Bombus flavifrons* (with a medium-length tongue) is increased with increasing corolla length. Hummingbirds likewise exhibit an inverse relationship between corolla length (of artificial flowers) and the rate of nectar intake (Hainsworth and Wolf, 1976).

Floral preferences depend in part on the relationship between pollinator morphology and floral morphology. The flowers of cow vetch (*Vicia cracca*) vary in corolla length. Morse (1978) captured bumble bees of one species foraging on various-size flowers and found a positive relationship between corolla length and tongue length. Similarly, the tongue lengths of *B. flavifrons* workers foraging on the long-corolla larkspur, *Delphinium,* and the short-corolla *Mertensia* had significantly different tongue lengths (Inouye, 1980). Presumably, these patterns emerge from the pollinators enhancing their individual foraging returns by making choices which minimize handling time.

Handling costs likely depend on relationships other than tongue length and corolla length; however, I know of no other systematic efforts to quantify other possibilities. This is another area for further, and likely fruitful, investigation.

Floral choices are dependent on factors not yet considered in foraging models. For example, they may depend on interactions with other individuals, yet current models ignore this fact. The recruits in species of bees which communicate (honey bees and stingless bees) may not make individual foraging decisions at all (Lindauer, 1967); instead, they respond to a signal based on a scout's decisions. Additionally, competitive interactions are known to influence floral choice in some situations. Inouye (1978) studied the behavior of two species of bumble bees foraging on two floral species in the Rocky Mountains; each species of bumble bee had an apparent preference for one floral species. However, when nearly all of the individuals of either bumble bee species were removed from a local patch, individuals of the remaining bumble bee species would visit the other (previously less preferred) floral species more frequently. Similarly, Morse (1977) found that *Bombus terricola* workers spatially displace *B. ternarius,* so that the latter forage on the distal florets of goldenrod (*Solidago*) clusters. The introduction of honey bees into an area in Central America dominated by stingless bees resulted in a localized reduction of foraging by the stingless bees. The stingless bees presumably shifted to different plants (Roubik, 1978). Johnson and Hubbell's (1974) work clearly shows the importance of aggression on the foraging behavior of several species of stingless bees. Furthermore, they find that the degree of aggression (and potentially the magnitude of the affect of behavior) is positively related to the quality of the food source (Johnson and Hubbell, 1974).

Pollinators' foraging behavior may be biased from the start by certain inherent preferences. That is, preferences for flowers are, at least in part, built into the genome. Honeybees prefer certain shapes (Zerrahn, 1934) and colors, and hummingbirds may have a spontaneous preference for red. Mackensen and Nye (1969) have shown that the tendency is inherited for honeybees to collect pollen from alfalfa (*Medicago sativa*) in preference to other plants. Some other bees exhibit strong genetically determined preferences for pollen (Lindsley and MacSwain, 1957); these bees are called "oligolectic." Some adaptations associated with oligolecty include the ability to locate a pollen source, synchronization of activity with flowering of the usual source, and morphological adaptation to the flower (Eickwort and Ginsberg, 1980). Strickler (1979) found that the oligolectic mason bee (*Hoplitus anthocopoides*) produces offspring more efficiently by feeding its brood its preferred plant than do some polylectic species; thus, there may be strong selection for the mason bee to maintain its inherent floral preference.

The spatial and temporal distribution of flowers is an important factor influencing the sequence of visits among flowers. Plants often grow in single-species stands. Since pollinator flights are usually to neighboring flowers, the species choice is certain. Even when plants grow intermingled, there may be considerable floral species fidelity. One reason is that for a particular pollinator, one species may usually be perceived as best in terms of energetics. Second, it was found by Levin and Kerster (1973) and Faulkner (1976) that bees exhibit horizontal flight paths between flowers (they tend to forage at the same height on successive visits). With species of different heights, this behavior would promote successive visits to the same species (Waddington, 1979c). These and related topics are discussed in more detail by Rathcke in Chapter 12, this volume. They may be among the reasons why bees have been called "flower constant" and why other pollinators are often observed exhibiting high floral species fidelity.

VII. Summary

The purpose of this chapter has been to summarize information on pollinator foraging behavior. It is argued that complex foraging behavior can best be understood, and behavioral patterns best predicted, by first understanding the simpler physiological and behavioral mechanisms; these mechanisms comprise the foundation for understanding the foraging animal. Thus, this chapter starts with descriptions of sensory inputs, integration of this sensory information, and decisions based on it, and ends with a discussion of the resultant complex behavioral patterns. After considering the most recent literature on optimal foraging theory, decisions and behavioral patterns are discussed in the context of energetic constraints.

Throughout the chapter, I have noted gaps in our knowledge on pollinator foraging behavior, and there are many. As these gaps are filled, I hope that attention will be paid to integrating knowledge at several levels. Too often, behavioral ecologists interested in predicting behavior using mathematical models ignore important details on the physiological and behavioral mechanisms of the animal in question. The results are models founded on incorrect assumptions yielding inaccurate predictions. Post-hoc arguments are then made to explain the disparity between a prediction and an observed result, with little assessment of the validity of the model's initial set of assumptions (Maynard Smith, 1978). Likewise, physiologists often lose sight of the importance of their findings for potentially understanding complex behavioral patterns. Information on sensory systems and mechanisms of behavior should be collected with an eye toward piecing it together in order to understand better complex behavioral patterns in the field.

The pollinator–flower system has served us well in the quest for establishing general concepts in animal foraging behavior. In addition, information on pollinator foraging behavior may be important in studies of floral structure and plant population and community structure. The potential for fruitful lines of research on the foraging behavior of pollinators remains unlimited.

Acknowledgments

My work on pollinator foraging behavior and the writing of this chapter were supported in part by grants from the National Science Foundation (BNS 8004537 and DEB 81-19280). Randy Breitwisch made helpful comments on a draft of the manuscript. This is contribution No. 76 from the Program in Ecology, Behavior and Tropical Biology, the University of Miami.

References

Beling, I. (1929). Über das Zeitgedachtnis der Bienen. *Z. Vergl. Physiol.* **9,** 259–338.

Bene, F. (1941). Experiments on the color preference of black-chinned hummingbirds. *Condor* **43,** 237–242.

Bennett, M. F., and Renner, M. (1963). The collecting performance of honey bees under laboratory conditions. *Biol. Bull. (Woods Hole, Mass.)* **125,** 416–430.

Best, S. L., and Bierzychudek, P. (1982). Pollinator foraging on foxglove (*Digitalis purpurea*): A test of a new model. *Evolution (Lawrence, Kans.)* **36,** 70–79.

Brown, J. L. (1964). The evolution of diversity in avian territorial systems. *Wilson Bull.* **76,** 160–169.

Butler, C. G., Jeffree, E. P., and Kalmus, H. (1943). The behaviour of a population of honeybees on an artificial and on a natural crop. *J. Exp. Biol.* **20,** 65–73.

Carpenter, F. L., and MacMillen, R. E. (1976). Threshold model of feeding territoriality and test with a Hawaiian honeycreeper. *Science* **194,** 639–642.

Charnov, E. L. (1976). Optimal foraging: Attack strategy of a mantid. *Am. Nat.* **110,** 141–151.

Darwin, C. R. (1883). "The Effects of Cross and Self Fertilization in the Vegetable Kingdom." Appleton, New York.

Daumer, K. (1958). Blumenfarben, wie sie die Bienen sehen. *Z. Vergl. Physiol.* **41,** 49–110.

Eckert, J. E. (1933). The flight range of the honey bee. *J. Agric. Res. (Washington, D.C.)* **47,** 257–285.

Eickwort, G. C., and Ginsberg, H. S. (1980). Foraging and mating behavior in apoidea. *Annu. Rev. Entomol.* **25,** 421–446.

Emlen, J. M. (1966). The role of time and energy in food preferences. *Am. Nat.* **100,** 611–617.

Faulkner, G. J. (1976). Honeybee behaviour as affected by plant height and general colour in Brussels sprouts. *J. Apic. Res.* **15,** 15–18.

Feinsinger, P., and Chaplin, S. B. (1975). On the relationship between wing disc loading and foraging strategy in hummingbirds. *Am. Nat.* **109,** 217–224.

Frankie, G. W. (1976). Pollination of widely dispersed trees by animals in Central America, with an emphasis on bee pollination systems. *In* "Tropical Trees: Variation, Breeding and Conservation" (J. Burley and B. T. Styles, eds.), pp. 151–159. Academic Press, New York.

Frankie, G. W., and Baker, H. G. (1974). The importance of pollinator behaviour in the reproductive biology of tropical trees. *Anal. Inst. Biol.* (*Botanica*), No. 45, 1–10.

Frankie, G. W., Vinson, S. B., and Coville, R. E. (1980). Territorial behavior of *Centris adani* and its reproductive function in the Costa Rican dry forest (Hymenoptera: Anthohoridae). *J. Kans. Entomol. Soc.* **53,** 837–857.

Free, J. B. (1966). The pollinating efficiency of honeybee visits to apple flowers. *J. Hortic. Sci.* **41,** 91–94.

Gary, N. E., Witherell, P. C., Lorenzen, K., and Marston, J. M. (1977). Area fidelity and intra-field distribution of honey bees during the pollination of onions. *Environ. Entomol.* **6,** 303–310.

Gill, F. B., and Wolf, L. L. (1975a). Economics of feeding territoriality in the golden-winged sunbird. *Ecology* **56,** 333–345.

Gill, F. B., and Wolf, L. L. (1975b). Foraging strategies and energetics of east African sunbirds at mistletoe flowers. *Am. Nat.* **109,** 491–510.

Gill, F. B., and Wolf, L. L. (1977). Nonrandom foraging by sunbirds in a patchy environment. *Ecology* **58,** 1284–1296.

Gould, J. L. (1982). "Ethology: The Mechanisms and Evolution of Behavior." Norton, New York.

Grant, K. A., and Grant, V. (1968). "Hummingbirds and Their Flowers." Columbia Univ. Press, New York.

Grant, V. (1950). The flower constancy of bees. *Bot. Rev.* **16,** 379–398.

Grant, V. (1963). "The Origin of Adaptations." Columbia Univ. Press, New York.

Hainsworth, F. R., and Wolf, L. L. (1976). Nectar characteristics and food selection by hummingbirds. *Oecologia* **25,** 101–113.

Heinrich, B. (1975). Energetics of pollination. *Annu. Rev. Ecol. Syst.* **6,** 139–170.

Heinrich, B. (1976). The foraging specializations of individual bumblebees. *Ecol. Monogr.* **46,** 105–128.

Heinrich, B. (1979a). Resource heterogeneity and patterns of movement in foraging bumblebees. *Oecologia* **40,** 235–246.

Heinrich, B. (1979b). "Majoring" and "minoring" by foraging bumblebees, *Bombus vagans:* An experimental analysis. *Ecology* **60,** 245–255.

Heinrich, B., and Raven, P. H. (1972). Energetics and pollination ecology. *Science* **176,** 597–602.

Heinrich, B., Mudge, P. R., and Deringis, P. G. (1977). Laboratory analysis of flower constancy in foraging bumblebees: *Bombus ternarius* and *B. terricola. Behav. Ecol. Sociobiol.* **2,** 247–265.

Hodges, C. M., and Miller, R. B. (1981). Pollinator flight directionality and the assessment of pollen returns. *Oecologia* **50,** 376–379.

Inouye, D. W. (1978). Resource partitioning in bumblebees: experimental studies of foraging behavior. *Ecology* **59,** 672–678.

Inouye, D. W. (1980). The effect of proboscis and corolla tube lengths on patterns and rates of flower visitation by bumblebees. *Oecologia* **45,** 197–201.

Jamieson, C. A., and Austin, G. H. (1958). Preferences of honeybees for sugar solutions. *Proc. Int. Congr. Entomol.* **4,** 1059–1062.

Jander, R. (1975). Ecological aspects of spatial orientation. *Annu. Rev. Ecol. Syst.* **6,** 171–188.

Janzen, D. (1971). Euglossine bees as long-distance pollinators of tropical plants. *Science* **171,** 203–205.

Johnson, L. K., and Hubbell, S. P. (1974). Aggression and competition among stingless bees: Field studies. *Ecology* **55,** 120–127.

Kamil, A., and Sargeant, T., eds. (1981). "Foraging behavior: Ecological, Ethological and Psychological Approaches." Garland STPM Press, New York.

Knuth, P. (1906). "Handbook of Flower Pollination," Vol. 1. Oxford Univ. Press (Clarendon), London and New York (transl. J. R. Ainsworth Davis).

Kugler, H. (1950). Der Blutenbesuch der Schlammfliege (*Eristalomyia tenax*). *Z. Vergl. Physiol.* **32,** 328–347.

Kugler, H. (1951). Blutenokologische Untersuchungen mit Goldfliegen (Lucilien). *Ber. Dtsch. Bot. Ges.* **64,** 327–341.

Kugler, H. (1956). Über die optische Wirkung von Fliegenblumen auf Fliegen. *Ber. Dtsch. Bot. Ges.* **69,** 387–398.

Levin, D. A. (1972). Low frequency disadvantage in the exploitation of pollinators by corolla variants in *Phlox*. *Am. Nat.* **106,** 453–459.

Levin, D. A., and Kerster, H. W. (1968). Local gene dispersal in *Phlox*. *Evolution (Lawrence, Kans.)* **22,** 130–139.

Levin, D. A., and Kerster, H. W. (1969). The dependence of bee-mediated pollen and gene dispersal upon plant density. *Evolution (Lawrence, Kans.)* **23,** 560–571.

Levin, D. A., and Kerster, H. W. (1973). Assortative pollination for stature in *Lythrum salicaria*. *Evolution (Lawrence, Kans.)* **27,** 144–152.

Levin, D. A., Kerster, H. W., and Niedzlek, M. (1971). Pollinator flight directionality and its effect on pollen flow. *Evolution (Lawrence, Kans.)* **25,** 113–118.

Lindauer, M. (1967). "Communication among Social Bees." Atheneum Press, New York.

Lindsley, E. G., and MacSwain, J. W. (1957). The nesting habits, flower relationships, and parasites of some North American species of *Diadasia*. *Wasmann J. Biol.* **15,** 119–235.

Linhart, Y. B. (1973). Ecological and behavioral determinants of pollen dispersal in hummingbird-pollinated *Heliconia*. *Am. Nat.* **107,** 511–523.

Linsley, E. G., and MacSwain, J. W. (1957). The nesting habits, flower relationships, and parasites of some North American species of *Diadasia*. *Wasmann J. Biol.* **15,** 199–235.

Lovell, J. (1918). "The Flower and the Bee." Charles Scribner's Sons, New York.

Lyerly, S. B., Riess, B. F., and Ross, S. (1950). Color preference in the Mexican Violet-eared hummingbird, *Calibri t. thalassinus* (Swainson). *Behaviour* **2,** 237–248.

MacArthur, R. H., and Pianka, E. R. (1966). An optimal use of a patchy environment. *Am. Nat.* **100,** 603–609.

McCabe, R. A. (1961). The selection of colored nest boxes by house wrens. *Condor* **63,** 322–329.

Macior, L. W. (1970). The pollination ecology of *Pedicularis* in Colorado. *Am. J. Bot.* **57,** 716–728.

Mackensen, O., and Nye, W. P. (1969). Selective breeding of honeybees for alfalfa pollen collection: sixth generation and outcrosses. *J. Apic. Res.* **8,** 9–12.

Marden, J. H., and Waddington, K. D. (1981). Floral choices by honeybees in relation to the relative distances to flowers. *Physiol. Entomol.* **6,** 431–435.

Maynard Smith, J. (1978). Optimization theory in evolution. *Ann. Rev. Ecol. Syst.* **9,** 31–56.

Menzel, R., and Erber, J. (1978). Learning memory in honey bees. *Sci. Am.* **239,** 102–111.

Menzel, R., Erber, J., and Masuhr, T. (1974). Learning and memory in the honeybee. *In* "Experimental Analysis of Insect Behaviour" (L. Barton Browne, ed.), pp. 195–217. Springer-Verlag, Berlin and New York.

Moffett, J. O., Stith, L. S., Burkhardt, C. C., and Shipman, C. W. (1976). Nectar secretion in cotton flowers and its relation to floral visits by honey bees. *Am. Bee J.* **116,** 32, 34, 36.

Morse, D. H. (1977). Resource partitioning in bumble bees: The role of behavioral factors. *Science* **19,** 678–680.

Morse, D. H. (1978). Size-related foraging differences of bumble bee workers. *Ecol. Entomol.* **3,** 189–192.

Morse, D. H. (1979). Foraging rate, foraging position, and worker size in bumble bee workers. *Proc. 4th Int. Symp. on Pollination, Md. Agric. Exp. Stn., Spec. Misc. Publ.* **1,** 447–452.

Opfinger, E. (1931). Über die Orientierung der Biene an der Futterquelle. *Z. Vergl. Physiol.* **1,** 431–487.

Oster, G., and Heinrich, B. (1976). Why do bumblebees major? A mathematical model. *Ecol. Monogr.* **46,** 129–133.

Oster, G. F., and Wilson, E. O. (1978). "Caste and Ecology in the Social Insects." Princeton Univ. Press, Princeton, New Jersey.

Plateau, F. (1899). Nouvelles recherches sur les rapports entre Les insectes et les fleurs. II. Le croix des couleurs par les insectes. *Mem. Soc. Zool. Fr.* **12,** 336–370.

Pleasants, J. M., and Zimmerman, M. (1979). Patchiness in the dispersion of nectar resources: evidence for hot and cold spots. *Oecologia* **41,** 283–288.

Proctor, M., and Yeo, P. (1972). "The Pollination of Flowers." Taplinger Publ., New York.

Pyke, G. H. (1978a). Optimal foraging: Movement patterns of bumblebees between inflorescences. *Theor. Pop. Biol.* **13,** 72–98.

Pyke, G. H. (1978b). Optimal foraging in bumblebees and coevolution with their plants. *Oecologia* **36,** 281–293.

Pyke, G. H. (1978c). Optimal foraging in hummingbirds: testing the marginal value theorem. *Am. Zool.* **18,** 739–752.

Pyke, G. H. (1979a). The economics of territory size and time budget in the golden-winged sunbird. *Am. Nat.* **114,** 131–145.

Pyke, G. H. (1979b). Optimal foraging in bumblebees: rule of movement between flowers within inflorescences. *Anim. Behav.* **27,** 1167–1181.

Pyke, G. H. (1981). Optimal foraging in hummingbirds: Rule of movement between inflorescences. *Anim. Behav.* **29,** 889–896.

Real, L. (1980a). Fitness, uncertainty and the role of diversification in evolution and behavior. *Am. Nat.* **11,** 623–638.

Real, L. (1980b). On uncertainty and the law of diminishing returns in evolution and behavior. *In* "Limits to Action: The Allocation of Individual Behavior" (J. E. R. Staddon, ed.), pp. 37–64. Academic Press, New York.

Real, L. (1981). Uncertainty and pollinator-plant interactions: The foraging behavior of bees and wasps on artificial flowers. *Ecology* **62,** 20–26.

Real, L., Otte, J., and Silverfine, E. (1983). On the trade-off between the mean and variance in foraging: an experimental analysis with bumblebees. *Ecology* **63,** 1617–1623.

Roubik, D. W. (1978). Competitive interactions between neotropical pollinators and Africanized honey bees. *Science* **201,** 1030–1032.

Singh, S. (1950). Behavior studies of honeybees in gathering nectar and pollen. *N. Y. Agric. Exp. Stn. (Ithaca)* **288,** 1–57.

Strickler, K. (1979). Specialization and foraging efficiency of solitary bees. *Ecology* **60,** 998–1009.

Thomson, J. D. (1981). Spatial and temporal components of resource assessments by flower-feeding insects. *J. Anim. Ecol.* **50,** 49–59.

Vansell, G. H. (1942). Factors affecting the usefulness of honeybees in pollination. *Circ. U.S. Dept. Agric.*, No. 650.

von Frisch, K. (1967). "The Dance Language and Orientation of Bees." Harvard Univ. Press, Cambridge, Massachusetts.

Waddington, K. D. (1979a). Flight patterns of three species of sweat bees (Halictidae) foraging at *Convolvulus arvensis*. *J. Kans. Entomol. Soc.* **52,** 751–758.

Waddington, K. D. (1979b). Quantification of the movement patterns of bees: A novel method. *Am. Midl. Nat.* **101,** 278–285.

Waddington, K. D. (1979c). Divergence in inflorescence height: An evolutionary response to pollinator fidelity. *Oecologia* **40,** 43–50.

Waddington, K. D. (1980). Flight patterns of foraging bees relative to density of artificial flowers and distribution of nectar. *Oecologia* **44,** 199–204.

Waddington, K. D. (1981). Factors influencing pollen flow in bumblebee-pollinated *Delphinium virescens*. *Oikos* **37,** 153–159.

Waddington, K. D. (1982a). Optimal diet theory: Sequential and simultaneous encounter models. *Oikos* **39,** 279–280.

Waddington, K. D. (1982b). Information used in foraging. *In* "The Biology of Social Insects" (M. D. Breed, C. D. Michener, and H. E. Evans, eds.), pp. 24–27. Westview Press, Boulder, Colorado.

Waddington, K. D. (1983). Floral-visitation-sequences by bees: Models and experiments. *In* "Handbook of Experimental Pollination Biology" (E. Jones and R. Little, eds.), pp. 461–473. Van Nostrand-Reinhold, New York.

Waddington, K. D., and Heinrich, B. (1979). The foraging movements of bumblebees on vertical inflorescences: an experimental analysis. *J. Comp. Physiol.* **134,** 113–117.

Waddington, K. D., and Heinrich, B. (1981). Patterns of movement and floral choice by foraging bees. *In* "Foraging Behavior: Ecological, Ethological, and Psychological Approaches" (A. Kamil and T. Sargent, eds.), pp. 215–230. Garland STPM Press, New York.

Waddington, K. D., and Holden, L. R. (1979). Optimal foraging: On flower selection by bees. *Am. Nat.* **114,** 179–196.

Waddington, K. D., Allen, T., and Heinrich, B. (1981). Floral preferences of bumblebees (*Bombus edwardsii*) in relation to intermittent versus continuous rewards. *Anim. Behav.* **29,** 779–784.

Wagner, H. (1946). Food and feeding habits of Mexican hummingbirds. *Wilson Bull.* **58,** 69–93.

Wahl, O. (1932). Neue Untersuchungen über das Zeitgedachtnis der Bienen. *Z. Vergl. Physiol.* **16,** 529–589.

Waller, G. D. (1972). Evaluating responses of honeybees to sugar solutions using an artificial-flower feeder. *Ann. Entomol. Soc. Am.* **65,** 857–862.

Weaver, N. (1957). The foraging behavior of honeybees on hairy vetch. II. The foraging area and foraging speed. *Insectes Sociaux* **4,** 43–57.

Zerrahn, G. (1934). Formdressur und Formunterscheidung bei der Honigbiene. *Z. Vergl. Physiol.* **20,** 117–150.

Zimmerman, M. (1979). Optimal foraging: A case for random movement. *Oecologia* **43,** 261–267.

Zimmerman, M. (1982). Optimal foraging: Random movement by pollen collecting bumblebees. *Oecologia* **53,** 394–398.

CHAPTER 10

The Adaptive Nature of Floral Traits: Ideas and Evidence

NICKOLAS M. WASER
Department of Biology
University of California
Riverside, California
and
Rocky Mountain Biological Laboratory
Crested Butte, Colorado

If we look at genetics, or general physiology, we find that a decisive advance has been made there, after the investigators had greatly simplified their problems and taken their stand upon the firm basis of experimental methods (Gause, 1934, p. 1).

POLLINATION BIOLOGY

ISBN 0-12-583980-4 (hardcover)
ISBN 0-12-583982-0 (paperback)

I. Introduction

In their textbook of general botany, Jensen and Salisbury (1972) ask what a visitor from outer space might deduce about Earth's vegetation as his space-ship approached our planet. I wonder, in addition, how the alien would react to a really close look at angiosperm reproductive structures! He might well exclaim over the tremendous variety of ways in which flowers are shaped, colored, and arranged into inflorescences, and he soon would notice equivalent variation in other floral traits. How could we respond were such an alien (or any other interested party) to ask what has led to this elaboration of floral expression?

We could begin by ascribing many types of floral features to the need for successful pollination. This explanation is not new. It can be traced back at least two centuries to authors who praised the cleverness of a wise Creator (e.g., Kölreuter, 1761; Sprengel, 1793). The more modern view, which considers evolution by natural selection as the force responsible for floral attributes, goes back to Darwin (1859, 1876) and his contemporaries (e.g., Müller, 1871, 1883; Knuth, 1906). Drawing on this legacy, we could tell our alien visitor that flowering plants seem to have evolved and diversified in large part because of their reliance on animals as agents of gamete transfer (Baker, 1963; Baker and Hurd, 1968; Mulcahy, 1979; Regal, 1977; Stebbins, 1981). We could further stress that the colorful conspicuousness and shape of flowers and their provision of food or other useful substances have induced certain animals to adopt ways of life in which they pick up and deliver pollen as they search for floral rewards (Baker and Hurd, 1968; Stebbins, 1970; Faegri and van der Pijl, 1971; Proctor and Yeo, 1972). To what extent, however, are we in a position to go beyond these vague statements and to provide rigorous adaptive explanations for floral traits? If such explanations are not at hand, what are the prospects for, and inherent difficulties in, obtaining them?

A profitable way to examine the adaptive significance of any floral trait is to detail as completely as possible the processes by which it can affect plant fitness. It can do so in two ways. To begin with, a trait can influence pollinator visitation behavior, which in turn can be divided into three sequential components. First, a certain set of animal species approaches the plant to visit its flowers, having just visited a set of other plants. Second, once at the plant, these animals visit a certain number of its flowers in some sequence. Finally, the visitors leave the plant and travel to another set of plants. A trait can influence any of these parts of the visitation sequence (Pyke, 1981b; Pyke and Waser, 1981).

In addition, floral traits can influence the exact mechanics of where and how pollen is removed from and applied to the pollinator's body, even if they do not influence behavior. This may determine the efficiency of any pollinator as well as the exact pattern of pollen transfer among flowers (Levin and Berube, 1972;

Silander and Primack, 1978; Parker, 1981; Tepedino, 1981; Motten *et al.*, 1981; Bertin, 1982a; Lertzman and Gass, 1983; Morse, 1982; Price and Waser, 1982). In one respect, these types of effects are subsumed under those of a trait on behavior, since much of pollen transfer mechanics will depend on variations in what the pollinator does. However, this is not always the case, so that the distinction drawn here seems to me a useful one.

The effect a floral trait has on visitor behavior or pollen transfer mechanics feeds back on the reproductive success of the plant as a female (pollen-receiving) or male (pollen-donating) parent (cf. Charnov, 1979; Willson, 1979; Waser and Price, 1983a). This feedback can be complex, and it may often be useful to simplify the problem by considering that traits ultimately influence plant fitness through their effect on only two things: the quantity and the quality of pollen transfer. ''Quantity'' refers to the amount of pollen reaching stigmas and being carried away from anthers to appropriate destinations. ''Quality'' refers to the extent to which desirable genetic properties of pollinations are achieved—for example, optimal genetic congruence of mates (Mulcahy, 1971; Hogenboom, 1975; Price and Waser, 1979; Lertzman, 1981; Waser and Price, 1983a), absolute genetic excellence of pollen deposited (Janzen, 1977; Willson, 1979; Charnov, 1979; Stephenson and Bertin, Chapter 6, this volume), or optimal mixture of pollen genotypes deposited (Mulcahy, 1971, 1979; Janzen *et al.*, 1980; Waser and Price, unpublished). ''Quality'' may also refer to the extent to which conspecific pollen is admixed with heterospecific pollen in cases in which mixing has a reproductive effect (see Waser, 1978a,b; Sukhada and Jayachandra, 1980; Thomson *et al.*, 1981).

What methods are appropriate for exploring the effects a given trait has on pollinator behavior, pollen transfer, and ultimately plant fitness? An obvious first step is simply careful observation of pollination and eventual inference of the value of a trait (Faegri and van der Pijl, 1971; see also Williams, 1966). This approach has a serious methodological drawback, however, in allowing one only to propose *a posteriori* adaptive hypotheses (which, unfortunately, often become incorporated into a ''mythology'' of untested adaptive explanations; Gould and Lewontin, 1979). Mere observation can help to generate hypotheses but cannot serve as a means for testing them, which is logically necessary for demonstrating adaptation (Platt, 1964; Curio, 1973; Maynard Smith, 1978; Krebs and Davies, 1981).

Two possible forms of hypothesis testing can be recognized; both rely on the existence of some trait variation. In the comparative method, this variation occurs naturally in time or space, One looks for times or places (whether within or between populations) in which expression of a trait varies and asks how variation alters pollinator behavior, pollen transfer mechanics, and plant reproductive success (in actual studies, one of these steps is often left out). This gives insight into the relative fitness values of trait expressions observed if one knows or assumes that trait variation has a substantial genetic basis.

The second method of hypothesis testing is through manipulative experiments. Proponents of comparative studies may refer to temporal or spatial trait variation as providing a "natural experiment." But this is not the case, since factors other than the trait in question do not remain tightly controlled (Connell, 1975). Manipulations allow one, in principle, to vary only the trait of interest, and to test hypotheses about how this trait will influence the reproductive success of the plant through its influence on the pollination process. If reproductive success is reduced by manipulation, one assumes that the observed trait expression indeed represents an optimum of some sort. This type of experiment employs, explicitly or implicitly, an evolutionarily stable strategy (ESS) approach (Maynard Smith, 1976) in which one tries to determine whether the observed expression could be invaded by a mutant expression. The "mutant" in this case is produced by the experimenter. This brings up a potential problem: The experimenter must be sure to manipulate in a way that does not cause bizarre alterations in the natural order and thus lead to predicted results for the wrong reasons. It seems surprising that manipulative experiments remain the rarest of the three methods for arguing that a floral trait is adaptive, since they are often relatively straightforward. As discussed by Connell (1980) and Waser (1983), combinations of comparative and experimental methods may be especially useful.

In this chapter, I will discuss several traits—the color, form, and distribution in time and space of flowers—and will consider how they may influence pollinator behavior, pollination mechanics, and plant fitness. The discussion is organized into sections that treat the three sequential components of visitation mentioned earlier: the attraction of pollinators to the plant, their behavior while at the plant, and their behavior on leaving the plant. I will emphasize temperate, hermaphroditic, animal-pollinated systems (especially hummingbird-pollinated ones); my own work; and other studies with which I am familiar, especially those employing experiments. The chapter will largely ignore certain topics, including sexual systems of plants (Bawa and Beach, 1981; Wyatt, Chapter 4, this volume), community-level patterns (Frankie *et al.*, 1974; Feinsinger, 1978; Waser, 1983; Rathcke, Chapter 12, this volume), population structure (Levin, 1978; Waser and Price, 1983a; Handel, Chapter 8, this volume), and foraging economics from the point of view of pollinators (Wolf and Hainsworth, 1971; Gass and Montgomerie, 1981; Pyke, 1981b; Pleasants, 1981; Waddington, Chapter, 9, this volume; Real, Chapter 11, this volume). A discussion of the genetics of variation in specific traits also lies outside the scope of this chapter. Finally, I intend for simplicity to treat floral traits individually, ignoring potential genetic and ontogenetic links among them. This method has inherent drawbacks (Gould and Lewontin, 1979), but it provides the only reasonable starting point and the only obvious means by which adaptive hypotheses can be posed.

II. What Species Visit a Plant, and Where Do They Come From?

A. Flowering Time as a Determinant of Visitation

Flowering time dictates the identities and relative availabilities of flower visitors, and therefore may often represent an adaptation for matching periods of abundance of specific visitors, especially those whose activity is limited by external constraints. Daily flowering time has been regarded commonly as an adaptation for attracting specific pollinators (Baker, 1961; Faegri and van der Pijl, 1971). Similarly, seasonal flowering patterns sometimes seem to match seasonal cycles in pollinator activity. For example, the ephemeral flowering of some desert annuals appears to match closely the equally ephemeral life cycles of their specialized insect visitors (Linsley, 1958; Baker and Hurd, 1968), and the spring flowering of some North American shrubs and herbs appears to correspond closely to the migratory passage of their hummingbird pollinators (Grant and Grant, 1968; Austin, 1975; Simpson, 1977; Waser, 1979; Bertin, 1982b).

Most studies of temporal synchrony between flowers and visitors have been of an observational nature (Robertson, 1895, is an excellent early example). A few, however, have made comparisons among years, locations, or individual plants. For example, Bertin (1982b) found that peak flowering of jewelweed (*Impatiens biflora*) matched peak migration of ruby-throated hummingbirds (*Archilochus colubris*) at an Illinois site over 3 years, and he summarized records suggesting a reasonable match in other geographic areas as well. It appears that both insects and hummingbirds exert selective pressure on jewelweed flowering time, and Bertin suggests that flowering time, in turn, has had much to do with the timing of ruby-throat migration. In a similar vein, I studied flowering time and seed set of marked ocotillo plants (*Fouquieria splendens*), and availability of several migratory hummingbird species and of nonmigatory carpenter bees (*Xylocopa californica*), at two Arizona sites over 4 years (Waser, 1979). The temporal match between migration and peak flowering varied among plants and years, and this was reflected in seed set. Based on this natural experiment, I argued that there is selection on ocotillo flowering to match hummingbird migration through deserts each spring. In this case, the timing of migration appears to be relatively constant, perhaps because birds are penalized if they do not arrive at distant breeding sites in time to anticipate the flowering of other important food plants (Waser, 1976, 1979; Calder *et al.*, 1983).

Although the results from such comparative studies suggest the adaptive nature of flowering time, they lack the decisiveness of a true manipulation. In 1974, I attempted to manipulate flowering time of ocotillo (Waser, 1979 and unpublished). This plant has a strange "candelabrum" growth form of multiple woody stems. I removed pairs of stems from plants at three sites near Tucson, Arizona; these were at the north and south of a latitudinal gradient of 80 km and at the bottom and top of an elevational gradient of 380 m at the southern limit of the latitudinal gradient. One of each pair of stems was transplanted next to the plant from which it was taken, whereas the other was removed to the other end of the appropriate gradient. Of a total of 90 stems, 29 produced some buds or flowers one season after transplantation, although many of these aborted before maturity. Regardless of the site of transplantation, flowering stems retained the timing of their ancestral populations. Those stems moved to a new elevation or latitude flowered asynchronously with surrounding populations, since ocotillo flowering progresses from south to north and from low to high elevations each spring in southern Arizona. Asynchrony was on the order of 1 week in all cases. I suspect, but cannot confirm because of infrequent censuses, that I was also varying synchrony of flowering with northward migration of hummingbirds. Of the few stems that matured fruits, those transplanted to a new site set fewer seeds per flower than controls taken from surrounding populations; however, sample sizes are too small for firm conclusions to be drawn. An unusual freeze in 1976 killed most of the transplants before further results could be obtained. I outline the experiment here mainly to emphasize the feasibility of manipulating flowering time. One problem, which I did not deal with adequately in the ocotillo study, is the difficulty of varying asynchrony with pollinators, aside from asynchrony with flowering of surrounding members of the same and other species. This can produce effects that will complicate the interpretation of experimental results (Waser, 1979).

B. Color as an Attractant: A Classic View

The spectral reflectances of flowers and associated plant parts are exceedingly variable between about 300–800 nm, although not all colors are equally represented on a local or global scale (Weevers, 1952; Kevan, 1978). What dictates variation in color? One idea, which has had a long history and which survives in the pollination syndrome concept, is that flower visitors have relatively fixed color preferences (van der Pijl, 1960; Faegri and van der Pijl, 1971; Procter and Yeo, 1972; Stebbins, 1970). If this is true, diversity of visitors and of color preferences could promote diversity of flower colors. This view assumes that color itself is an attractant of pollinators, along with food and other floral rewards.

A logical consequence of the strict form of this classic view of flower color is that there will be specific color associations for each type of flower visitor. This prediction is not borne out by observations. The rule, rather than the exception, is that flowers of any given color are approached regularly by a wide range of animals (Baker, 1961, 1963; Grant and Grant, 1965; Baker and Hurd, 1968; O'Brien, 1980; Waser, 1978a, 1979, 1982; Bertin, 1982a). Furthermore, most types of flower visitors approach a range of colors. For example, there is no evidence that hummingbirds consistently visit red flowers and avoid those of any other color, either in temperate or tropical climates (V. Grant and K. A. Grant, 1966, 1967; K. A. Grant and V. Grant, 1967, 1968; Stiles, 1976; Feinsinger, 1978; see also Table I). Innate color preference thus does not seem to account for the common red coloration of hummingbird flowers. This example is not unique; bees also tend to be catholic in color choice (Bennett, 1883; Plateau, 1899; Pederson, 1967; Heinrich, 1975a; Hannan, 1981; Pleasants, 1981).

The available experimental studies of behavior and neurophysiology of flower visitors support the observation that color preferences are not pronounced. Hummingbirds have a visual spectrum that ranges from deep red to near ultraviolet (UV) (Goldsmith, 1980), and they can be trained rapidly to associate any color within this range with a food reward (Bené, 1941; Collias and Collias, 1968; Miller and Miller, 1971; Stiles, 1976; Goldsmith and Goldsmith, 1979). Broad spectral sensitivity also seems to occur with some butterflies (Swihart, 1970; Bernard, 1979), and Swihart (1970) showed that the color preference of a swallowtail (*Papilio troilus*) was very plastic. Honeybees (*Apis mellifera*) and bumblebees (*Bombus* spp.) are sensitive from red-orange or red to near UV (Goldsmith and Bernard, 1974; Menzel *et al.*, 1974; Kevan, 1978; and references therein) and can be trained to choose different colors within this range (von Frisch, 1914; Kugler, 1943; Menzel *et al.*, 1974). The speed of training does

TABLE I

Some Characteristics of Flowers Visited by Broad-tailed Hummingbirds (*Selasphorus platycercus*) at the Rocky Mountain Biological Laboratory in Colorado

Plant species	Flower color	Flower shape	Hummingbird visitation	Pollination[a]
Ipomopsis aggregata	Red	Tube	Common	Yes
Castilleja miniata	Red	Tube	Moderate	Yes
Aquilegia elegantula	Red	Tube	Moderate	—
Epilobium angustifolium	Purple	Open	Moderate	—
Hydrophyllum fendleri	Pale purple	Cup	Moderate	—
Delphinium nelsonii	Blue	Tube	Common	Yes
Delphinium barbeyi	Blue	Tube	Common	—
Aconitum columbianum	Blue	Complex	Rare	—
Mertensia fusiformes	Pale blue	Cup	Moderate	—
Mertensia ciliata	Pale blue	Cup	Moderate	—
Iris missouriensis	Pale blue	Complex	Rare	—
Aquilegia coerulea	Pale blue/white	Tube	Rare	—
Pedicularis bracteosa	Yellow	Tube	Moderate	—
Lonicera involucrata	Yellow	Tube	Moderate	—
Castilleja sulphurea	Yellow	Tube	Moderate	—
Erythronium grandiflorum	Yellow	Open	Rare	—
Frasera speciosa	Pale green	Open	Moderate	—

[a]Experimental verification of pollination was obtained by Waser (1978a and unpublished) and Waser and Price (1981).

vary: Bees appear to be especially good at discriminating colors containing blue and near-UV wavelengths, and have more difficulty with longer wavelengths (Kugler, 1943; Menzel *et al.*, 1974) and perhaps also with colors that are white to humans (Heinrich *et al.*, 1977; Clement, 1965; but see Pederson, 1967; Mogford, 1974; Horovitz, 1969; Waser and Price, 1981). Bees can be trained to red colors (Kugler, 1943; Menzel *et al.*, 1974), and they visit red flowers in nature—even some that should have evolved as much as possible to be inconspicuous to them, for example, those typical hummingbird flowers that some bumblebees systematically rob of nectar (personal observations; see also Waser, 1979, 1982).

C. Color as an Advertisement, and Economics as an Ultimate Attractant

From the above discussion, it seems that flower visitors show little innate aversion to any color falling within their visual spectrum, although varying sensitivity across the spectrum can influence discrimination ability. The fact that all flower visitors studied have broad spectral sensitivities may be no accident: Their food comes in packages that vary in color over space and time (Feinsinger, 1983). From the plants' point of view, broad sensitivity of visitors must inadvertently create a desirable situation in many cases. Flowers of any given color have the potential to obtain visits from a mixture of animals, and experiments may show that the services provided by alternative visitors are beneficial in terms of quantity and quality of pollen transfer (Feinsinger, 1983; Waser and Price, 1983a).

This leaves unexplained the original observation that flower color varies. The ultimate explanation may come from considering color as a cue that visitors can associate with food or other rewards provided by flowers and with the likelihood or difficulty of their removal (Heinrich, 1975a; Baker and Hurd, 1968; Stiles, 1976; Waser and Price, 1981). In my terminology, then, the general adaptive explanation of color must be that it provides "advertisement" to pollinators over relatively long distances, whereas the ultimate "attractant" is floral reward. Stiles (1981) makes a similar distinction (as do many others), but he uses the terms "attraction unit" and "reward unit," respectively.

The hypothesis that pollinator choice is primarily based on reward economics implies that the primary criterion for flower color of any species is that it falls within the spectral sensitivity of desirable visitors and contrasts to background vegetation. In many cases, for reasons that I will bring up later in discussing flower constancy (Section II,E), a color that contrasts maximally to those of other flowers in the community may also provide substantial reproductive benefits. Indeed, Kevan (1978) provides intriguing evidence that flowers of co-occurring insect-pollinated species are separated more widely when plotted on a trichromaticity diagram based on bee vision than when plotted on a similar diagram based on human vision. This holds true for plant communities in several geographical regions and suggests that flowers within each community have evolved to be visually distinct. Extending this approach to southern California mountains, W. Sawyer (personal communication) found a wide separation of flowers on an insect trichromaticity diagram, and showed that this separation is maintained throughout an entire flowering season despite rapid turnover of individual species. An examination of floras of several countries by Weevers (1952) also supports the idea that flower colors have somehow evolved to maximize the coverage of pollinator visual spectra. Weevers found no substantial overall variation among countries in the proportions of species with flowers of various colors, at least in the human visual spectrum, despite the fact that his survey included temperate and tropical regions that might be expected to differ substantially in the makeup of their pollinator faunas. Differences did occur, however, in comparing low and high elevations; the same may be true when comparing arctic and temperate latitudes (Kevan, 1978). Finally, the economics-as-attractant hypothesis implies that flower color within a single species may vary geographically in a seemingly arbitrary way as a function of the local color environment. Table II, taken from my surveys of *Ipomopsis aggregata,* presents some information that bears on this suggestion. This herbaceous monocarpic perennial is widespread in the western United States, occupying mountain ranges that often are isolated from one another by intervening deserts. *Ipomopsis aggregata* is indeed variable in flower color, corolla length, and nectar concentrations (see also Wherry, 1961; Grant and Grant, 1965), as well as in flowering time (Waser, 1983). Judging from Table II, the former three traits do not vary concordantly. The range of corolla length in populations with red flowers includes those in populations with pink, purple, and orange flowers, and nectar concentrations show no obvious relationships with color or corolla length. These patterns certainly are more suggestive of color being a relatively arbitrary cue than they are of specific colors being linked with other floral traits to form constant syndromes that might be associated with geographic variation in major pollinators.

Direct support for the hypothesis that color acts as an advertisement and that visitor choice depends on reward comes from observations of pollinator foraging within single communities. For example, Heinrich (1976) and Pleasants (1981) have both shown that bumblebees utilize flowers of different species in densities proportional to their nectar rewards and unrelated to their colors. This is a pattern expected by logical extension of the "ideal free distribution" model of optimal habitat choice theory (Fretwell and Lucas, 1969). McDade (1983) similarly found that nectar rewards alone explained choice by tropical hummingbirds among differently colored flowers of a single species. Mulligan and Kevan (1973) found that insect visitation frequency was not consistently related to any particular combination of UV, blue, or yellow reflectance of flowers of Canadian weeds, so long as flowers reflected strongly. Marden and Waddington (1981) presented honeybees with pairwise choices of equally rewarding blue and yellow

TABLE II

CHARACTERISTICS OF *IPOMOPSIS AGGREGATA* FLOWERS FROM VARIOUS PARTS OF THE WESTERN UNITED STATES

Location	Color[a]	Corolla length[b]	Nectar concentration[c]
Pinaleño Mtns., Graham Co., Arizona, 2700 m	Red	37.8 ± 0.53 (39)	15.6 ± 0.78 (11)
Wheeler Peak, White Pine Co., Nevada, 2700 m	Red	31.2 ± 1.10 (10)	—
La Sal Mtns., San Juan Co., Utah, 2400 m	Red	27.8 ± 0.36 (10)	—
Elk Mtns., Gunnison Co., Colorado, 2900 m	Red	26.6 ± 0.29 (76)	24.7 ± 0.37 (202)
Pinaleño Mtns., Graham Co., Arizona, 2300 m	Red	26.2 ± 0.70 (17)	18.9 ± 0.38 (5)
Aquarius Plateau, Garfield Co., Utah, 2000 m	Red	22.3 ± 0.41 (15)	28.3 ± 2.32 (6)
Wasatch Mtns., Salt Lake Co., Utah, 1500 m	Red	21.5 ± 0.33 (8)	—
Teton Mtns., Teton Co., Wyoming, 2100 m	Red	20.3 ± 0.20 (90)	—
Telescope Peak, Inyo Co., California, 3000 m	Red	17.4 ± 0.16 (10)	—
Vermilion Cliffs, Kane Co., Utah, 1800 m	Red	17.0 ± 0.23 (20)	28.1 ± 1.51 (9)
Mt. Trumbull, Coconino Co., Arizona, 1700 m	Red	16.8 ± 0.47 (10)	—
Jacob Lake, Coconino Co., Arizona, 2100 m	Red	15.3 ± 0.34 (10)	—
Abajo Mtns., San Juan Co., Utah, 2700 m	Purple	29.9 ± 0.39 (20)	24.0
White Mtns., Apache Co., Arizona, 2600 m	Pale purple	34.3 ± 0.25 (10)	—
San Juan Mtns., San Miguel Co., Colorado, 2700 m	Red orange	26.8 ± 0.27 (80)	—
White Mtns., Navajo Co., Arizona, 1900 m	Orange	23.9 ± 0.55 (7)	—
La Sal Mtns., Montrose Co., Colorado, 2300 m	Pale pink	36.8 ± 0.47 (30)	26.0 ± 2.00 (2)

[a]Basic flower color was judged as closely as possible from field notes or photographs. There was also substantial variation in some details of coloration, for example in spotting at the corolla mouth; this was ignored.

[b]Corolla lengths are in mm and are expressed as mean ± 1 SE (n).

[c]Nectar concentrations, where available, are in percent of sucrose-equivalent sugars weight/weight and are expressed in the same way as corolla lengths.

TABLE III[a]

HANDLING TIMES OF REPRESENTATIVE BUMBLEBEE AND HUMMINGBIRD POLLINATORS WHILE FLYING BETWEEN SUCCESSIVE FLOWERS ON A SINGLE PLANT (BETWEEN FLOWERS) AND WHILE EXTRACTING NECTAR FROM SINGLE FLOWERS (WITHIN FLOWERS) DURING VISITS TO BLUE- AND ALBINO-FLOWERED *DELPHINIUM NELSONII*[b]

Pollinator and flower type	Handling time	
	Between flowers	Within flowers
Basic experiments		
Bombus appositus		
Blue flowers	1.86 ± 0.14 (54)	3.36 ± 0.19 (106)
Albino flowers	2.05 ± 0.17 (36)	3.97 ± 0.38 (62)
Selasphorus platycerus		
Blue flowers	0.38 ± 0.02 (59)	.65 ± 0.05 (96)
Albino flowers	0.41 ± 0.02 (57)	.93 ± 0.10 (82)
Flower painting experiments		
Bombus appositus (before painting)		
Blue flowers	2.35 ± 0.19 (19)	3.50 ± 0.37 (30)
Albino flowers	2.62 ± 0.22 (22)	3.79 ± 0.40 (40)
Bombus appositus (after painting)		
Blue flowers	1.84 ± 0.11 (15)	3.20 ± 0.48 (34)
Albino flowers	1.75 ± 0.12 (27)	2.96 ± 0.37 (49)
Selasphorus platycercus (before painting)		
Blue flowers	0.52 ± 0.08 (16)	0.94 ± 0.16 (46)
Albino flowers	0.61 ± 0.07 (25)	1.82 ± 0.23 (35)
Selasphorus platycercus (after painting)		
Blue flowers	0.36 ± 0.05 (14)	0.62 ± 0.05 (56)
Albino flowers	0.33 ± 0.04 (30)	0.59 ± 0.05 (69)

[a]Adapted by permission from *Nature*, Vol. 302, pp. 422–424. Copyright © 1983, Macmillan Journals Limited.

[b]Values are from representative flight cage experiments, as discussed further in Fig. 3; they are in seconds and are expressed as are corolla lengths in Table II. Bumblebees used were queens, and hummingbirds were males. Notice that the handling times for albinos exceed those for blue flowers and plants in all cases except the "after painting" treatments of flower-painting experiments.

artificial flowers. They found no overall preferences unless flowers were placed at different distances from bees, in which case bees always chose the closest flower, regardless of its color. Real (1981) found that both bumblebees and wasps preferred artificial flowers of a color that contrasted well to the background over those that contrasted poorly when all flowers held equal rewards. Finally, Waser and Price (1981) found that although hummingbirds and bumblebees will visit both the normal blue flowers and rare "albino" flowers of *Delphinium nelsonii* in nature, albinos are undervisited slightly. Recent experiments (Waser and Price, 1983b; see Table III) show that handling times are longer for pollinators visiting albino flowers and plants. Longer handling times and associated greater energetic costs of dealing with albinos could explain discrimination against them, if we assume that birds and bees are behaving as optimal foragers (Pyke *et al.,* 1977). Under this assumption, pollinators might be expected to undervisit albinos, even though their nectar production rate and concentration equal those of blue flowers, since undervisitation would allow accumulation of sufficient extra reward to repay extra handling costs. We are now exploring this hypothesis further, but the evidence seems excellent that choice is based on reward alone. In this case, the economics themselves seem, in turn, to be related to color—not the overall color of a flower but instead the details of its color patterns that may act as nectar guides and thus influence harvest efficiency (see Section III,A).

D. Experiments with Advertisement Intensity

With some elegant manipulations of glass and color paper cylinders in a natural plant population, Kugler (1943) showed that approach behavior in bumblebees is elicited by the sight of an inflorescence or object of similar shape, and that it depends on color rather than on exact shape and odor. Having deduced that one function of flower color is to serve as a long-distance advertisement, Kugler next explored the effects of advertisement intensity. He constructed artificial flowers that differed only in size, and placed them in a room containing bees. The mean and maximum distances at which bees initiated direct flights toward artificial flowers increased directly with target size, and the relationships suggested that an approach reaction depends on the angle subtended by a flower in the visual field of a potential visitor. Kugler also noted that the total number of approaches to an artificial flower increased with its size. This could come about because a randomly flying bee would be more likely to enter a ''sphere of recognition'' around a large flower than around a small one. It could also be explained if bees were cognizant of all flowers, could judge their absolute size at a distance, and could actively choose the largest. The latter effect is necessary for the evolution of some floral traits that I will discuss shortly. The general conclusion that visitation rate depends more on target size than on exact color is supported by the observations and experiments of Mulligan and Kevan (1973), Willson and Price (1977), Thomson (1981b), Geber (1982), and Augspurger (1980).

Given that the size of a colored target can influence pollinator approach rate and that approach rate may be important to plant fitness, one might expect selection to favor the largest possible flowers and inflorescences. The situation is not so simple, however, since pollinator behavior during and after a visit to a plant will also influence fitness. One illustration of this may be provided by an experiment in which I used surveyor's flagging to increase artificially the target

TABLE IV

SEED SETS OF OCOTILLO FLOWERS WITH AND WITHOUT ENHANCEMENT OF THE TARGET SIZE OF INFLORESCENCES CONTAINING THEM

Individual	Treatment[a]			
	Unmanipulated		Flagged	
Plant 1 (1974)	11.6	(10)[b]	10.7	(10)
Plant 2 (1974)	9.1	(10)	10.4	(10)
Plant 3 (1975)	5.8 ± 0.38	(64)	3.0 ± 0.27	(57)
Plant 4 (1975)	4.7 ± 0.45	(22)	6.6 ± 0.45	(59)

[a]The basic method consisted of choosing two stalks of a single plant, each of which carried a terminal inflorescence of equivalent size. Red-orange surveyor's flagging was wound around one of each pair of stalks at the base of the inflorescence, to approximately double the linear extent of the colored target. Flagging matched flower color closely for a human observer.

[b]Seed sets are expressed as are corolla lengths in Table II. Standard errors were not obtained in 1974 for reasons explained by Waser (1979).

sizes of ocotillo inflorescences (see Section II,A: Waser, 1979). Because of the relative rarity of visits by hummingbird and solitary bee pollinators, I did not monitor their behavioral response to my manipulation, but instead measured seed set. No systematic effect on seed set was produced by augmenting the target size of inflorescences (Table IV). It may be that my crude manipulation did not effectively augment the already overpowering advertisement of ocotillo inflorescences. On the other hand, the experiment changed the advertisement independent of reward, and thus altered the ratio of these two variables. This could have had some effect on the behavior of pollinators once they reached an inflorescence, and ultimately on seed set.

This example illustrates again the need to consider effects of any plant trait on all components of pollinator visitation before attempting to assess how the trait influences fitness. Gori (1983) makes a related point in his review of postpollination and age-related color changes of flowers. Whereas flowers of most plants simply dehisce their petals as they age, a number of species in scattered families retain old flowers or those already pollinated. Such flowers often undergo a change in color pattern, but not in basic color. They therefore contribute to the overall advertisement display of the plant. Gori (1983) and C. E. Jones (personal communication), using *Lupinus argenteus* and *Lotus scoparius,* respectively, found that retention of color-changed flowers increased the pollinator approach rate per plant relative to that at plants from which such flowers were removed. Gori (1983) makes the critical point, however, that increased pollinator approach to plants will not by itself explain a change in flower color pattern. Rather, such a change must function to alter in beneficial ways the behavior of pollinators once they actually reach the plant (see Section III,A).

Increased pollinator approach to large floral displays, and the consequent selective advantage of retaining spent flowers, could result from active choice or from an essentially passive sphere of recognition effect as discussed above. In the case of *L. argenteus,* the effect appears to be passive: Approach rate is directly proportional to the total number of flowers, regardless of their color pattern, in manipulated or unmanipulated inflorescences (Gori, personal communication). Active choice, on the other hand, is vital to the arguments of Schaffer and Schaffer (1977, 1979) for the evolution of semelparous reproduction in plants. They show that semelparity will be selected for if the trade-off between immediate and future reproduction is concave. One factor that could lead to concavity is strong pollinator preference for plants with large inflorescences. Schaffer and Schaffer discuss evolutionary and ecological reasons why pollinators might show preference of this sort, and provide some evidence for preference on the part of insects visiting semelparous members of the agave family (but see conflicting evidence in Udovic 1981; Aker, 1982). Preference is especially strong when levels of floral reward are increased experimentally, and Schaffer and Schaffer (1979) use this to explain the high incidence of semelparity in nectar-rich agaves relative to nectar-poor yuccas. Unfortunately, their measurements of visitation rate were mostly indirect, and they did not assess actual reproductive effects of visitation.

E. Flower Color, Shape, and the Constancy of Pollinators

What has been said so far suggests that flower color serves to signal to a variety of animals the existence of a certain floral reward. As measured by the biologist, this reward seems to have some mean value and variance, but the effective reward may well differ for different types of animals. There are several reasons for this. One, which I will not discuss at length, is that the match between the morphologies of flower and visitor determines whether there is access to the entire reward, or part of it, or none at all (Grant, 1949, 1952; Hobbs *et al.*, 1961; Sprague, 1962; Stebbins, 1970; Hainsworth and Wolf, 1976; Schemske, 1976; Whitham, 1977; Inouye, 1980). This relative accessibility will determine the effective mean and variance in the reward for each potential visitor, and these parameters, in turn, will jointly determine whether the flower is a profitable one for the animal to include in its foraging itinerary (Real, 1981 and Chapter 11, this volume; Waddington *et al.*, 1981). Thus, accessibility will strongly influence the spectrum of flower visitors. One driving force in the evolution of floral morphology and reward production, therefore, has probably been some sort of selective advantage to being more or less specific in pollinator affinities. If some subset of all available visitors, taken together, provides a given plant with an optimal quantity and quality of pollen transfer, then selection should favor, where possible, the evolution of an effective reward that attracts exactly these visitors. Recent studies suggest that it will be useful to investigate visitor affinities carefully with this hypothesis in mind (Primack and Silander, 1975; Motten *et al.*, 1981; Waser and Price, 1981, 1983a; Waser, 1982; Bertin, 1982a; see also Feinsinger, 1983).

A second determinant of effective reward has to do with the possibility that some flower visitors that would be expected on morphological grounds to have access to a given flower are excluded because of behavioral constraints. Such constraints are probably needed to explain the behavior known as "flower constancy." "Constancy" refers to the tendency of an animal to restrict its visits to flowers of a single species, even when these occur intermixed with other equally accessible and rewarding flowers. Reports on the degree of constancy of a given type of animal are often conflicting (Waddington, 1983). There are at least two reasons for this. First, there is a conceptual, and even a greater practical, potential for confusion between constancy and preference based on differences in effective reward (Bateman, 1951). Animals may well choose flowers on the basis of rewards, but it does not seem reasonable to invoke some special term such as "constancy" to describe this behavior. Second, some methods for measuring constancy are indirect and can give spurious results. With bees, for example, constancy is sometimes judged by the amount of mixing of loads carried in pollen baskets (Clements and Long, 1923; Brittain and Newton, 1933). This will be inaccurate if bees do not collect pollen from all flowers visited, if different flowers produce different amounts of pollen, or if the actual mixture of available flowers is unknown (Grant, 1950; Thomson, 1981a; Waddington, 1983). A more accurate method is actually to observe floral choices in nature. This approach is now becoming relatively common again, after a lapse of almost a century (Bennett, 1883; Christy, 1883; Bateman, 1951; McNaughton and Harper, 1960; Heinrich, 1976; Jones, 1978). Its drawback is that it is often difficult, given complex spatial mixing of plants in nature, to generate an expectation of the sequence of flower visits that would be observed under the null expectation of no constancy. A final general method for evaluating constancy that circumvents this last drawback is to present an experimentally controlled mixture of flowers. This approach was pioneered by Clements and Long (1923), and Thomson (1981a) recently used a modification of their "floral bouquet" experiment to measure constancy in bumblebees. An alternative experimental method is to expose pollinators to arrays either of artificial or real flowers (Bateman, 1951; Manning, 1957; Faulkner, 1976; Heinrich *et al.*, 1977; Waddington, 1983). Two examples of this from my own work are described later. Despite these variations in method, the degree of constancy does seem to differ among flower visitors, with honeybees appearing to be more constant than bumblebees, which in turn seem to be more constant than solitary bees (Brittain and Newton, 1933; Grant, 1950; Manning, 1957; Free, 1963, 1966, 1970;

Heinrich, 1976; Jones, 1978; Waddington, 1983). The lepidopterans and hummingbirds appear to exhibit even less constancy (Bennett, 1883; Bateman, 1951; Kislev *et al.*, 1972; Waddington, 1983; and references therein; Waser, 1983; and references therein).

Constancy can be very important to plant fitness. If the floral traits of a plant serve to attract an appropriate type of pollinator (social bees, for example), and somehow to induce constancy in these animals, we may hypothesize that the traits have evolved in part because of the benefits of constancy. These benefits have to do with where pollinators come from as they approach the plant, and where they go after they leave. If pollinators are completely constant, the plants they visit will be spared any detrimental effects of interspecific pollen transfer (Waser, 1978b; Feinsinger, 1978). These include effects of receiving foreign pollen that can actively disrupt or passively block stigmatic surfaces or lead to formation of seeds of low viability (Char, 1977; Waser, 1977, 1978a; Ockendon and Currah, 1977; Sukhada and Jayachandra, 1980; Campbell and Motten, 1981; Thomson *et al.*, 1981; McCrea, 1981), as well as effects of losing pollen picked up by the pollinator and carried to inappropriate destinations. The evidence is growing that these reproductive effects can be substantial even with a relatively small amount of interspecific pollen transfer.

Other than perhaps producing an effective reward that attracts appropriate visitors, how can floral traits evolve to induce constancy? This reintroduces the question of why some pollinators should be constant at all. Darwin (1876, p. 419) was the first to propose the explanation that I think is most likely:

> That insects should visit the flowers of the same species for as long as they can, is of great importance to the plant, as it favours the cross-fertilisation of distinct individuals of the same species; but no one will suppose that insects act in this manner for the good of the plant. The cause probably lies in insects being thus enabled to work quicker; they have just learned how to stand in the best position on the flower, and how far and in what direction to insert their proboscides.

Notice that this hypothesis invokes behavioral constraints. Darwin is suggesting that constancy is due to an inability of flower visitors to remember simultaneously how to handle several different flowers of complex morphologies. The proposal is that such constrained visitors have the highest harvest rate if they keep to one type of flower. Darwin's hypotheses has been widely accepted (Grant, 1950; Proctor and Yeo, 1972), but surprisingly enough appears never to have been tested experimentally.

Cast in a more detailed and modern form, Darwin's idea can be thought of as follows. An animal not subjected to memory constraints, and thus able to act as an optimal forager in the sense of simple models (MacArthur and Pianka, 1966; Pyke *et al.*, 1977), should not exhibit constancy. This is because constancy, except in certain restricted types of plant mixtures, requires a forager to pass over perfectly rewarding flowers. An optimal forager is expected to include all flowers with rewards above a certain threshold (and certainly all equally rewarding flowers) in its foraging itinerary, since such an itinerary minimizes travel costs and thus maximizes net energy gain (Waser, 1983). If an animal is unable to remember more than one or a few sets of floral cues at a time, however, the savings in travel costs of a mixed itinerary may be more than offset by increased handling costs stemming from the need to relearn repeatedly how to deal with different flowers (Manning, 1956; Laverty, 1980).

An alternative explanation of constancy that should be mentioned is that it is in fact a very efficient foraging tactic, rather than a less than maximally efficient one imposed by behavioral constraints. One version of this explanation proposes that constancy accompanies increased intelligence or specialization of flower visitors (Faegri and van der Pijl, 1971). This version seems consistent with the trends noted above regarding the degree of constancy of different types of bees, but is much less consistent with the overall pattern if other intelligent and specialized pollinators such as lepidopterans and hummingbirds are included. In any case, this version provides no useful evidence on why or whether constancy is efficient. Another, more sophisticated version of the explanation of constancy as efficient behavior is given in a theoretical paper by Oster and Heinrich (1976). The weakness of their approach and similar ones (Levin, 1978), I believe, is that they fail to consider any potential savings in travel costs for a pollinator that includes a mixture of rewarding flowers in its itinerary rather than skipping all but those of a single species.

In light of these comments, I believe the most likely available hypothesis to be that constancy is a result of memory constraints. Darwin proposed that difficulties in remembering how to deal with flowers would have to do with their morphological complexity. The evidence at hand indeed suggests that complexity causes learning problems for pollinators such as bees (Laverty, 1980; Heinrich, 1976) and that constancy is often associated with different, and complex, morphologies (Heinrich, 1976). Also consistent with this idea are observations that bees sometimes forage with little or no constancy in mixtures of flowers that differ only in color (Darwin, 1876; Bennett, 1883; Christy, 1883; Plateau, 1899; Manning, 1957; Lewis, 1961; Macior, 1971). However, other studies suggest that color differences alone, or such differences associated with only slight differences in morphology, are sufficient to induce constancy (Grant, 1949; Bateman, 1956; Manning, 1956; McNaughton and Harper, 1960; Levin, 1968; Levin and Schaal, 1970; Faulkner, 1976; Heinrich *et al.*, 1977; Jones, 1978; Thomson, 1981a).

In the examples from my own work, constancy appeared to be enhanced by differences in both morphology and color (Table V). An experiment conducted in 1981 in the mountains of Costa Rica showed that bumblebees foraged with slightly greater constancy in a mixture of flowers differing only in morphology

TABLE V

Frequencies of Different Types of Flower-to-Flower Movements of Bees Foraging in Artificial Arrays of Flowers

(Same color, different morphology)		Movements to:		
		D	*F*	Constancy
Movements	*D*	139	47	index
from:	*F*	42	31	= 0.19
(Different color, same morphology)		Movements to:		
		D(P)	*D(W)*	Constancy
Movements	*D(P)*	141	104	index
from:	*D(W)*	105	125	= 0.12
(Same color, same morphology)		Movements to:		
		G	*H*	Constancy
Movements	*G*	25	18	index
from:	*H*	19	36	= 0.24
(Different color, same morphology)		Movements to:		
		G	*C*	Constancy
Movements	*G*	95	17	index
from:	*C*	16	73	= 0.67

Top: Movements of bumblebees (*Bombus ephippiatus*) in two arrays, one containing equal numbers of purple flowers of *Digitalis purpurea* (D) and a *Fuchsia* sp. (F), the other equal numbers of purple and white flowers of *D. purpurea*, D(P) and D(W). Values for Bateman's (1951) constancy index, which factors out different preferences for flower types and ranges from 0 to 1, are shown. Bottom; Movements of solitary bees (*Diadasia* sp.) in two arrays, one containing equal numbers of yellow flowers of *Gaillardia arizonica* (G) and *Haplopappus spinulosis* (H), the other equal numbers of yellow flowers of the first species and white flowers of *Chaenactis stevioides* (C).

than in one of flowers differing only in color. Mixtures in this case consisted of 6 × 4 rectangular arrays of vases in which flowers of two types were alternated; vases were separated by about 15 cm and arrays by about 3 m. Another such experiment conducted in 1979 in the southern Arizona desert showed solitary bees to be more constant in a mixture of morphologically similar composite flowers differing in color than in one of flowers of a single color. Here I constructed square arrays 1 m on a side and separated from one another by 1 m; each array contained 100 regularly spaced vases in which flowers of two types were alternated. Taken together, these results and those of the other authors discussed above suggest that memory constraints of some pollinators extend to floral advertisement cues as well as to morphological features that actually dictate effective rewards. In some cases, then, the evolution of all these floral traits may have been influenced by beneficial effects of constancy on plant reproductive success (see also Heinrich, 1975a).

F. Comments on Flower Nectar and Pollinator Attraction

In discussing how floral traits may be adaptations that influence which animals approach a plant and where they come from, I have stressed the apparent roles of floral reward. Given what I have said and the available reviews on pollinator foraging, I will limit myself here to a few additional comments. These have to do with the possibilities for experimental studies of one particular reward, flower nectar.

Flowers of animal-pollinated plant species vary greatly in the amounts and concentrations of nectar they produce. If one ignores the general paucity of experiments showing what pollinates what, and accepts an estimate of major pollinators based on pollination syndromes, the following general patterns emerge: Sugar concentrations of nectars generally decrease, and nectar production rates increase, with pollinator body mass (Pyke, 1981b; Pyke and Waser, 1981). The adaptive significance of these intriguing patterns has received almost no experimental attention.

Taking a pollination system that by now is familiar, several authors have recently suggested reasons that the relatively dilute nectars of flowers pollinated by hummingbirds may be adaptive (Baker, 1975; Calder, 1979; Bolten and Feinsinger, 1978). Potentially serious problems with these explanations were discussed by Pyke and Waser (1981), who pointed out that the explanations might be taken to imply that nectar concentrations have evolved for the benefit of pollinators. In addition, Pyke and I summarized experimental evidence of our own and others showing that concentrated nectars are actually the ones optimal for hummingbirds, and that birds prefer these nectars given a choice. Bees also seem to prefer concentrated nectars (Waller, 1972; Bachman and Waller, 1977), although they will forage at hummingbird flowers with dilute nectars as well (Pleasants, 1983; Waser, 1982; see Section II,B). In the absence of experimental studies, Pyke and I were unable to generate an alternative adaptive explanation for dilute hummingbird nectars, much less one for general trends in nectar concentration according to pollinator type. We were able to outline, however, the sort of (admittedly difficult) experimental protocol that will be needed for a really rigorous exploration of this floral trait (see also Pyke, 1981b).

Compared to the situation with nectar concentration, experimental studies of nectar production rate are relatively advanced. For example, both Pyke

(1981b,c) and Pleasants (1983, personal communication) have examined nectar production in *Ipomopsis aggregata* (see Section II,C). Both studies employed elements of an ESS approach. Whereas Pyke attempted to make quantitative predictions of the optimal nectar production rate based on considerations of both male and female components of plant fitness, Pleasants dealt with a qualitative prediction concerning variation in optimal nectar production rate as a function of inflorescence and plant size. Both studies indicate some strengths of experimental studies of adaptive value, and both illustrate some difficulties.

A clear antecedent to the studies just discussed comes from the well-known observations of Heinrich (1975b) and Heinrich and Raven (1972) on relationships between pollinator energetics and floral rewards. Heinrich and Raven advanced a proposal, now widely accepted, that floral rewards are constrained by the need to be sufficiently rich to sustain the metabolism of the desired pollinator, yet not rich enough to satiate it and keep it from moving to other conspecifics. The problem I see with this hypothesis, other than lack of experimental verification, is that it seems to set only minimal conditions rather than to explain interspecific and geographical details of variation in rewards such as nectar. This must be in part because the hypothesis does not consider, as an ESS approach would, how the behavioral response of pollinators toward a given plant will depend on rewards of other plants in the surrounding community. How can Heinrich and Raven's hypothesis alone explain that tropical flowers visited by "high-energy" hummingbirds often produce 100 μl of nectar a day, whereas temperate counterparts visited by hummingbirds of equivalent size produce one or two orders of magnitude less (Feinsinger, 1978; Stiles, 1976; Waser, 1978a)?

III. How Do Flower Visitors Behave Once at the Plant?

A. Floral Color Pattern and Pollinator Behavior

Despite the emphasis on color in Section II, one more related topic bears discussion, and that is how pollinators respond to exact patterns of flower color once they reach a plant. Sprengel (1793) proposed that certain kinds of color patterns act as *Saftmale,* or nectar guides, that facilitate discovery of rewards by visitors. It is easier to construct an explicit ESS scenerio for the evolution of this floral trait than it is for some others. Suppose, for example, that the presence of a nectar guide decreases the caloric cost and time required for a pollinator to handle a flower; that pollinators choose flowers based on their expected value in terms of net rate of caloric yield; and that pollinators can learn to associate guides with expected reward and to discriminate against flowers lacking guides, at least once they have approached flowers closely. These conditions seem realistic (Pyke, 1981b; Waser and Price, 1981; Pleasants, 1981), and they should often favor the spread of a rare mutant whose flowers have guides but are otherwise unchanged. If such flowers are distinguishable from a distance, for example, they may increase the rate at which pollinators approach the plant, and this may increase plant reproductive success. Even if guides are visible only after pollinators have reached the plant, they could influence behavior in several ways. When mutants with guides are rare, pollinators that arrive at them may visit more flowers before leaving than they do on plants without guides. This prediction follows from consideration of simple foraging models that predict optimal tenure on a plant (or in any other ''patch'' of food; Charnov, 1976; Hodges, 1981; Best and Bierzychudek, 1982). In cases in which this extra pollinator tenure increases plant fitness, mutants should spread until at some point their expected rewards dominate the ''gain curve'' for the plant mixture in general and thus determine by themselves the maximum possible net rate of caloric intake for a pollinator. At this point, plants without guides may even be dropped from the optimal foraging itinerary. Figure 1 summarizes these ideas graphically, following the notation of Charnov (1976).

There is a strong experimental tradition in nectar guide studies. After reviewing earlier inconclusive studies, Kugler (1943) reported on his extensive observations of bumblebees visiting artificial feeders. Feeders consisted of glass tubes filled with sugar water and placed in holes drilled into a piece of cardboard. Kugler distinguished several different types of color pattern that might act as nectar guides, and used variously colored papers to construct guides of each kind that were then placed near feeders. The effect on bees was striking: Feeders with guides elicited extension of the proboscis and feeding far more commonly than those without guides. Daumer (1958) continued such experiments, after conducting the first systematic survey of UV relectance of flowers and finding many patterns that looked like nectar guides. Using feeders similar to Kugler's and surrounding them with ligules of ray florets from different composite species, he showed that honeybees treated the UV-absorbing portions of the ligules as guides. When ligules were arranged as in a normal composite head, with UV-absorbing portions toward the center, bees alighted and walked to the center of the artificial flower. When ligules were turned around, however, bees walked along them to the edge of the artificial flower, extended their proboscises, and obtained no reward. Several elaborations of these experiments yielded similar results.

Experimental manipulations of color pattern have also been attempted under more natural conditions. When Scora (1964) presented bees and wasps with a mixture of normal *Mondarda punctata* flowers, which have spotted lower corolla lips, and colchicine-induced mutants, which lack spotting, mutants were approached but then always ignored. When mutants were presented by themselves in a relatively large stand, bees and wasps landed on them but then failed to act in a normal and purposeful way and soon left. Jones and Buchmann (1974) carried

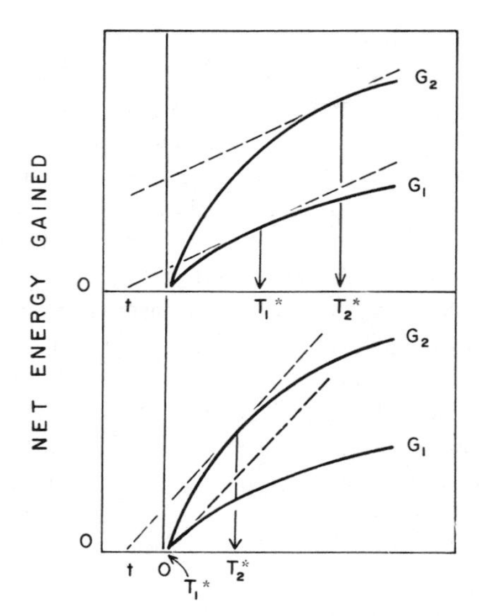

Fig. 1. Hypothetical effects of a mutation producing flower nectar guides on the time spent by pollinators visiting a plant. Gain curves relating net energy gained to time spent at a plant are G_1 and G_2 for normal plants and mutants, respectively, and there is a fixed average travel time between plants, t.
Top: Mutants are rare, so that effective rewards of normal plants determine the maximum net rate of energy intake in the plant mixture (given by the slope of straight lines tangent to the gain curves); this, in turn, determines optimal tenure times T_1* and T_2* for pollinators visiting normal and mutant plants, respectively.
Bottom: Mutants are common, and their effective rewards determine the maximum net rate of energy intake and optimal tenure times. In this example, gain curves are drawn sufficiently divergent so that normal plants are dropped from the foraging itinerary ($T_1* = 0$).

out similar experiments with two legumes in Costa Rica. Flowers of each species have banner petals that contrast to the rest of the flower by absorbing UV light. Jones and Buchmann used simple techniques to alter the orientation of the banner petal and its contrast with other petals. If contrast was eliminated, social and solitary bees usually approached flowers but then flew away without landing. If orientation was changed, bees tended to adjust their own orientation before or after landing, so that their heads always faced the banner petal.

All these findings indicate that color patterns can strongly influence behavior once pollinators reach flowers. They do not tell us what effect this has on plant fitness, however. In a series of experiments alluded to earlier (Section II,C) with hummingbird and bumblebee pollination of *Delphinium nelsonii,* Waser and Price (1981) determined that rare plants with albino flowers are undervisited. The ultimate reason appears to be that pollinators take longer to handle and extract nectar from albino flowers because these lack the strongly contrasting nectar guides found on normal blue flowers. Albinism is a cue that pollinators should be able to associate with lack of guides and thus use in adjusting their rate of approach to albino plants (Waser and Price, 1981, 1983b). Support for this hypothesis comes from experiments in which we painted albino flowers to match the normal color pattern, which includes a contrasting nectar guide, and presented these along with appropriately painted controls to hummingbirds and bumblebees in a flight cage. Once a nectar guide was provided, pollinators no longer discriminated against albinos (Fig. 2) or took longer to handle them than blue flowers (Table V). This suggests that nectar guides in *D. nelsonii* have a selective advantage because they increase the efficiency with which pollinators can extract nectar and thereby increase the rate of approach to plants, at least when a long-distance cue such as overall flower color is associated with them.

In the example with *D. nelsonii,* nectar guides serve, through their association with a color difference that acts as an advertisement cue, to increase the number of times pollinators approach and visit a plant. A different potential benefit of color patterns is that they may increase pollinator tenure on a plant. This sort of effect is discussed by Gori (1983; see Section II,D), who proposes that color change may provide a short-distance cue to flowers that contain no rewards. Changed flowers would still contribute to overall plant advertisement, as I have discussed. Once at the plant, however, a pollinator would realize a greater foraging efficiency if it could avoid old or already pollinated flowers. Indeed, Gori gives experimental evidence for *Lupinus argenteus* that bumblebees visit significantly more flowering stalks before leaving a plant if color pattern is accurately associated with absence of reward than if it is not. Although these differences in pollinator tenure seemed to have no detectable effect on seed set, it is possible, as Gori says, that they influenced male reproductive success in terms of amounts of pollen removed and disseminated. Several other possible selective advantages to plants of change in color or color pattern are proposed by Gori (1983); these do not seem to be important for *L. argenteus* but may be in other systems.

B. Some Effects of Inflorescence Size and Reward Distribution

Once a pollinator reaches a plant, it visits a certain number of flowers in some order before leaving. This has important effects on the amounts of pollen deposited (which, along with pollen quality, influences female reproductive success; see Section II), as well as on the amounts picked up (which, along with the quality of future transport, influences male reproductive success; see Section IV). The determinants of within-plant behavior of pollinators are extremely complex at one level; they include the total number of flowers on the plant, their spatial arrangement, and the rewards they contain, in addition to the corresponding features and overall density of other nearby plants. Once again, a way to simplify the situation that appears to have worked fairly well is to consider that behavior is dictated by reward economics and to attempt to understand details of

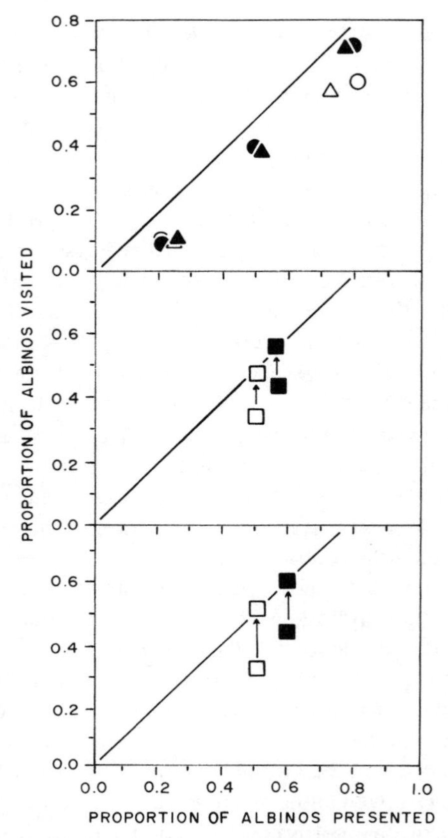

Fig. 2. Graphic representations of pollinator discrimination in flight cage experiments. In each experiment, a single hummingbird or several bumblebee queens were presented with either a 2:8, 8:2, or 5:5 mixture of blue- and albino-flowered potted plants that were matched for inflorescence height and flower number. Proportions of blue and albino flowers, however, deviated slightly from proportions of plants in each replicate. The straight line in all graphs is the line of no preference. *Top*: Open circles show mean proportions of plants visited by bees relative to proportions presented, whereas open triangles show the same for bees and flowers (from four experiments yielding 365 visits to plants and 823 visits to flowers). Closed circles and triangles show results for birds and plants or flowers, respectively (from five experiments yielding 779 visits to plants and 1376 visits to flowers). *Middle*: Open squares connected by an arrow show proportions of plants visited by bees before and after all flowers were painted to match a normal color pattern, whereas closed squares show proportions of flowers visited (from a single replicate yielding 224 visits to plants and 506 visits to flowers). *Bottom*: Symbols are the same as those in the middle graph, but show the results for a single bird (from a single replicate yielding 165 visits to plants and 286 visits to flowers).

pollinator movement in reference to the predictions of optimal foraging theory. This approach has been treated in some recent reviews, especially from the point of view of pollinators (Levin, 1978; Pyke, 1981b; Waddington and Heinrich, 1981). Here I will discuss how a few particular plant traits seem to influence pollinator behavior and/or plant fitness, using foraging economics as an organizing principle.

One interesting example has to do with the arrangement of rewards within inflorescences. In relatively small plants, at least, regular spatial patterns of rewards may act to direct pollinator movements in ways beneficial to plant fitness. This seems to be the case for a series of herbaceous wildflowers studied by Pyke (1978b, 1979), Hodges (1981), and Best and Bierzychudek (1982). Flowers in all but one of the species considered by these authors (Pyke, 1978b) open sequentially from the bottom to the top of inflorescences, and nectar contents apparently increase directly with flower age. Bumblebee pollinators tend to approach bottom flowers first, to work their way upward by flying between successive flowers, and to leave inflorescences before visiting all flowers. Several components of this pattern are in qualitative accord with predictions of optimal foraging theory, including movement down a gradient of floral reward, rarity of skipping or revisiting flowers, and departure from inflorescences before all flowers are visited. Indeed, Pyke (1979), Hodges (1981), and Best and Bierzychudek (1982) show that there is good agreement with some quantitative predictions of optimal foraging theory as well, especially with predictions of optimal tenure on an inflorescence derived from the marginal value theorem of Charnov (1976).

The point has been raised that any tendency of insects to move upward on vertical inflorescences may be due in part to factors other than distribution of nectar in flowers. Waddington and Heinrich (1979) studied movements of bumblebees on an artificial inflorescence consisting of five "flowers," and found upward movement to be the rule whether the bottom two flowers were most rewarding, all flowers were equally rewarding, or the top two were most rewarding. They did find, however, that bees quickly learned to begin foraging at the first flowers, and cease at the last flowers, that contained any rewards. Corbet *et al.* (1981) used a natural analog of these laboratory experiments to examine movements of bees and wasps (see also Heinrich, 1979). Both types of flower visitor tended to move upward on inflorescences of *Scrophularia aquatica,* even though there was no consistent relationship between flower position and nectar contents. Nectar contents increased from bottom to top on inflorescences of *Linaria vulgaris,* apparently as a result of robbing by *Bombus terrestris*. Bumblebees of this species that faced downward while robbing showed a slight tendency to move down, whereas the equivalent number that faced upward while robbing moved mostly up. Legitimately foraging *B. hortorum* faced upward and tended strongly to move up inflorescences. Corbet and her co-workers concluded that movement direction is strongly influenced by the direction insects must face while feeding, which in turn is a function of flower orientation. This conclusion does not necessarily contradict those of Pyke and others. Instead, it seems that the distribution of floral rewards, together with an innate or learned response to a common flower orientation, may lead bees and other insects to work upward on inflorescences of many species.

In general terms at least, the reproductive benefits of such movement patterns seems obvious for the plants. Most species involved in the studies mentioned above have protandrous flowers, so that by moving up inflorescences pollinators first deposit the pollen they bring with them on receptive stigmas, and then pick up pollen which they carry to further plants. Thus, it appears that the distribution of floral rewards, along with flower orientation, may directly induce pollination in a way that enhances outcrossing, or at least that the distribution of sexual stages within inflorescences may capitalize on innate movement tendencies of pollinators to the same end. In none of the studies discussed so far, however, have the reproductive consequences of within-plant movement behavior been assessed directly. Thus, we do not know to what extent floral traits are inducing pollinator behavior that is optimal for plants. There is obviously room for much more work.

Plants with small, regularly arranged inflorescences are probably much rarer than those whose inflorescences exhibit no apparent order in terms of rewards or sexual states of flowers. Pyke (1981c) and Pleasants (1983, personal communication) have argued, however, that there is always a number of flowers to be visited that is optimal for a plant in terms of amounts of outcrossed pollen deposited and self pollen picked up. In their studies of *Ipomopsis aggregata,* both authors explored nectar production rates and other floral properties that might induce the optimal number of visits by hummingbirds (see Section II,F). The arguments used are sometimes complex, but at least these studies point the way for future experimental research on how plant traits can lead to pollinator movements within inflorescences that enhance plant fitness through both male and female sexual functions. It remains to be seen, however, whether this approach will prove to be of much use with much larger plants such as tropical trees (Frankie *et al.,* 1974).

Several other workers have taken a somewhat different approach to the general question of adaptive significance of inflorescence size. For example, Willson and Rathcke (1974) measured reproductive success in the self-incompatible milkweed *Asclepias syriaca,* in terms of both pollinia removal (related to male success) and fruit set (a measure of female success). Using experimental reduction of flower number and other methods to vary inflorescence size, they found that inflorescences containing about 30 flowers were most efficient in terms of achieving maximal fruit set per flower, and that those containing about 40 flowers achieved the highest pollinia removal per flower. From this they argued that selection would be expected to favor allocation of flowers of a plant among multiple inflorescences so that each contained about 30–40 flowers, since this would maximize the efficiency of each inflorescence and absolute reproductive success over the entire plant. Inflorescence size in nature actually averaged about 80 flowers, however. One explanation for this discrepancy might be that by removing flowers experimentally, Willson and Rathcke altered the competition within inflorescences for scarce resources needed to mature fruits. If this interpretation is correct, the maximum efficiency of fruit maturation with relatively small inflorescence size may be an artifact, and large inflorescences may actually be most efficient.

Willson and Price (1977) went on to measure components of reproductive success as a function of total flower output per plant in three species of milkweeds. They found that total fruit production increased with the number of flowers per inflorescence in some but not all cases, and that the pollinia removal rate always increased. As a result, they proposed that male reproductive success would provide a selective benefit to large plants and inflorescences even if female reproductive success did not (cf. Gori, 1983; Section III,A). The most common plants and inflorescences in nature actually were fairly small, from which Willson and Price argued that there must be trade-offs between immediate and future reproductive success that limit flower production in each season. Willson *et al.* (1979) and Schemske (1980) proposed similar constraints for two temperate herbs and a tropical orchid, respectively, after finding again that the largest plants or inflorescences produced the greatest numbers of seeds and had (in the case of the orchid) the greatest pollinia removal, whereas most plants actually were small. In an independent study of one of the milkweeds used by Willson and her co-workers, Wyatt (1980) stressed that it can be misleading to express reproductive success in terms of efficiency on a per-flower basis, since selection responds only to total reproductive output of a plant. This is correct up to a point, but patterns of efficiency probably are crucial in determining the evolution of the overall reproductive life history of a plant (Schaffer and Schaffer, 1979; Section II,D) and thus are important to understand if one wants to make predictions about trade-offs between present and future flower production along the lines proposed by Willson *et al.* (1979) and Schemske (1980). In any case, Wyatt failed again to explain why plants producing large numbers of flowers were rare, yet had the highest total reproductive success of any in his populations.

In addition to the possiblity of trade-offs among multiple reproductive episodes, one obvious problem with these attempts to explain inflorescence and plant size is that variation in these traits may reflect plant age or microsite rather than heritable genetic variation (but see Willson and Rathcke, 1974, on heritability of inflorescence size). Another problem is that only one of the studies discussed above (Willson *et al.*, 1979) included even a minimal attempt to observe the effects of floral display size on pollinator behavior, in terms either of attraction or of movements while at the plant. As I have argued throughout, an approach to the study of adaptation of floral traits that considers their detailed effects on pollinator behavior and pollen transfer mechanics is likely to be much more profitible than one that does not.

One study of total flower number per plant that explicitly considers pollinator behavior is that of Geber (1982). She found that bumblebees approached large plants of the partially self-compatible herb *Mertensia ciliata* more often than they did small plants, and that these pollinators visited only slightly more flowers per foraging bout on large than on small plants. These behavioral differences turned out to compensate exactly for flower number, so that seed set per flower was essentially constant across plant sizes. This result is not especially helpful in explaining the observed distribution of flower number in *M. ciliata* plants, for the same reasons discussed above in regard to other studies. However, Geber's study serves as a useful model in that it combines careful observation of pollinator behavior, experimentation, and measurement of at least one component of plant reproductive success.

C. Flower Shape and the Mechanics of Pollen Transfer

As I argued in Section I, an understanding of the link between floral trait expression and plant fitness will depend partly on knowing how traits affect the mechanics of deposition of pollen on, and later recovery from, a pollen pool on the body of the pollinator. Although the exploration of pollen transfer mechanics is still in its infancy, there are a few examples with which to illustrate the value of this approach.

The first example concerns the relative efficiencies of different visitors in transferring pollen to stigmas of a given species. Several recent studies have addressed this point, with the general finding that efficient transfer is often provided by an unexpected variety of visitors. In my experiments with ocotillo (Waser, 1979), for example, I found that major pollinators in southern Arizona included solitary bees along with hummingbirds, even though the red tubular flowers might seem primarily adapted to the birds. Hummingbird visits were actually relatively rare, but they contributed substantially to seed set. Several solitary bee species were more common visitors, and all contributed to pollen transfer and seed set. The most common visitor overall was the carpenter bee *Xylocopa californica,* which pollinates by scraping the underside of its body across the brush of exserted sexual parts of flowers in the process of robbing them of nectar. Bertin (1982a) carried out detailed studies of another species with red tubular flowers, *Campsis radicans*. In one experiment, he exposed flowers with virgin stigmas to single visits by the hummingbirds and bees that frequented the plants, and found that birds deposited pollen loads an order of magnitude larger than did bees. An opposite trend in the relative efficiencies of bees and hummingbirds occurs with the blue-flowered larkspur *Delphinium nelsonii*

(Waser and Price, 1981; see Sections II,C and III,A). Seed set in this species is linearly related to the number of visits by each pollinator type, but the slope of the relationship is much steeper for bees. Because hummingbirds visit more flowers per unit time, however, the contributions of each pollinator type to female fecundity appear to be equivalent. Overall equivalence of pollen donation was also reported by Motten *et al.* (1981) and by Tepedino (1981) for generalist and specialist visitors of *Claytonia virginica* and *Cucurbita pepo,* respectively, whereas differences were reported by Primack and Silander (1975), O'Brien (1980), and Parker (1981) in pollen loads and/or pollen donation of various visitors to primroses, pavement plain wildflowers, and sunflowers, respectively. Similarities and differences in each of these cases appear to be complex functions of pollinator morphology and behavior.

A second example concerns recent work on how floral traits can influence pollen carryover. The distribution of distances over which pollen is carried away from a flower may be very important to plant fitness, as Price and Waser (1979) and Waser and Price (1983a), among others, have argued. Waser (1982) and Waser and Price (1983b) have discussed the likelihood that individual selection acting on floral traits will have relatively little direct influence on pollinator flight distances, but it may be that indirect influence is possible through variation in amount of carryover. Thomson and Plowright (1980) have recently shown that the amount of pollen deposited on individual flowers of *Diervilla lonicera* increases as a function of their nectar reward. Thus, it seems possible that reward could influence pollen transfer mechanics in ways that affect carryover. In addition, carryover may be sensitive to the arrangement of sexual parts of flowers. Consider, for example, a flower such as that of ocotillo, with a brush of exserted anthers that deposits pollen over a wide area on a pollinator's body. If the stigma is relatively small and contacts any part of this pollen pool with equal likelihood, the pool can be considered to be divided into n parts each of which has the area of the stigma. In this case, it can be shown that carryover follows an exponential decay pattern, with the mean flower reached by pollen being the nth past the one from which it is picked up (Fig. 3). Thus, floral variability of this kind might enhance carryover. The result is misleading, however, in that it does not take into account pollen from the original flower being layered over or diluted by that from subsequently visited flowers. It is difficult, in fact, to decide how to incorporate such effects into an analytical model, and simulations are likely to be the most profitable means of exploring the problem further. Lertzman (1981) has made an excellent beginning by simulating some of the effects of variability in anther and stigma placement on carryover using model parameters that mimic a hummingbird-pollinated system. Under the assumptions of his model, which include perfect layering of pollen from successively visited flowers and bivariate normal distributions of placement of anthers and stigmas, carryover appeared to be longest with an intermediate degree of floral variability (Lertzman, 1981; Fig.

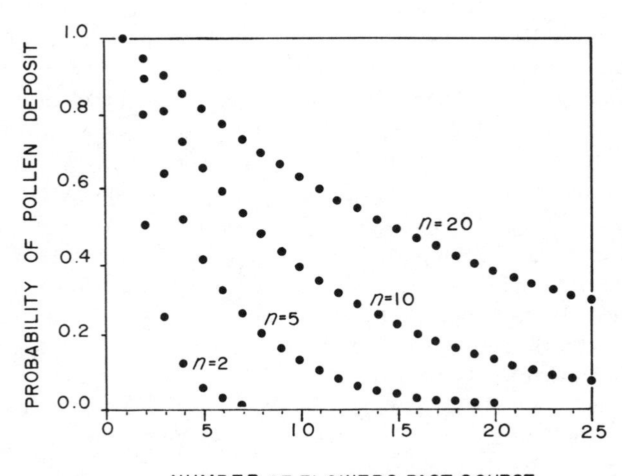

Fig. 3. Probability distributions for pollen carryover with different values of *n*, the ratio of pollen pool area to stigma area. Simplifying assumptions are explained in the text.

21 and p. 103 cited therein). The reasons for this are complex; they may also be very sensitive to model assumptions. Much remains to be done with such simulations, and they should serve as an extremely useful tool for focusing the design of experimental studies with real systems (cf. Price and Waser, 1982).

IV. What Do Flower Visitors Do after Leaving the Plant?

A. Pollinator Directionality and Flight Distances

In the previous two sections, I have attempted to discuss some effects floral traits have on pollinator behavior before and during a visit to a plant. However, topics have crept in here and there, such as pollen carryover, that concern a third component of visitation: the behavior of pollinators after they leave the plant. Here I wish to elaborate briefly on this final component of visitation, stressing the directionality and distances of interplant movements of pollinators.

In a seminal paper, Pyke (1978a) considered optimal movement patterns from a theoretical standpoint. In a uniform environment of infinite size, an optimal pollinator would be expected to show perfect directionality (see Levin *et al.*, 1971, for definition), that is, to move in a straight line. This is because such behavior eliminates revisitation of flowers that have just been emptied. Using computer simulations, Pyke calculated optimal degrees of directionality in uniform environments of finite sizes. It turns out, however, that directionalities actually recorded for pollinators and other animals do not overlap at all those predicted by simulations. Pyke (1978a,c) concluded that animals do not behave like the optimal "harvesters" of his models in terms of minimizing revisitation. The reasons are not obvious but may include behavioral constraints arising from the difficulty of remembering previous directions of movement. What pollinators actually do on leaving a plant can be best explained by assuming that they scan a sector of certain angular width in front of them and then move to the nearest or apparently nearest plant within the sector (Pyke, 1978c, 1981a).

Table VI summarizes much of the information on pollinator directionality that is available at present. Directionality appears to be high in many cases (i.e., successive movements tend to be in the same direction), but in other cases it is essentially zero (i.e., directions of successive movements are independent of one another). Zimmerman (1979) suggested that lack of directionality is most likely in systems where pollinators visit only a small proportion of the flowers on each inflorescence, so that the chances of encountering empty flowers upon revisiting an inflorescence are small; this indeed seems consistent with most of the results shown in Table VI. Relative detriments of revisitation might also be reduced under some circumstances if simultaneous foraging by other pollinators lowers the value of inflorescences near the one just visited by the pollinator in question. By themselves, however, these explanations are not sufficient; there must also be some direct benefit of reduced directionality. One very reasonable benefit becomes apparent if one considers that floral resources are seldom evenly distributed, as was assumed for simplicity in Pyke's simulations. By scanning a wide sector of its visual environment, a pollinator will increase the probability of actually finding the absolutely nearest or the apparently nearest inflorescence; in both cases, choice of this inflorescence will reduce travel costs (Pyke, 1978c). A different explanation for lack of directionality was proposed by Hodges and Miller (1981) for bees collecting pollen. It rests on the reasonable assumption that assessment of the value of individual flowers as pollen sources is difficult or impossible. It is hard to see how this alone will influence directionality, however, since selection could favor a behavior pattern that minimizes revisitation even if assessment of immediate return is impossible.

A clear benefit of momentary reduction in directionality occurs when rewards are patchy, since increased turning rate keeps a pollinator in a locally rewarding patch (Pyke, 1978c). Indeed, Heinrich (1979) found that bees exhibited less directionality when foraging in patches of clover that had been allowed to accumulate nectar than in those that were depleted. He also found that the bees flew shorter distances between inflorescences in the former case, as expected if they were scanning a wider visual sector. Waddington (1981) reported similar flight distance results for bees visiting larkspurs in which reward was varied experimentally, although he did not measure directionality changes. Finally, Pyke (1978c) reported on differences in directionality as a function of inferred rewards in larkspur and monkshood flowers. However, he did not measure rewards

TABLE VI

CHARACTERISTICS OF SOME POLLINATION SYSTEMS FOR WHICH DEGREE OF POLLINATOR DIRECTIONALITY HAS BEEN REPORTED

Pollinator	Plant	Reward	Inflorescence size[a]	Proportion used[b]	Directionality[c]	Source
Bumblebees	*Delphinium*	Nectar	Small	Large	High	Pyke (1978c)
Bumblebees	*Aconitum*	Nectar	Small	Large	High	Pyke (1978c)
Bumblebees	*Trifolium*	Nectar	Small	Large	High	Heinrich (1979)
Honey bees	Artificial flowers	Nectar	One	Large	High	Waddington (1980)
Solitary bees	*Convolvulus*	Pollen, nectar	One	Large	High	Waddington (1979)
Mixed bees	*Lythrum*	Nectar	Small?	Large?	High	Levin *et al.* (1971)
Mixed butterflies	*Lythrum*	Nectar	Small?	Large?	High	Levin *et al.* (1971)
Bumblebees	*Mertensia*	Nectar	Large	Small	None	Geber (1982)
Bumblebees	*Polemonium*	Nectar	Large	Small	None	Zimmerman (1979)
Bumblebees	*Aquilegia*	Pollen	Small	Small	None	Hodges and Miller (1981)
Hummingbirds	*Ipomopsis*	Nectar	Moderate	Moderate	None	Pyke (1981c)
Sunbirds	*Leonotis*	Nectar	Large	Small	None	Gill and Wolf (1977)

[a]Inflorescence size refers to the number of flowers.

[b]Proportion used refers to the proportion of all flowers visited before the pollinator leaves an inflorescence.

[c]Directionality refers to movements between inflorescences.

directly, and Waddington *et al.* (1981) point out other potential explanations for his results.

The extent to which, and the scale over which, rewards are actually patchy are largely unknown. Pleasants and Zimmerman (1979), Zimmerman (1981), and Brink (1982) provide evidence that nectar is distributed spatially to form a pattern of rich and poor patches in meadows of *Delphinium nelsonii,* and that this pattern builds up as a result of foraging of bees. In contrast, I found no clear evidence of rich and poor patches at a study site near theirs (see Table VII), either in the morning when pollinator foraging began or later in the day. In addition, hummingbirds and bumblebees at my study site showed no clear tendency to exhibit more variable interplant flight distances or tenure on plants as the day progressed, as would be expected if the meadow were becoming increasingly divided into rich and poor patches (Table VIII). The spatial scale on which I measured patchiness was slightly larger than that of Pleasants and Zimmerman, and my criteria for accepting evidence of patchiness were more strict than theirs.

The other general aspect of pollinators' behavior once they leave a plant is the

TABLE VII

Some Measures of Spatial Heterogeneity in Nectar Contents of *Delphinium nelsonii* Flowers at Two Different Times of Day

	Rank as "neighbors" along linear transect[a]									
	1st	2nd	3rd	4th	5th	6th	7th	8th	9th	10th
Morning										
r^b	−.01	−.07	−.12	−.28	−.04	+.02	+.12	−.01	−.18	+.26
$\bar{x}_d{}^c$	.44	.49	.48	.57	.51	.48	.51	.46	.46	.51
s	.38	.35	.38	.35	.33	.35	.29	.29	.37	.40
n	31	30	29	28	27	26	25	24	23	22
Evening										
r	−.01	−.04	−.16	−.27	+.02	−.25	+.10	+.18	−.09	−.15
$\bar{x}_d$	.51	.52	.69	.73	.62	.61	.60	.43	.63	.56
s	.49	.41	.52	.50	.54	.48	.45	.35	.49	.49
n	24	23	22	21	20	19	18	17	16	15

[a]Single study plants were chosen at intervals of approximately 1 m along a 30-m transect line. First "neighbors" are all pairs of adjacent study plants; second "neighbors" are all pairs separated by one other study plant, and so on. Actual nearest-neighbor distances are less than 1 m for most plants in the meadow. Nectar contents of bottom female-stage flowers of study plants were measured at 7 PM ("evening") by destructive sampling; new plants along the same transect were measured at 8 AM the next morning ("morning").

[b]These are correlation coefficients of nectar contents of bottom flowers of all first, second, . . . tenth neighbors.

[c]These are mean absolute deviations, in microliters, of nectar contents. Standard deviations and sample sizes are also shown.

TABLE VIII

SOME MEASURES OF FORAGING BEHAVIOR OF POLLINATORS OF *DELPHINIUM NELSONII* AT TWO DIFFERENT TIMES OF DAY

	Interplant distance moved[a] ($\bar{x}$, *cv*, *n*)	Flowers visited per plant ($\bar{x}$, *cv*, *n*)
Morning		
Hummingbirds	2.28, 294%, 44	1.91, 48%, 45
Bumblebees	0.71, 359%, 235	1.55, 55%, 235
Evening		
Hummingbirds	1.62, 200%, 50	2.18, 60%, 51
Bumblebees	0.96, 278%, 109	1.48, 56%, 109

[a]Distances are in meters and are for pollinators foraging at the same times and in the same meadow from which nectar data given in Table VII were taken.

distance they travel to the next plant visited. As I have already intimated, pollinators engaged solely in flower visitation (and not in other simultaneous behavioral tasks) tend to minimize flight distances, at least within constraints imposed by such things as directionality (see the studies mentioned above; also see Linhart, 1973; Schmitt, 1980; Waser and Price, 1983a; Waser, 1982). Such behavior is as one would expect if pollinators are sensitive to foraging economics (Waser, 1982).

B. EVOLUTIONARY EFFECTS ON PLANT TRAITS

What I have just said is meant to supplement the many recent reviews of pollinator foraging behavior, most of which are firmly grounded in considerations of optimal foraging theory. Somewhat less consideration has been given to the effects of pollinator behavior on the evolution of plant traits. Levin *et al.* (1971), among others, pointed out some of the ways in which directionality, flight distances, and pollen carryover will influence pollen flow, and Price and Waser (1979) and Waser and Price (1983a) have discussed ways in which this factor, in turn, may affect plant fitness. It seems certain that there will be many subtle ways in which pollen flow patterns influence the quantity and quality of pollen transfer away from and to a plant, and that some optimal pollen flow pattern will therefore exist for the plant. What does not seem as certain (at least to me) is the extent to which selection acting on plant traits will ultimately influence pollinator behavior and pollen flow patterns (Waser and Price, 1983a). To take the example at hand, we may assume, if Zimmerman (1979) and Hodges and Miller (1981) are correct, that pollinator directionality and flight distances will be influenced by inflorescence size and the type of floral reward. However, I suspect that expression of these plant traits in most cases is determined by other selective forces that are stronger than the effects of changes in flight parameters. Likewise, flight behavior may be influenced by patchy distribution of rewards. However, it is hard to see how anything other than a uniform spatial distribution of rewards can be produced by individual selection alone, without the action of edaphic heterogeneity or pollinator foraging, unless optimal nectar production rate is related to plant age or size (Pleasants, 1983). Thus, unlike some other workers in the field, I suspect that the evolution of floral traits is explained relatively rarely in terms of their influence on the behavior of pollinators that have just left the plant.

V. Conclusions

In planning this chapter, I considered several possible approaches to the huge topic at hand. First, I thought of simply reviewing the literature on selected topics (as some friends pointed out, this approach might be useful but would make for very tedious reading). Second, I thought about advancing my own views on the relative value of detailed, process-oriented approaches to the study of adaptation, and then concentrating on examples of this approach. What I have actually done is a little of both. I certainly have advanced my own views, but I have also strayed far and wide in reviewing what is known and inferred about the adaptive value of certain floral traits. One reason for this mixed approach is that there are still relatively few experimental studies to draw on. Thus, I could not realistically limit my discussion to process-oriented experiments.

The continued avoidance of experimental approaches by those who discuss the evolution of floral traits is both surprising and unsurprising. In a large number of situations, experiments are completely straightforward, albeit time-consuming. This is true, for example, when trait expressions found in nature or produced by manipulations involve little or no difference in costs for the plants, so that it suffices in principle to concentrate only on reproductive benefits of alternative expressions. A case in point is flower color expression in *Delphinium nelsonii* (Waser and Price, 1981, 1983b; Sections II,C and III,A): The savings in energy and resources enjoyed by albino plants by not providing their flowers with anthocyanin pigments can probably be safely ignored in calculating net reproductive benefits of different colors or color patterns. Other examples might include differences in flowering time and flower shape, and some alterations in the distribution of flowers among inflorescences. It may be reasonable to focus solely on reproductive benefits of various expressions of traits such as these.

Measuring reproductive benefits may be far from easy, of course. Although it is relatively straightforward to measure seed set, this represents only one component of fitness. It is far more challenging to measure the value of the seeds produced as vehicles of eventual gene transfer, in terms of their germination and survival to maturity under natural conditions. In this regard, the most desirable approach is to follow the demography of seedlings planted in a situation they would normally be expected to encounter (i.e., in the parental environment), and to do so in as controlled a fashion as possible. Such an approach, especially as part of studies of floral traits, appears to be very rare (Waser and Price, 1983a). Similarly, it is extremely difficult to devise ways to measure the reproductive success attributable to the floral traits of an individual plant from its donation of pollen to conspecifics, even though this represents on average one-half of all possibilities for gene transfer in outbred species. Pollination ecologists traditionally have concentrated on seed set and other components of female reproductive success, although it has become more common recently to ascribe adaptive benefits of floral traits to male success through pollen donation (Willson and Price, 1977; Gori, 1983; see Sections II,D and III,A, B). The only study of which I am aware that actually attempts experimentally to measure male (as well as female) reproductive success is that of Pyke (1981c) on nectar production rate in *Ipomopsis aggregata* (see Sections II,F and III,B). Although the methods used are crude, this study seems to me to pioneer a difficult territory, and it should serve as a starting point for future efforts.

The task of the experimenter becomes even more difficult when floral traits are considered for which there seem to be major trade-offs between benefits and costs of alternative expressions. For example, variations in size or architecture of inflorescences or in reward quality may elicit changes in pollinator behavior or pollen transfer mechanics that will be expected to lead to different reproductive outcomes for the plant (see, e.g., Thomson and Plowright, 1980; Geber, 1982; Gori, 1983). At the same time, substantially different costs in terms of energy or nutrients expended may be associated with different trait expressions. A mutant plant with increased nectar production rate, for instance, must pay some increased caloric cost (of as yet unknown magnitude); this cost may reasonably be expected to appear in terms of reduced ability to provision seeds or produce pollen. The problem is that benefits and costs are expressed in different currencies—the former in terms of fitness and the latter in terms of energy—and it will be very difficult to determine the degree to which these currencies are interchangeable and the exchange rate between them. In one respect, these problems may not be too severe. If one only attempts to assess the overall effects of different trait variants on all components of fitness (including fecundity, offspring viability, and output of the plant during future reproductive episodes), one can be confident in principle of measuring a quantity that includes all variation in costs. On the other hand, it is essential to determine trade-offs if one wishes to construct any sort of quantitative model that provides *a priori* predictions of optimal trait expression. Again, Pyke (1981c) has pioneered such an approach in his study of nectar production (for an attempt at predictions that are similar but qualitative, see Pleasants, 1983). Pyke used bomb calorimetry of seeds to express their value in terms that could be easily compared to the energetic costs of nectar. This, of course, is simplistic, but it provides a starting point.

Let me return to our hypothetical visitor from outer space, to whom we owe an explanation of selective forces that maintain the remarkable diversity of floral trait expression seen on our planet. In terms of rigorous experimental results, we might not be able to tell this visitor much at present, but he, in turn, might not be overly critical of our efforts, given the sorts of difficulties I have outlined. In mentioning these difficulties and in reviewing selectively what we do know or infer, I hope at least to have succeeded in pointing out some profitable directions for future workers to pursue.

Acknowledgments

Numerous friends and colleagues contributed to the ideas and research projects I have discussed in this chapter. For their inspiration and help, I would like to single out W. A. Calder III, P. Feinsinger, M. V. Price, H. R. Pulliam, G. H. Pyke, and J. D. Thomson. The manuscript was read in various drafts and greatly improved by M. Geber and M. V. Price. Financial support for some of the studies came from Sigma Xi, the Chapman Memorial Fund, the American Philosophical Society, the University of Arizona Foundation, the National Science Foundation (Grant DEB 81-02774), the University of Utah (through a teaching postdoctoral fellowship), and the University of California at Riverside (through several Intramural Grants).

References

Aker, C. L. (1982). Spatial and temporal dispersion patterns of pollinators and their relationship to the flowering strategy of *Yucca whipplei* (Agavaceae). *Oecologia* **54,** 243–252.

Augspurger, C. K. (1980). Mass-flowering of a tropical shrub (*Hybanthus prunifolius*): Influence on pollinator attraction and movement. *Evolution* (*Lawrence, Kans.*) **34,** 475–488.

Austin, D. F. (1975). Bird flowers in the eastern United States. *Fla. Sci.* **38,** 1–12.

Bachman, W. W., and Waller, G. D. (1977). Honeybee responses to sugar solutions of different compositions. *J. Apic. Res.* **16,** 165–173.

Baker, H. G. (1961). The adaptation of flowering plants to nocturnal and crepuscular pollinators. *Q. Rev. Biol.* **36,** 64–73.

Baker, H. G. (1963). Evolutionary mechanisms in pollination biology. *Science* **139,** 877–883.

Baker, H. G. (1975). Sugar concentration in nectars from hummingbird flowers. *Biotropica* **7,** 37–41.

Baker, H. G., and Hurd, P. D. (1968). Intrafloral ecology. *Annu. Rev. Entomol.* **13,** 385–414.

Bateman, A. J. (1951). The taxonomic discrimination of bees. *Heredity* **5,** 271–278.

Bateman, A. J. (1956). Cryptic self-incompatibility in the wall flower: *Cheiranthus cheiri* L. *Heredity* **10,** 257–261.

Bawa, K. S., and Beach, J. H. (1981). Evolution of sexual systems in flowering plants. *Ann. M. Bot. Gard.* **68,** 254–274.

Bené, F. (1941). Experiments on the color preference of black-chinned hummingbirds. *Condor* **43,** 237–323.

Bennett, A. W. (1883). On the constancy of insects in their visits to flowers. *Zool. J. Linn. Soc.* **17,** 175–185.

Bernard, G. D. (1979). Red-absorbing visual pigment of butterflies. *Science* **203,** 1125–1127.

Bertin, R. I. (1982a). Floral biology, hummingbird pollination, and fruit production of trumpet creeper (*Campsis radicans,* Bignoniaceae). *Am. J. Bot.* **69,** 122–134.

Bertin, R. I. (1982b). The ruby-throated hummingbird and its major food plants: Ranges, flowering phenology, and migration. *Can. J. Zool.* **60,** 210–219.

Best, L. S., and Bierzychudek, P. (1982). Pollinator foraging on foxglove (*Digitalis purpurea*): A test of a new model. *Evolution* (*Lawrence, Kans.*) **36,** 70–79.

Bolten, A. B., and Feinsinger, P. (1978). Why do hummingbird flowers secrete dilute nectar? *Biotropica* **10,** 307–309.

Brink, D. (1982). A bonanza-blank pollinator reward schedule in *Delphinium nelsonii* (Ranunculaceae). *Oecologia* **52,** 292–294.

Brittain, W. H., and Newton, D. E. (1933). A study in the relative constancy of hive bees in pollen gathering. *Can. J. Res.* **9,** 334–349.

Calder, W. A. (1979). On the temperature-dependency of optimal nectar concentration for birds. *J. Theor. Biol.* **78,** 185–196.

Calder, W. A., Waser, N. M., Hiebert, S. M., Inouye, D. W., and Miller, S. (1982). Site-fidelity, longevity, and population dynamics of broad-tailed hummingbirds: A ten year study. *Oecologia* **56,** 359–369.

Campbell, D. R., and Motten, A. F. (1981). Competition for pollination in two spring wildflowers. *Bull. Ecol. Soc. Am.* **63,** 99.

Char, M. B. S. (1977). Pollen allelopathy. *Naturwissenschaften* **64,** 489–490.

Charnov, E. L. (1976). Optimal foraging, the marginal value theorem. *Theor. Pop. Biol.* **9,** 129–136.

Charnov. E. L. (1979). Simultaneous hermaphroditism and sexual selection. *Proc. Natl. Acad. Sci. U.S.A.* **76,** 2480–2484.

Christy, R. M. (1883). On the methodic habits of insects when visiting flowers. *Zool. J. Linn. Soc.* **17,** 186–194.

Clements, F. E., and Long, F. L. (1923). "Experimental pollination: An outline of the ecology of flowers and insects." *Carnegie Inst. Wash. Publ.* No. 336.

Clement, W. M. (1965). Flower color, a factor in attractiveness of alfalfa clones for honey bees. *Crop Sci.* **5,** 267–268.

Collias, N. E., and Collias, E. C. (1968). Anna's hummingbirds trained to select different colors in feeding. *Condor* **70,** 78–86.

Connell, J. H. (1975). Some mechanisms producing structure in natural communities: A model and evidence from field experiments. *In* "Ecology and Evolution of Communities" (M. L. Cody and J. M. Diamond, eds.), pp. 460–490. Belknap Press of Harvard Univ. Press, Cambridge, Massachusetts.

Connell, J. H. (1980). Diversity and the coevolution of competitors, or the ghost of competition past. *Oikos* **35,** 131–138.

Corbet, S. A., Cuthill, I., Fallows, M., Harrison, T., and Hartley, G. (1981). Why do nectar-foraging bees and wasps work upward on inflorescences? *Oecologia* **51,** 79–83.

Curio, E. (1973). Towards a methodology of teleonomy. *Experientia* **29,** 1045–1180.

Darwin, C. H. (1859). "On the Origin of Species by Means of Natural Selection." Murray, London.

Darwin, C. H. (1876). "On the Effects of Cross and Self Fertilisation in the Vegetable Kingdom." Murray, London.

Daumer, K. (1958). Blumenfarben, wie sie die Bienen sehen. *Z. Vergl. Physiol.* **41,** 49–110.

Faegri, K., and van der Pijl, L. (1971). "The Principles of Pollination Ecology," 2nd revised ed. Pergamon, Oxford.

Faulkner, G. J. (1976). Honeybee behavior as affected by plant height and flower colour in brussels sprouts. *J. Apic. Res.* **15,** 15–18.

Feinsinger, P. (1978). Ecological interactions between plants and hummingbirds in a successional tropical community. *Ecol. Monogr.* **48,** 269–287.

Feinsinger, P. (1983). Coevolution and pollination. *In* "Coevolution" (D. J. Futuyma and M. Slatkin, eds.), pp. 282–310. Sinauer, Sunderland, Massachusetts.

Frankie, G. W., Baker, H. G., and Opler, P. A. (1974). Comparative phenological studies of trees in tropical wet and dry forests in the lowlands of Costa Rica. *J. Ecol.* **62,** 881–913.

Free, J. B. (1963). The flower constancy of honeybees. *J. Anim. Ecol.* **32,** 119–131.

Free, J. B. (1966). The foraging behavior of bees and its effect on the isolation and speciation of plants. *In* "Reproductive Biology and Taxonomy of Vascular Plants" (J. G. Hawkes, ed.), pp. 76–91. Pergamon, Oxford.

Free, J. B. (1970). The flower constancy of bumblebees. *J. Anim. Ecol.* **39,** 395–402.

Fretwell, S. D., and Lucas, H. L. (1969). On territorial behavior and other factors influencing habitat distribution in birds. I. Theoretical development. *Acta Biotheor.* **19,** 16–36.

Gass, C. L., and Montgomerie, R. D. (1981). Hummingbird foraging behavior: Decision-making and energy regulation. *In* "Foraging Behavior: Ecological, Ethological, and Psychological Approaches" (A. C. Kamil and T. D. Sargent, eds.), pp. 159–194. Garland STPM Press, New York.

Gause, G. F. (1934). "The Struggle for Existence." Williams and Wilkins, Baltimore, Maryland.

Geber, M. (1982). Architecture, size, and reproduction in plants: A pollination study of *Mertensia ciliata* (James) G. Don. Thesis, Oregon State Univ., Corvallis.

Gill, F. B., and Wolf, L. L. (1977). Nonrandom foraging by sunbirds in a patchy environment. *Ecology* **40,** 235–246.

Goldsmith, T. H. (1980). Hummingbirds see near ultraviolet light. *Science* **207,** 786–788.

Goldsmith, T. H., and Bernard, G. D. (1974). The visual system of insects. *In* "The Physiology of Insecta" (M. Rockstein, ed.), 2nd ed. pp. 165–272. Academic Press, New York.

Goldsmith, T. H., and Goldsmith, K. M. (1979). Discrimination of colors by the black-chinned hummingbird, *Archilochus alexandri. J. Comp. Physiol.* **130,** 209–220.

Gori, D. F. (1983). Post-pollination phenomena and adaptive floral changes. *In* "Handbook of Experimental Pollination Biology" (C. E. Jones and R. J. Little, eds.), pp. 31–45. Van Nostrand-Reinhold, New York.

Gould, S. J., and Lewontin, R. C. (1979). The spandrels of San Marco and the Panglossian paradigm: A critique of the adaptationist programme. *Proc. R. Soc. London, Ser. B* **205,** 581–598.

Grant, K. A., and Grant, V. (1967). Records of hummingbird pollination in the western American flora. III. Arizona records. *Aliso* **6,** 107–220.

Grant, K. A., and Grant, V. (1968). "Hummingbirds and their Flowers." Columbia Univ. Press, New York.

Grant, V. (1949). Pollination systems as isolating mechanisms in angiosperms. *Evolution (Lawrence, Kans.)* **3,** 82–97.

Grant, V. (1950). The flower constancy of bees. *Bot. Rev.* **16,** 379–398.

Grant, V. (1952). Isolation and hybridization between *Aquilegia formosa* and *A. pubescens. Aliso* **2,** 341–360.

Grant, V., and Grant, K. A. (1965). "Flower Pollination in the Phlox Family." Columbia Univ. Press, New York.

Grant, V., and Grant, K. A. (1966). Records of hummingbird pollination in the western American flora. I. some California plant species. *Aliso* **6,** 51–66.

Grant, V., and Grant, K. A. (1967). Records of hummingbird pollination in the western American flora. II. additional California records. *Aliso* **6,** 103–105.

Hainsworth, F. R., and Wolf, L. L. (1976). Nectar characteristics and food selection by hummingbirds. *Oecologia* **25,** 101–113.

Hannan, G. L. (1981). Flower color polymorphism and pollination biology of *Platystemon californicus* Benth. (Papaveraceae). *Am. J. Bot.* **68,** 233–243.

Heinrich, B. (1975a). Bee flowers: A hypothesis on flower variety and blooming times. *Evolution (Lawrence, Kans.)* **29,** 325–334.

Heinrich, B. (1975b). Energetics of pollination. *Annu. Rev. Ecol. System.* **6,** 139–170.

Heinrich, B. (1976). The foraging specializations of individual bumblebees. *Ecol. Monogr.* **46,** 105–128.

Heinrich, B. (1979). Resource heterogeneity and patterns of movement in foraging bumblebees. *Oecologia* **40,** 235–246.

Heinrich, B., and Raven, P. H. (1972). Energetics and pollination ecology. *Science* **176,** 597–602.

Heinrich, B., Mudge, P. R., and Deringis, P. G. (1977). Laboratory analysis of flower constancy in foraging bumblebees: *Bombus ternarius* and *B. terricola. Behav. Ecol. Sociobiol.* **2,** 247–265.

Hobbs, G. A., Nummi, W. O., and Virostek, J. F. (1961). Food-gathering behavior of honey, bumble, and leaf-cutter bees (*Hymenoptera*: Apoidea) in Alberta. *Can. Entomol.* **93,** 409–419.

Hodges, C. M. (1981). Optimal foraging in bumblebees: Hunting by expectation. *Anim. Behav.* **29,** 1166–1171.

Hodges, C. M., and Miller, R. B. (1981). Pollinator flight directionality and the assessment of pollen returns. *Oecologia* **50,** 376–379.

Hogenboom, N. G. (1975). Incompatibility and incongruity: Two different mechanisms for the non-functioning of intimate partner relationships. *Proc. R. Soc. London, Ser. B* **188,** 361–375.

Horovitz, A. (1969). Effect of flower color variations on the mating system in some forms of *Lupinus nanus* Dougl. (ex Benth.). Dissertation, Univ. of California, Davis.

Inouye, D. W. (1980). The effect of proboscis and corolla tube lengths on patterns and rates of flower visitation by bumblebees. *Oecologia* **45,** 197–201.

Janzen, D. H. (1977). A note on optimal mate selection by plants. *Am. Nat.* **111,** 365–371.

Janzen, D. H., DeVries, P., Gladstone, D. E., Higgins, M. L., and Lewinsohn, T. M. (1980). Self-and cross-pollination of *Encyclia cordigera* (Orchidaceae) in Santa Rosa National Park, Costa Rica. *Biotropica* **12,** 72–74.

Jensen, W. A., and Salisbury, F. D. (1972). "Botany, an Ecological Approach." Wadsworth, Belmont, California.

Jones, C. E. (1978). Pollinator constancy as a pre-pollination isolating mechanism between sympatric species of *Cercidium. Evolution (Lawrence, Kans.)* **32,** 189–198.

Jones, C. E., and Buchmann, S. L. (1974). Ultraviolet floral patterns as functional orientation cues in hymenopterous pollination systems. *Anim. Behav.* **22,** 481–485.

Kevan, P. G. (1978). Floral coloration, its colorimetric analysis and significance in anthecology. *In* "The Pollination of Flowers by Insects" (A. J. Richards, ed.), pp. 51–78. Academic Press, New York.

Kislev, M. E., Kraviz, Z., and Lorch, J. (1972). A study of hawkmoth pollination by a palynological analysis of the proboscis. *Isr. J. Bot.* **21,** 57–75.

Knuth, P. (1906). "Handbook of Flower Pollination," Vol. 1. Oxford Univ. Press (Clarendon), Oxford.

Kölreuter, J. G. (1761). "Vorläufige Nachrichten von einigen das Geschlecht der Pflanzen betreffenden Versuchen und Beobachtungen." Gleditschischen Handlung, Leipzig.

Krebs, J. R., and Davies N. B. (1981). "An Introduction to Behavioral Ecology." Sinauer, Sunderland, Massachusetts.

Kugler, H. (1943). Hummeln als Blütenbesucher. *Ergeb. Biol.* **19,** 143–323.

Laverty, T. M. (1980). The flower-visiting behavior of bumblebees: Floral complexity and learning. *Can. J. Zool.* **58,** 1324–1335.

Lertzman, K. P. (1981). Pollen transfer: Processes and consequences. Thesis, Univ. of British Columbia, Vancouver.

Lertzman, K. P., and C. L. Gass (1983). Alternative models of pollen transfer. *In* "Handbook of Experimental Pollination Biology" (C. E. Jones and R. J. Little, eds.), pp. 474–489. Van Nostrand-Reinhold, New York.

Levin, D. A. (1968). The effect of corolla color and outline on interspecific pollen flow in *Phlox*. *Evolution (Lawrence, Kans.)* **23,** 444–455.

Levin, D. A. (1978). Pollinator behavior and the breeding structure of plant populations. *In* "The Pollination of Flowers by Insects" (A. J. Richards, ed.), pp. 133–150, Academic Press, New York.

Levin, D. A., and Berube, D. E. (1972). *Phlox* and *Colias*: The efficiency of a pollination system. *Evolution (Lawrence, Kans.)* **26,** 242–250.

Levin, D. A., and Schaal, B. A. (1970). Corolla color as an inhibitor of interspecific hybridization in *Phlox*. *Am. Nat.* **104,** 273–283.

Levin, D. A., Kerster, H. W., and Niedzlek, M. (1971). Pollinator flight directionality and its effect on pollen flow. *Evolution (Lawrence, Kans.)* **25,** 113–118.

Lewis, H. (1961). Experimental sympatric populations of *Clarkia*. *Am. Nat.* **95,** 155–168.

Linhart, Y. B. (1973). Ecological and behavioral determinants of pollen dispersal in hummingbird-pollinated *Heliconia*. *Am. Nat.* **107,** 511–523.

Linsley, E. G. (1958). The ecology of solitary bees. *Hilgardia* **27,** 543–599.

MacArthur, R. H., and Pianka, E. R. (1966). On optimal use of a patchy environment. *Am. Nat.* **100,** 603–609.

McCrea, K. D. (1981). Ultraviolet floral patterning, reproductive isolation, and character displacement in the genus *Rudbeckia* (Compositae). Dissertation, Purdue Univ., Lafayette, Indiana.

McDade, L. A. (1983). Long-tailed hermit hummingbird visits to inflorescence color morphs of *Heliconia irrasa*. *Condor* (in press).

Macior, L. W. (1971). Co-evolution of plants and animals: Systematic insights from plant-insect interactions. *Taxon* **20,** 17–28.

McNaughton, I. H., and Harper, J. L. (1960). The comparative biology of closely related species living in the same area. I. External breeding barriers between *Papaver* species. *New Phytol.* **59,** 15–26.

Manning, A. (1956). Some aspects of the foraging behavior of bumblebees. *Behaviour* **9,** 164–201.

Manning, A. (1957). Some evolutionary aspects of the flower constancy of bees. *Proc. R. Soc. Edinburgh* **25,** 67–71.

Marden, J. H., and Waddington, K. D. (1981). Floral choices by honeybees in relation to the relative distances to flowers. *Physiol. Entomol.* **6,** 431–435.

Maynard Smith, J. (1976). Evolution and the theory of games. *Am. Sci.* **64,** 41–45.

Maynard Smith, J. (1978). Optimization theory in evolution. *Annu. Rev. Ecol. Syst.* **9,** 31–56.

Menzel, R., Erber, J., and Masuhr, T. (1974). Learning and memory in the honeybee. *In* "Experimental Analysis of Insect Behaviour" (L. B. Browne, ed.), pp. 195–217. Springer-Verlag, Berlin and New York.

Miller, R. S., and Miller, R. E. (1971). Feeding activity and color preference of ruby-throated hummingbirds. *Condor* **73,** 309–313.

Mogford, D. J. (1974). Flower colour polymorphism in *Cirsium palustre*. II. Pollination. *Heredity* **33,** 257–263.

Morse, D. H. (1982). The turnover of milkweed pollinia on bumble bees, and implications for outcrossing. *Oecologia* **53,** 187–196.

Motten, A. F., Campbell, D. R., Alexander, D. E., and Miller, H. L. (1981). Pollination effectiveness of specialist and generalist visitors to a North Carolina population of *Claytonia virginica*. *Ecology* **62,** 1278–1287.

Mulcahy, D. L. (1971). A correlation between gametophytic and sporophytic characteristics in *Zea mays* L. *Science* **171,** 1155–1156.

Mulcahy, D. L. (1979). Rise of the angiosperms: A genecological factor. *Science* **206,** 20–23.
Müller, H. (1871). "Application of the Darwinian theory to flowers and the insects which visit them. *Am. Nat.* **5,** 271–279.
Müller, H. (1883). "The Fertilisation of Flowers." Macmillan, London.
Mulligan, G. A., and Kevan, P. G. (1973). Color, brightness, and other floral characteristics attracting insects to the blossoms of some Canadian weeds. *Can. J. Bot.* **51,** 1939–1952.
O'Brien, M. H. (1980). The pollination biology of a pavement plain: Pollinator visitation patterns. *Oecologia* **47,** 213–218.
Ockendon, D. J., and Currah, L. (1977). Self-pollen reduces the number of cross-pollen tubes in the styles of *Brassica oleracea* L. *New Phytol.* **78,** 675–680.
Oster, G., and Heinrich, B. (1976). Why do bumblebees major? A mathematical model. *Ecol. Monogr.* **46,** 129–133.
Parker, F. D. (1981). How efficient are bees in pollinating sunflowers? *J. Kans. Entomol. Soc.* **54,** 61–67.
Pederson, M. W. (1967). Cross-pollination studies involving three purple-flowered alfalfas, one white-flowered line, and two pollinator species. *Crop Sci.* **7,** 59–62.
Plateau, F. (1899). Nouvelles recherches sur les rapports entre les insectes et les fleurs. II. Le choix des couleurs par les insectes. *Mem. Soc. Zool. Fr.* **12,** 336–370.
Platt, J. R. (1964). Strong inference. *Science* **146,** 347–353.
Pleasants, J. M. (1981). Bumblebee responses to variation in nectar availability. *Ecology* **62,** 1648–1661.
Pleasants, J. M. (1983). Nectar production patterns in *Ipomopsis aggregata. Am. J. Bot.* (in press).
Pleasants, J. M., and Zimmerman, M. (1979). Patchiness in the dispersion of nectar resources: Evidence for hot and cold spots. *Oecologia* **41,** 283–288.
Price, M. V., and Waser, N. M. (1979). Pollen dispersal and optimal outcrossing in *Delphinium nelsoni. Nature (London)* **277,** 294–296.
Price, M. V., and Waser, N. M. (1982). Experimental studies of pollen carryover: Hummingbirds and *Ipomopsis aggregata. Oecologia* **54,** 353–358.
Primack, R. B., and Silander, J. A. (1975). Measuring the relative importance of different pollinators to plants. *Nature (London)* **255,** 143–144.
Procter, M., and Yeo, P. (1972). "The Pollination of Flowers." Taplinger, New York.
Pyke, G. H. (1978a). Are animals efficient harvesters? *Anim. Behav.* **26,** 241–250.
Pyke, G. H. (1978b). Optimal foraging in bumblebees and coevolution with their plants. *Oecologia* **36,** 281–293.
Pyke, G. H. (1978c). Optimal foraging: Movement patterns of bumblebees between inflorescences. *Theor. Pop. Biol.* **13,** 72–98.
Pyke, G. H. (1979). Optimal foraging foraging in bumblebees: Rule of movement between flowers within inflorescences. *Anim. Behav.* **27,** 1167–1181.
Pyke, G. H. (1981a). Optimal foraging in hummingbirds: Rule of movement between inflorescences. *Anim. Behav.* **29,** 889–896.
Pyke, G. H. (1981b). Optimal foraging in nectar-feeding animals and coevolution with their plants. *In* "Foraging Behavior: Ecological, Ethological, and Psychological Approaches" (A. C. Kamil and T. D. Sargent, eds.), pp. 19–48. Garland STPM Press, New York.
Pyke, G. H. (1981c). Optimal nectar production in a hummingbird pollinated plant. *Theor. Pop. Biol.* **20,** 326–343.
Pyke, G. H., and Waser, N. M. (1981). The production of dilute nectars by hummingbird and honeyeater flowers. *Biotropica* **13,** 260–270.
Pyke, G. H., Pulliam, H. R., and Charnov, E. L. (1977). Optimal foraging: A selective review of theory and tests. *Q. Rev. Biol.* **52,** 137–154.
Real, L. A. (1981). Uncertainty and plant-pollinator interaction: The foraging behavior of bees and wasps on artificial flowers. *Ecology* **62,** 20–26.

Regal, P. J. (1977). Ecology and evolution of flowering plant dominance. *Science* **196,** 622–629.

Robertson, C. (1895). The philosophy of flower seasons, and the phaenological relations of the entomophilous flora and the anthophilous insect fauna. *Am. Nat.* **29,** 97–117.

Schaffer, W. M., and Schaffer, M. V. (1977). The adaptive significance of variation in reproductive habit in Agavaceae. *In* "Evolutionary Ecology" (B. Stonehouse and C. M. Perrins, eds.), pp. 261–276. Macmillan, New York.

Schaffer, W. M., and Schaffer, M. V. (1979). The adaptive significance of variations in reproductive habit in the Agavaceae II: Pollinator foraging behavior and selection for increased reproductive expenditure. *Ecology* **60,** 1051–1069.

Schemske, D. W. (1976). Pollinator specificity in *Lantana camara* and *L. trifolia* (Verbenaceae). *Biotropica* **8,** 260–264.

Schemske, D. W. (1980). Evolution of floral display in the orchid *Brassavola nodosa. Evolution (Lawrence, Kans.)* **34,** 489–493.

Schmitt, J. (1980). Pollinator foraging behavior and gene dispersal in *Senecio* (Compositae). *Evolution (Lawrence, Kans.)* **34,** 934–943.

Scora, R. W. (1964). Dependency of pollination on patterns in *Monarda* (Labiatae). *Nature (London)* **204,** 1011–1012.

Silander, J. A., and Primack, R. B. (1978). Pollination intensity and seed set in the evening primrose (*Oenothera fruticosa*). *Am. Midl. Nat.* **100,** 213–216.

Simpson, B. B. (1977). Breeding systems of dominant perennial plants of two disjunct warm deserts. *Oecologia* **27,** 203–226.

Sprague, E. F. (1962). Pollination and evolution in *Pedicularis* (Scrophulariaceae). *Aliso* **5,** 181–209.

Sprengel, C. K. (1793). "Das entdeckte Geheimniss der Natur im Bau und in der Befruchtung der Blumen." Vieweg, Berlin.

Stebbins, G. L. (1970). Adaptive radiation of reproductive characteristics in angiosperms, I: Pollination mechanisms. *Annu. Rev. Ecol. Syst.* **1,** 307–326.

Stebbins, G. L. (1981). Why are there so many species of flowering plants? *BioScience* **31,** 573–576.

Stiles, F. G. (1976). Ecology, flowering phenology, and pollination of some Costa Rican *Heliconia* species. *Ecology* **56,** 285–301.

Stiles, F. G. (1981). Geographical aspects of bird-flower coevolution, with particular reference to Central America. *Ann. M. Bot. Gard.* **68,** 323–351.

Sukhada, K., and Jayachandra (1980). Pollen allelopathy: A new phenomenon. *New Phytol.* **84,** 739–746.

Swihart, S. L. (1970). The neural basis of colour vision in the butterfly *Papilio troilus. J. Insect Physiol.* **16,** 1623–1636.

Tepedino, V. J. (1981). The pollination efficiency of the squash bee (*Peponapis pruinosa*) and the honey bee (*Apis mellifera*) on summer squash (*Cucurbita pepo*). *J. Kans. Entomol. Soc.* **54,** 359–377.

Thomson, J. D. (1981a). Field measures of flower constancy in bumblebees. *Am. Midl. Nat.* **105,** 377–380.

Thomson, J. D. (1981b). Spatial and temporal components of resource assessment by flower-feeding insects. *J. Anim. Ecol.* **50,** 49–59.

Thomson, J. D., and Plowright, R. C. (1980). Pollen carry-over, nectar rewards, and pollinator behavior with special reference to *Diervilla lonicera. Oecologia* **46,** 68–74.

Thomson, J. D., Andrews, B. J., and Plowright, R. C. (1981). The effect of a foreign pollen on ovule development in *Diervilla lonicera* (Caprifoliaceae). *New Phytol.* **90,** 777–783.

Udovic, D. (1981). Determinants of fruit set in *Yucca whipplei*: Reproductive expenditure vs. pollinator availability. *Oecologia* **48,** 389–399.

van der Pijl, L. (1960). Ecological aspects of flower evolution. I. Phyletic evolution. *Evolution (Lawrence, Kans.)* **14,** 403–416.

von Frisch, K. (1914). Der Farbensinn und Formensinn der Bienen. *Zool. Jahrb., Abt. Allg. Zool. Physiol. Tiere* **35,** 1–188.

Waddington, K. D. (1979). Flight patterns of three species of sweat bees (*Halictidae*) foraging at *Convolvulus arvensis. J. Kans. Entomol. Soc.* **52,** 751–758.

Waddington, K. D. (1980). Flight patterns of foraging bees relative to density to artificial flower and distribution of nectar. *Oecologia* **44,** 199–204.

Waddington, K. D. (1981). Factors influencing pollen flow in bumblebee-pollinated *Delphinium virescens. Oikos* **37,** 153–159.

Waddington, K. D. (1983). Floral-visitation-sequences by bees: Models and experiments. *In* "Handbook of Experimental Pollination Biology" (C. E. Jones and R. J. Little, eds.), pp. 461–473. Van Nostrand-Reinhold, New York.

Waddington, K. D., and Heinrich, B. (1979). The foraging movements of bumblebees on vertical "inflorescences:" An experimental analysis. *J. Comp. Physiol.* **134,** 113–117.

Waddington, K. D., Allen T., and Heinrich, B. (1981). Floral preferences of bumblebees (*Bombus edwardsii*) in relation to intermittant versus continuous rewards. *Anim. Behav.* **29,** 779–784.

Waller, G. D. (1972). Evaluating responses of honey bees to sugar solutions using an artificial-flower feeder. *Ann. Entomol. Soc. Am.* **65,** 857–862.

Waser, N. M. (1976). Food supply and nest timing of broad-tailed hummingbirds in the Rocky Mountains. *Condor* **78,** 133–135.

Waser, N. M. (1977). Competition for pollination and the evolution of flowering time. Dissertation, Univ. of Arizona, Tucson, Arizona.

Waser, N. M. (1978a). Competition for hummingbird pollination and sequential flowering in two Colorado wildflowers. *Ecology* **59,** 934–944.

Waser, N. M. (1978b). Interspecific pollen transfer and competition between co-occurring plant species. *Oecologia* **36,** 223–236.

Waser, N. M. (1979). Pollinator availability as a determinant of flowering time in ocotillo (*Fouquieria splendens*). *Oecologia* **39,** 107–121.

Waser, N. M. (1982). A comparison of distances flown by different visitors to flowers of the same species. *Oecologia* **55,** 251–257.

Waser, N. M. (1983). Competition for pollination and floral character differences among sympatric plant species: A review of evidence. *In* "Handbook of Experimental Pollination Biology" (C. E. Jones and R. J. Little, eds.), pp. 277–293. Van Nostrand-Reinhold, New York.

Waser, N. M., and Price, M. V. (1981). Pollinator choice and stabilizing selection for flower color in *Delphinium nelsonii. Evolution (Lawrence, Kans.)* **35,** 376–390.

Waser, N. M., and Price, M. V. (1983a). Optimal and actual outcrossing in plants, and the nature of plant-pollinator interaction. *In* "Handbook of Experimental Pollination Biology" (C. E. Jones and R. J. Little, eds.), pp. 341-359. Van Nostrand-Reinhold, New York.

Waser, N. M., and Price, M. V. (1983b). Pollinator behavior and natural selection for flower colour in *Delphinium nelsonii. Nature (London)* **302,** 422–424.

Weevers, T. (1952). Flower colours and their frequency. *Acta Bot. Neerl.* **1,** 81–92.

Wherry, E. T. (1961). Remarks on the *Ipomopsis aggregata* group. *Aliso* **5,** 5–8.

Whitham, T. G. (1977). Coevolution of foraging in *Bombus* and nectar dispensing in *Chilopsis*: A last dreg theory. *Science* **197,** 593–596.

Williams, G. C. (1966). "Adaptation and Natural Selection." Princeton Univ. Press, Princeton, New Jersey.

Willson, M. F. (1979). Sexual selection in plants. *Am. Nat.* **113,** 777–790.

Willson, M. F., and Price, P. W. (1977). The evolution of the inflorescence size in *Asclepias* (Asclepiadaceae). *Evolution (Lawrence, Kans.)* **31,** 495–511.

Willson, M. F., and Rathcke, B. J. (1974). Adaptive design of the floral display in *Asclepias syriaca* L. *Am. Midl. Nat.* **92,** 47–57.
Willson, M. F., and Miller, L. J., and Rathcke, B. J. (1979). Floral display in *Phlox* and *Geranium*: Adaptive aspects. *Evolution (Lawrence, Kans.)* **33,** 52–63.
Wolf, L. L., and Hainsworth, F. R. (1971). Time and energy budgets of territorial hummingbirds. *Ecology* **52,** 976–988.
Wyatt, R. (1980). The reproductive biology of *Asclepias tuberosa*: I. Flower number, arrangement, and fruit-set. *New Phytol.* **85,** 119–131.
Zimmerman, M. (1979). Optimal foraging: A case for random movement. *Oecologia* **43,** 261–267.
Zimmerman, M. (1981). Patchiness in the dispersion of nectar resources: Probable causes. *Oecologia* **49,** 154–157.

CHAPTER 11

Microbehavior and Macrostructure in Pollinator–Plant Interactions

LESLIE REAL
Department of Zoology
North Carolina State University
Raleigh, North Carolina

I. Microbehavior and Macrostructure

One of the goals of a theory of community structure should be to uncover the manner by which global properties of the community are generated from the individual behaviors and pursuits of its members. Community ecologists, in their admirable attempt at holism, sometimes are satisfied with descriptions of global patterns uncoupled from any mechanism that may account for the pattern, whereas behavioralists and autecologists with proclivities toward more reductionist perspectives often lose sight of the implications of individual behavior on more large-scale patterns. Neither approach is wholly satisfactory, and we must develop a much keener theoretical understanding of how community pattern (macrostructure) is derived from the underlying behavior of individuals in that community (microbehavior).

POLLINATION BIOLOGY

ISBN 0-12-583980-4 (hardcover)
ISBN 0-12-583982-0 (paperback)

Attempts at deriving macrostructure from microbehavior are in their infancy in biology but are standard fare for economists. Thomas Schelling's (1978) entertaining book "Micromotives and Macrobehavior" illustrates this kind of analysis. The stimulus for writing his book came during a lecture when Schelling noticed that everyone sat toward the back of the hall. At one level of analysis, one can be satisfied with simply describing this fact. One might even undertake an analysis to verify that, indeed, listeners were highly clumped and toward the rear. Alternatively, one can try to account for this pattern based on knowledge of the listeners' choices in selecting seats. In making this latter attempt, we wish to know how the listeners' microbehavior generated the audience's macrostructure.

There are at least two apparent mechanisms that will account for everyone's sitting at the back of the hall. Each illustrates an important point about the connection between microbehavior and macrostructure. One case assumes that each listener prefers to sit as far back as possible and simply sits where there is an available seat. The overall pattern merely reflects the individual pattern of choice by the listener. The pattern of the audience is the sum of individual preferences, and there is no real interaction between members of the audience. Alternatively, each listener actually wishes to sit as close to the speaker as possible but must also satisfy a strong impulse to sit behind someone else. If everyone in the audience wishes to sit behind someone else, then the audience as a whole inevitably will drift toward the back of the hall. The macrostructure results from the interaction among members and may lead individuals to less preferred seating arrangements. The interaction generates a pattern which is *not* in accord with each individual's preference for a seat near the front if that individual were alone. These mechanisms illustrate two important points: (1) microbehavior may or may not be interactive and, more importantly, (2) macrostructure may or may not reflect the preferences of individuals that comprise the community. By and large, the most interesting cases for the ecologist will be interactions. Consequently, we must carefully ascertain the differences between unconstrained individual preferences and the constrained behaviors observed after the equilibrium macrostructure has been attained.

There are already a few studies in pollination biology that examine the relationship between microbehavior and macrostructure. Numerous authors have explained territoriality of pollinators, especially hummingbirds, and their concomitant spatial distribution patterns, to be a result of individual behavior and competition over nectar sources (Kodric-Brown and Brown, 1978; Wolf, 1978; Gill and Wolf, 1975; Carpenter and MacMillan, 1976). Pimm (1978) similarly explained community structure and species diversity in hummingbird assemblages as resulting from individual responses to environmental predictability. In a series of elegant studies, Johnson and Hubbell (1974, 1975) accounted for the spatial arrangement among stingless bee nests within a tropical dry forest by understanding the hierarchy of aggressiveness among pairs of bees fighting over a common nectar source. Schaffer *et al.* (1979) explained the temporal pattern of bee visits to *Agave schottii* as a response by individuals of different species of bees to temporal variation in the nectar contents of open flowers. I have observed a similar temporal pattern in bees visiting tropical morning glories (Real, 1981b). Pollinator behavior and preference for particular resources, coupled with some form of competitive effect, account for the structure of hummingbird communities on islands (Feinsinger and Colwell, 1978) and tropical montane forests (Feinsinger, 1976).

In each of the above studies, the overall patterns of the communities (e.g., spatial and/or temporal distribution of species or individuals, community diversity, hierarchical arrangement of territories) were generated by the behavior of individual pollinators and the competitive interactions attendant upon these behaviors. Unfortunately, there has been little attempt to formulate a theoretical foundation for the exact mechanisms that generate macrostructure from behavior. What quantifiable aspects of behavior lead to the observed community patterns? In most cases, individual behaviors are not sufficiently quantified (or even articulated) to uncover these connections. In this chapter, I hope to demonstrate heuristically the form that such an investigation might take.

The ecological literature has seen a growing debate over whether plant species compete for pollinators. This debate was fueled in the late 1960s and early 1970s by observations on arctic and subarctic wildflowers in Canada. Hocking (1968) noticed that many insect-pollinated plant species had significant standing crops of nectar available to insect visitors throughout the day, but these resources seemed not to be exploited. Consequently, Hocking hypothesized that these plants must compete for visits and that pollinator visitation comprises a limiting resource. Extending this competition hypothesis, Mosquin (1971), also working with Canadian wildflowers, suggested that plants sharing pollinators should diverge in the time of their flowering so as to avoid competition. Mosquin's suggestion received immediate attention from many pollination biologists working in a variety of habitats. Gentry (1974) analyzed the flowering phenologies of six sympatric species of *Arrabidaea* (Bignoniaceae) occurring in Costa Rica and Panama. He suggested that these species showed different flowering peaks to avoid competition for shared bee pollinators. Flowering time divergence also has been suggested for plant phenologies in Maine bogs (Heinrich, 1975) and lowland tropical rain forests (Stiles, 1977).

Those who believe that competition occurs between simultaneously flowering plants distinguish two modes of its operation. First, simultaneously flowering plants must compete for the number of visits by pollinators. Two plants sharing a limited resource (pollinator visits) must surely do worse than one plant alone with a captive market. This "competition effect" will be enhanced if, in addition, one floral type is more attractive than the other. However, the effect exists even when the plants are equally attractive. The second form of competition suggests that plants will suffer reduced seed set when foreign pollen is deposited on the surface of the stigma. The limiting resource is then the stigma surface, and the number of seeds produced is related to the proportion of "correct" pollen grains deposited. The outcome of both types of competition eventually leads to the extinction of one of the species or to a divergence in the time of flowering for each or the use of alternative pollinators. This conclusion has received some convincing theoretical and empirical support (Levin and Anderson, 1970; Waser, 1978). However, in these examples the mechanism for competition is always stigma packing and the competition effect is rarely considered or supported. Recent authors have suggested that the competition effect can be offset by the increased attractiveness of plants flowering synchronously. Brown and Kodric-Brown (1979) examined the flowering of seven plant species with hummingbird-pollinated flowers in the Rocky Mountains. All seven species flowered at the same time, had the same pollinators, and looked very much alike. These authors suggested that stigma packing is avoided by differential placement of pollen on the birds and that the competition effect was counterbalanced by increased attractiveness. The flowers essentially acted as mutualists in attracting large numbers of foragers to their study area.

Rathcke (Chapter 12, this volume) and Waser (1983) reviewed the general evidence for and against these interpretations; their work will not be examined here. Instead, I will present a general theoretical framework for analyzing attractiveness (based upon pollinator foraging behavior) and its relation to convergence and divergence in flowering. I will be concerned with three main features of the study of pollinator–plant interactions:

1. Constructing a general model of pollinator foraging behavior based on sequential decision making by the pollinator.
2. Using this model to ascertain the general properties of a floral assemblage within and among plants in a habitat that will make that habitat attractive to pollinators.
3. Assessing the interaction between habitat and pollinator foraging as it affects (a) the evolution of convergence and divergence in flowering time and (b) its effect on the pollinators' foraging patterns.

The final remarks will relate this examination to traditional approaches in modeling pollinator–plant interactions.

II. A Sequential Foraging Model

I start with a pollinator foraging in a given habitat consisting of various floral patches with a fixed spatiotemporal location. These patches vary in the quality of rewards offered to the pollinator. Let the quality of energetic reward for a patch be designated by the random variable X_i with cumulative distribution function $F(X_i)$, with $E(X_i)<\infty$, and the X_i mutually independent. The pollinator is assumed to move sequentially from one patch to the next, assessing the quality of each patch as it moves. The random variable for patch quality is then indexed by i to designate the sequence of patches visited by the pollinator as it moves through the habitat, i.e., X_1 is the quality of the first patch visited, X_2 the second patch, etc. Assume that the pollinator searches among patches until it finds one with an energetic reward greater than some criteria level. When a patch meeting this criterion is found, the pollinator exploits the patch. The cost of moving from patch to patch is some constant c. Hence, the net energetic gain if the pollinator stops at the nth patch is

$$Y_n = \text{Max}\ (X_1 \ldots, X_n) - nc$$

For simplicity, let c be a constant; the argument can be easily extended to variable costs of moving between patches (Lippman and McCall, 1976). The pollinators' problem is to decide when to stop searching and to remain in a given patch. In other words, the pollinator maximizes $E(Y_n)$ where n is the random stopping time. The problem, as I have posed it, is identical to the job search problem in microeconomics theory. In this theory, an unemployed worker must decide whether to continue searching for jobs or accept a particular job with a fixed wage. Lippman and McCall (1976, 1979) provided the latest findings in economic search theory, and the notation I will be using throughout the text is theirs.

Let ξ be the criterion level for stopping search if the pollinator follows the best stopping rule. After Lippman and McCall (1976), any optimal decision rule by the pollinator then must satisfy the conditions

$$\text{Stay in patch if } X \geq \xi$$
$$\text{Continue search if } X < \xi$$

That is, if the energetic reward of the patch is equal to or greater than the expected gain from searching by the optimal rule, the pollinator should choose to forage in the patch.

Consider the first patch, X_1. For a pollinator following this simple decision rule, the expected energetic reward is given by

$$E \max\ (\xi, X_1) - c \tag{1}$$

Since ξ is defined as the expected energetic reward obtained through the best stopping role, the expected reward obtained must satisfy the condition

$$\xi = E \max\ (\xi, X_1) - c \tag{2}$$

The expected reward, $E \max\ (\xi, X_1)$, is the probability of the patch being below ξ

and the probability of the patch being above ξ weighted by their respective rewards:

$$
\begin{aligned}
E \max (\xi, X_1) &= \xi \int_0^\xi dF(X) + \int_\xi^\infty X dF(X) \\
&= \xi \int_0^\xi dF(X) + \xi \int_\xi^\infty dF(X) + \int_\xi^\infty X dF(X) - \xi \int_\xi^\infty dF(X) \\
&= \xi + \int_\xi^\infty (X - \xi) dF(X)
\end{aligned}
$$

So, equation (2) is identical with

$$
c = \int_\xi^\infty (X - \xi) dF(X) = H(\xi) \tag{3}
$$

where $H(X) = \int_X^\infty (y - X) dF(y)$

Equation (3) is the fundamental equation of search theory and has a simple interpretation. The critical value ξ is chosen to equate the cost of additional search c with the expected marginal return from one more observation of an alternative patch, $H(\xi)$. Note the similarities between this optimal stopping policy and the optimal giving-up policy advanced in the marginal value theorem (Charnov, 1975). These rules are compatible in that the former dictates when a pollinator should quit searching and start foraging in a given patch, whereas the latter dictates when a pollinator should quit a patch and resume searching.

It can be shown by taking successive derivatives that the function $H(x)$ is convex, nonnegative, and strictly decreasing (Fig. 1). We see immediately that the lower the search cost c the higher the acceptable reward level ξ. When search costs are low, pollinators will choose to search for longer durations and will require higher reward levels.

Given that we know $F(x)$ and c, equation (3) provides information on the level of acceptable rewards for different patches and a rule for calculating the acceptable reward levels that will induce a forager to stop searching and forage. We have seen that the propensity toward continued search is inversely related to search costs. Within the model's assumptions, we can ascertain the influence of several other ecological factors that promote search. These have been summarized by Kohn and Shavell (1974) and Stuart (1979) for the economic case. In addition to the stopping policy being influenced by search cost, it can be shown that:

1. An increase in the density of patches in the habitat increases the expected value of searching.

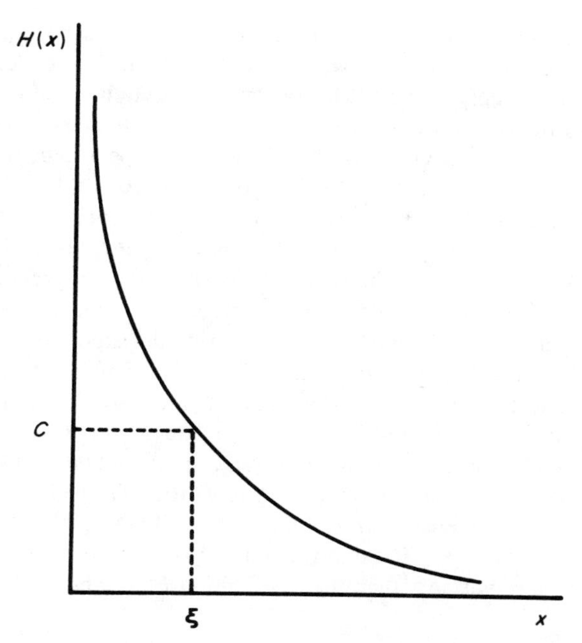

Fig. 1. The graph of the function $H(x) = \int_{\xi}^{\infty} (y - x)dF(y)$ representing the expected marginal return to the pollinator from one more observation of an alternative patch of flowers. The function $H(x)$ is convex, nonnegative, and strictly decreasing. A given search cost c fixes the acceptable level of reward ξ that will induce a forager to quit searching and begin foraging within a patch.

2. A translation of $F(x)$, the cumulative distribution of energetic rewards, to the right (i.e., an overall increase in reward at each patch) increases the expected value of a search.
3. An increase in the variability of rewards among patches increases the expected value of a search.
4. An increase in the forager's aversion to risk decreases the value of continued search.

The first two properties are easy to interpret. The greater the density of patches, the greater the opportunity to encounter a patch of higher quality. For instance, if there are two patches of flowers in the habitat that can be sampled, there is always a nonzero *probability* that the energetic rewards found in one of the patches will be less than the net expected energetic gain from sampling the other even though, at any particular instant, this probability may not be realized. This probability increases with the number and density of patches available for sampling. The second property, a translation of F to the right, indicates that some patches are now providing greater energetic rewards. Naturally, then, it may prove beneficial to continue searching since the expectation of a higher reward through search has increased. Conversely, when the quality of floral resources among patches is very low, we expect pollinators to be less selective in their choice of patches, and the general level of search should diminish. For habitats where floral resources are of high quality, pollinators engage in greater amounts of searching behavior and should therefore appear to be more selective. This point echoes the earlier conclusions on patch choice by MacArthur and Planka (1966) and Emlen (1973). Although the conclusions are similar to those reached in these earlier papers, the models used in previous analyses are very different from the model used here. I will discuss these differences later.

The third and fourth properties of the search model require more clarification. In order to examine the effects of increased variability among patches, we must establish (1) a measure of variability and (2) the manner in which the foragers generally respond to variability in shaping their foraging preferences.

Consider two habitats, G and F, each with characteristic cumulative distribution functions $G(X)$ and $F(X)$. What criterion can be applied to these habitats to determine if G is more variable than F in reward/patch distributions? For this model, I use a strengthened form of the Rothschild–Stiglitz (1970) condition advocated by Stuart (1979). The distribution $G(X)$ of rewards among patches will be considered more variable than the distribution $F(X)$ when

$$\int^{X} \{G(y) - F(y)\}\, dy = 0 \text{ if } G(X) = F(X) = 1 \quad (4a)$$

$$\int^{X} \{G(y) - F(y)\}\, dy > 0 \text{ if } G(X) > 0 \text{ and } F(X) < 1 \quad (4b)$$

Condition (4a) guarantees that the mean is preserved, and (4b) implies that distribution G has more weight in the tails. Hence, this criterion distinguishes the spread from the central tendency of the distribution.

In general, the forager can respond to spread (or what is usually called "risk" in the foraging literature) in one of three ways (Real, 1980a,b). Consider two distributions of rewards F and G with equal means but with the variance or spread of G greater than that of F. If the forager shows no preference for foraging from either F or G, then the foraging is termed "risk neutral." If the forager prefers F to G, then it is called "risk averse." If it prefers G to F, then it is called "risk accepting."

It is biologically very important to distinguish between these two aspects of a distribution of rewards. Mounting empirical evidence indicates that organisms respond to both the mean and the spread of reward distributions. Caraco (1980) and Caraco *et al*. (1980) showed that the response to risk, i.e., whether a bird is risk prone or risk averse, depends upon whether the expectation of the bird's foraging meets its daily energy requirements. This differential response was treated analytically by Stevens (1981). Real and co-workers (1981a; also see

Real *et al.* 1982) have shown that for risk-averse bumblebee foraging there is generally a linear trade-off between the means and the variances in reward distributions. Waddington *et al.* (1981) also showed that the level of risk sensitivity is itself variable among individual foragers. Some form of differential response to variability in reward distributions while foraging seems to be a general feature of animal behavior.

The search model articulated here and leading to property (3) assumes risk neutrality and a Rothschild–Stiglitz measure of variability on the part of foragers. In essence, this means that when there is no risk preference, increasing heterogenerty among patches will increase the expectation of continued search. However, when the forager is assumed to be risk averse, increasing heterogenerty decreases the expected value of search (property 4). Obviously, then, the final value to the forager of increasing variability among patches depends jointly on the magnitude of the variability and the magnitude of the forager's aversion to risk. Increasing variability will increase the probability of finding a better patch through more search, but this advantage is offset by greater orders of risk aversion.

A. A Nonintuitive Theorem on Patch Choice

In the move from an analysis of microbehavior to an understanding of macrostructure, we can begin by adding a greater degree of complexity to the habitat. The preceding model supposed that the foragers encountered a single patch of flowers with some determined reward level X_n with an associated search cost c at any one point in time, and this encounter process is repeated over successive time intervals. However, it is certainly possible that a forager can encounter more than one patch of flowers during a single search period. This situation can be rendered mathematically by letting the number T_i of patches encountered on the ith search move be a nonnegative random variable.

For example, we could consider a hummingbird foraging on different *Heliconia* species in a tropical forest. Some *Heliconia* species (e.g., *H. tortuosa*) occur as widely spaced, low-density patches, whereas other species (*H. latispatha*) occur as highly clumped, dense patches (Linhart, 1973). While foraging on *H. tortuosa,* the hummingbird might encounter only a single patch at a time and thus correspond to the preceding model of foraging. However, for a hummingbird utilizing *H. latispatha,* several patches might come into view at any one time; thus, it must choose from among many patches the single patch in which it will forage. Feinsinger and Colwell (1978) have shown that within hummingbird assemblages on islands, different hummingbird species will specialize on exploiting floral resources showing these kinds of differences in density and spatial distribution. Consequently, models describing the foraging behavior of each kind of specialist will have to incorporate different encounter rates.

For ease in comparison with the previous model (equation 3), assume that $E(T) = 1$, so that the expected cost per search remains c. The acceptable reward level ξ is then determined by the equation

$$\xi = -c + \sum_{k=0}^{\infty} \left(\xi F(\xi)^k + \int_{\xi}^{\infty} Xd\,[F(X)^k] \right) P(T = k) \tag{5}$$

This is equal to equation (3) when $P(T = 1) = 1$. Lippman and McCall (1976; appendix 1) demonstrate that equation (5) is equivalent to

$$c = -\xi + c^{\xi - 1} \tag{6}$$

Consequently, the fundamental equation (3) becomes,

$$c = \xi^2/2 - \xi + ½ \tag{7}$$

From equations (6) and (7), we see that the acceptable level is fixed by either $c > e^{-1}$ or $c > ½$, respectively. These conditions are graphically illustrated in Fig. 2, which shows that the forager would prefer to have only one patch become

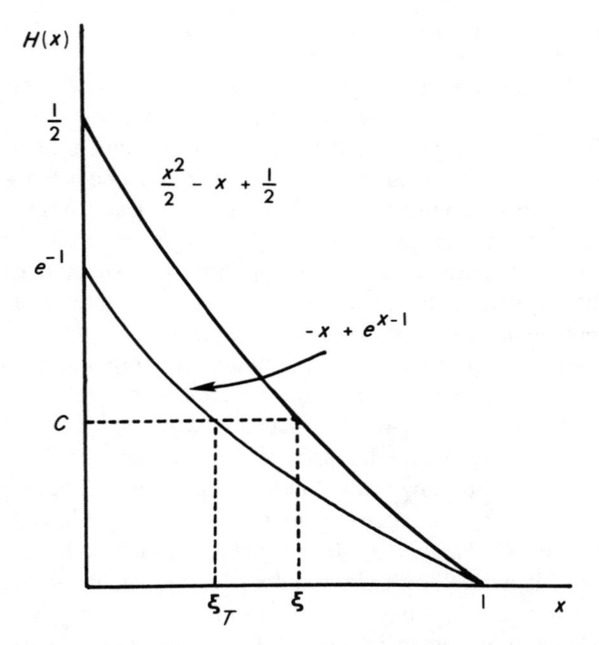

Fig. 2. The graphs of $H(x) = x^2/2 - x + ½$ and $H(x) = -x + e^{x-1}$ representing foraging when a single patch is available per search and when a random number T of patches become available per search, respectively. For the same search cost, the forager will quit searching and begin foraging at a lower acceptable level of reward when patches appear as a random number per search.

available at each search than a random number with a mean of 1, and that the acceptable reward level ξ will be lower when the number of patches appearing during a search is a random variable.

This result is far from intuitive and depends strictly upon the mathematical structure of the model. Yet there is a standard explanation for this result with some intuitive appeal. First, on any given search, in the random model, several acceptable patches can appear. Second, the forager can utilize only one of these acceptable patches at a time. Consequently, even though the per search cost is still c, the cost per utilized acceptable patch has increased. This conclusion is generalizable to a random distribution of patch numbers with any expectation $E(T_i) = n$. The pollinator will always prefer to encounter the fixed number $T_i = n$ rather than a random number with expectation n.

III. Habitat Selection and Flowering Time

The above model dictates how a forager's expected benefit from a search will be influenced by the conditions of a given habitat. The same arguments apply to choice of a habitat in which to forage. To analyze this, we make the following assumptions: (1) a forager uses several resources within a habitat, e.g., nectar, pollen, other food resources, nesting sites, refuges, etc.; (2) the forager experiences the same search costs for each of the resources it wishes to find, e.g., the differential costs for finding nectar and pollen are not appreciable; (3) the forager incurs a cost c_m to utilize habitat m, where m is a finite index of habitats; and (4) the forager can use only one habitat at a time. Let u_{im} denote the maximal utility attainable for resource i in habitat m. Caraco (1980) presents a general discussion of the biological meaning of utility. The utility of engaging in a particular activity or using a particular resource is associated with the long-term evolutionary fitness associated with using that resource; the higher the utility value, the higher the resulting fitness. The organism's perception of the utility values attached to different activities or resources is presumed to be shaped by natural selection.

The forager's decision is then to choose the habitat that maximizes

$$-c_m + \sum_{i=1}^{n} u_{\text{im}}$$

where $u_{\text{im}} = 0$ if resource i is not found in habitat m. This statementof the habitat selection problem and the preceding formulation is identical to the marketplace choice problem discussed by Stuart (1979). Stuart uses job search theory to address the question: When should sellers agglomerate, forming a single marketplace? The analogy to flowering plants being selected to occur in a single habitat and flowering simultaneously is direct.

For the preceding model, we see that one habitat will be more attractive to foragers than another if the following conditions hold:

1. There is a greater number of resources utilized by the forager in one of the habitats.
2. There is a general decrease in the level of search costs and/or travel costs within one of the habitats.
3. The perceived reward levels within one habitat are larger than those within another habitat.
4. Reward levels among patches within one habitat appear to be more variable than those within another habitat.
5. There is a greater number of discrete patches from which the pollinator can choose.

IV. Flowering Time Convergence and Divergence

The conditions just described dictate the manner in which foragers choose one habitat over another. If foragers are at all important in the plant's reproductive biology, then we can use these conditions to specify some of the selective forces acting on plant assemblages and generating their structure. What magnitude of attractiveness (based on these specified conditions) within a habitat must be achieved within the plant community to override the competitive effects of simultaneous flowering and promote convergence in flowering time? By determining this magnitude, we can specify the behavioral precursors to a very important community pattern, namely, community phenology.

The exact manner in which these conditions combine to determine attractiveness will depend in large part upon the peculiarities of the pollinators, their histories, and the plants they visit. Yet, we can make certain generalizations about the connection between foraging and community structure independent of specific conditions.

If convergence is to prove advantageous, then each individual plant must fare better when flowering collectively than when flowering alone. The conditions for this advantage can be stated explicitly.

Consider a simplified habitat where S_N plants (of the same or different species) provide nectar and S_P plants (of the same or different species) provide pollen for foragers in the habitat. Let the number of foragers, $H(S_N, S_P)$, be a function of the number of plants providing pollen and nectar. I assume that $H(S_N, S_P)$ is a monotonically increasing function of its arguments, i.e., the more plants providing nectar and/or pollen in an area, the more foragers there will be in that area. The increase in foragers can be due either to increased growth of established populations of foragers caused by the increased availability of their limiting resource (e.g., the number and size of bee colonies is correlated with the density of flowers in the habitat; Richards, 1975) or to the differential selection by mobile foragers of habitats showing a relative greater abundance of flowers (e.g., the number of hummingbirds in a Rocky Mountain habitat is correlated with floral density; Waser and Real, 1979). Considering nectar plants only, the average use of (or demand on) each individual nectar plant, Q_N, is given by

$$Q_N = \frac{H(S_N, S_p)}{S_N} \tag{8}$$

and

$$\frac{\partial Q_N}{\partial S_N} = \frac{1}{S_N}\left(\frac{\partial H}{\partial S_N} - \frac{H}{S_N}\right) \tag{9}$$

represents the change in individual demand in relation to a change in the number of plants providing nectar. For the change to be positive, the attraction effect must be greater than the competition effect, i.e.,

$$\frac{\partial H}{\partial S_N} > \frac{H}{S_N} \tag{10}$$

which implies that

$$\left(\frac{\partial H}{\partial S_N}\right) \Big/ \frac{H}{S_N} = \frac{\partial(\log H)}{\partial(\log S_N)} > 1 \tag{11}$$

The left-hand side of inequality (11) has a simple biological interpretation. The expression is the proportionate change in H, the number of foragers, relative to a proportionate change in S_N, the number of nectar-producing plants. This expression I call the "elasticity of attraction," after its economic analog, the "elasticity of demand" (Henderson and Quandt, 1971). For convergence to prove advantageous for the plant, the elasticity of attraction must be greater than 1; that is, a percentage change in the number of plants providing nectar must result in a more than proportionate change in the number of foragers in the area.

When the percentage change in floral number produces an equivalent change in pollinator numbers, there will be no advantage in convergence. Floral species will be indifferent (cf. Rapport, 1971) as to the time of flowering, and we would expect flowering times to be spread randomly throughout the possible flowering interval. Poole and Rathcke (1979) and Rabinowitz *et al.* (1981) have analyzed flowering sequences that suggest randomness, as they cannot be distinguished from a random array of flowering times drawn from a uniform distribution.

This leaves us with the major question: What biological conditions will generate an elasticity of attraction greater than 1?

Perhaps the most obvious (and most interesting) condition occurs when the per capita within-habitat growth rate of the pollinator population is greater than the per capita growth rate of the population of flowers they visit. This differential can result from either changes in birth and death rates or immigration of the pollinator into the habitat *relative* to the growth rate of the plants. Consequently, we expect to see flowering time convergence in habitats experiencing relatively rapid migration and/or growth in pollinator numbers. The existence of this differential can be determined empirically. We simply need to adjust artificially the number of plants in flower in a habitat by a certain percentage and then monitor percentage changes in pollinators utilizing these plants. For floral species presumed to be convergent, a percentage decrease in floral density should result in a greater than equal percentage decrease in pollinator density. The opposite would be true for plants presumed to be divergent due to the competition effect.

Floral convergence will also prove disproportionately attractive to pollinators if there are considerable nonadditive benefits to the pollinator which come about by the increased abundance of flowers. These benefits are manifested as decreases in searching and traveling between plants or an increases in harvesting efficiency due to learning or search image development. However, biotic interactions between foragers, e.g., territoriality or aggression, may reduce the likelihood of convergence since they may reduce the elasticity of attraction. Territoriality will prevent the plants from ever experiencing the benefits of increased attractiveness, since foragers drawn to the area may be turned back by dominant individuals already established in the area.

A third advantage to the pollinator occurs when the simultaneous flowering of species reduces the overall uncertainty of foraging in the habitat. As noted, both theoretical and empirical study suggests that pollinators will avoid risky foraging strategies (Real, 1980a,b, 1981; Real *et al.*, 1982; Caraco, 1980; Caraco *et al.*, 1980). The overall variance in nectar reward can be reduced through simultaneous flowering relative to the variance in reward for each species flowering separately. A simple numerical example will demonstrate this. Suppose there are two plant species, A and B, that share pollinators. In any given season, the individuals of each species produce uniform quantities of nectar per flower, but these qualities can vary from season to season. Let species A and B produce respectively floral rewards of 10 μl with a probability of ½ and 20 μl with a probability of ½. If the two species flowered separately, then the expected floral reward for any given season will be 15 μl with a variance equal to 25 (Table I). However, if the two plants flower synchronously, the probability of all flowers producing 10 μl is ¼ (Table II). In addition, there is now the combined probability that species A is producing 10 μl of flowers, whereas species B is producing 20 μl of flowers (and vice versa). The expected reward is still 15 μl/flower, but the variance is reduced to 12.5. This very artificial example indicates that combining the two distributions can reduce the temporal uncertainty for the

TABLE I

EXPECTED FORAGING REWARD WHEN FLOWERING ASYNCHRONOUSLY

Species A				Species B			
Nectar/flower		Probability		Nectar/flower		Probability	
10 μl		½		10 μl		½	
20 μl		½		20 μl		½	
$E(X_A)$	=	15 μl		$E(X_B)$	=	15 μl	
Var. (X_A)	=	25		Var. (X_B)	=	25	

foragers and may do so nonadditively. This nonadditive reduction in variance may generate the disproportionate attractiveness for a given time necessary to promote floral convergence.

Many insects separate their foraging using some plant species for pollen and others for nectar. What would be the impact of increasing the number of pollen plants on the use of individual nectar plants? The change in Q_N is given by

$$\frac{\partial Q_N}{\partial S_P} = \frac{1}{S_N}\left(\frac{\partial H}{\partial S_P}\right) \tag{12}$$

and is always positive. Intuitively, the use of individual nectar plants will always increase if they ococcur together with plants providing pollen. This is equally true for pollen plants. Consequently, we expect these plant types to show synchronous flowering.

Under conditions of unlimited nectar and/or pollen availability, the relation-

TABLE II

EXPECTED FORAGING REWARD WHEN FLOWERING SYNCHRONOUSLY

Species A + B		
Nectar/flower		Probability
10 μl		¼
15 μl		½
20 μl		¼
$E(X_{A+B})$	=	15 μl
Var. (X_{A+B})	=	12.5

ship between the number of plants providing these resources and the number of foragers utilizing them will be inelastic, i.e., a change in plant number will not result in a change in forager number. Since under these conditions the attraction effect $\partial H/\partial S_N$ will be zero but the dilution effect H/S_N will be increasing, we expect the plants to diverge.

Several issues have not been discussed. First, I have emphasized convergence in my arguments. Of course, antithetical arguments can be constructed that promote divergence. The important point to emphasize is that convergence, divergence, or neutrality under the forces of attraction depend upon the elasticity of attraction. Second, I have discussed only variation in the quantity and not in the quality of flowers in determining flowering time. The number of flowers in the habitat is surely not the only criterion used by pollinators in making their decisions as to when and where to forage. Variation in the quality of nectar and pollen will most certainly influence the evolution of flowering time, and future models should incorporate this variable.

All too often in analyses of population phenomena, we are constrained to descriptions of a single level of organization. In this chapter, I have tried to link aspects of community structure with the underlying behavior of individuals that compose an integral part of that community. The two levels of organization can now be viewed in interaction. I hope to have shown that at least some of the interesting macroscopic feature of communities can be deduced from the microscopic behaviors of their components. Uncovering the ties between microscopic behavior and macroscopic structure will undoubtedly help unify the diverse fields of population biology.

Acknowledgments

I thank Janis Antonvics, Beverly Rathcke, Jeff Walters, Roger Powell, Frank Benford and Tom Caraco for carefully reading the manuscript and making many helpful suggestions. I also thank Peter Feinsinger, Nick Waser, Bob Holte, Scott Gordan, and Hal Caswell for their thoughts on earlier versions of these ideas. This work was supported by a grant from the National Science Foundation (No. DEB-802089). This is paper number 8848 of the Journal Series of the North Carolina Agricultural Research Service, Raleigh, North Carolina.

References

Brown, J., and Kodric-Brown, A. (1979). Convergence, competition, and mimicry in a temperate community of humming-bird pollinated flowers. *Ecology* **60,** 1022–35.

Caraco, T. (1980). On foraging time allocation in a stochastic environment. *Ecology* **61,** 119–128.

Caraco, T., Martindale, S., and Whitman, T. S. (1980). An empirical demonstration of risk-sensitive foraging preferences. *Anim. Behav.* **28,** 820–830.

Carpenter, F. L., and MacMillan, R. E. (1976). Energetic cost of feeding territories in an Hawaiian honey creeper. *Oecologia* **26,** 213–223.

Charnov, E. (1975). Optimal foraging: The marginal value theorem. *Theor. Pop. Biol.* **9,** 129–136.

Emlen, J. M. (1973). "Ecology: An Evolutionary Approach." Addison-Wesley, New York.

Feinsinger, P. (1976). Organization of a tropical guild of nectarivorous birds. *Ecol. Monogr.* **46,** 257–291.

Feinsinger, P., and Colwell, R. K. (1978). Community organization among neotropical nectar-feeding birds. *Am. Zool.* **18,** 779–795.

Gentry, A. H. (1974). Flowering phenology and diversity in tropical Bignoniaceae. *Biotropica* **6,** 64–68.

Gill, F., and Wolf, L. (1975). Economics of territoriality in the golden-winged sunbird. *Ecology* **56,** 333–345.

Heinrich, B. (1975). Bee flowers: A hypothesis on flower variety and blooming times. *Evolution (Lawrence, Kans.)* **29,** 325–334.

Henderson, J. M., and Quandt, R. E. (1971). "Microeconomic Theory." McGraw-Hill, New York.

Hocking, B. (1968). Insect-plant associations in the high Arctic with special reference to nectar. *Oikos* **19,** 359–388.

Johnson, L. K., and Hubbell, S. P. (1974). Aggression and competitoin among stingless bees: Field studies. *Ecology* **55,** 120–127.

Johnson, L. K., and Hubbell, S. P. (1975). Contrasting foraging strategies and coexistance of two bee species on a single resource. *Ecology* **56,** 1398–1406.

Kodric-Brown, A., and Brown, J. (1978). Influence of economics, interspecific competition, and sexual dimorphism on territoriality of migrant rufous humming birds. *Ecology* **59,** 285–296.

Kohn, M. G., and Shavell, S. (1974). The theory of search. *J. Econ. Theor.* **9,** 93–124.

Levin, D., and Anderson, W. (1970). Competition for pollinators between simultaneously flowering species. *Am. Nat.* **104,** 455–67.

Linhart, Y. (1973). Ecological and behavioral determinants of pollen dispersal in hummingbird-pollinated *Heliconia. Am. Nat.* **107,** 511–523.

Lippman, S. A., and McCall, J. J. (1976). The economics of job search: A survey. *Econ. Inquiry* **14,** 155–189.

Lippman, S. A., and McCall, J. J. (1979). "Studies in the Economics of Search." North-Holland Publ., Amsterdam.

MacArthur, R., and Planka, E. (1966). Optimal use of a patchy environment. *Am. Nat.* **100,** 603–609.

Mosquin, T. (1971). Competition for pollinators as a stimulus for the evolution of flowering time. *Oikos* **22,** 398–402.

Pimm, S. L. (1978). An experimental approach to the effects of predictability on community structure. *Am. Zool.* **18,** 797–808.

Poole, R., and Rathcke, B. (1979). Regularity, randomness, and aggregation in flowering phenologies. *Science* **203,** 470–471.

Rabinowitz, D., Rapp, J. K., Sork, V. L., Rathke, B. J., Reese, G., and Weaver, J. (1981). Phelological properties of wind- and insect-pollinated prairie plants. *Ecology* **62,** 49–56.

Rapport, D. (1971). An optimization model of food selection. *Am. Nat.* **105,** 575–87.

Real, L. A. (1980a). Fitness, uncertainty, and the role of diversification in evolution and behavior. *Am. Nat.* **115,** 623–638.

Real, L. A. (1980b). On uncertainty and the law of diminishing returns in evolution and behavior. *In* "Limits to Action" (J. E. R. Sraddon, ed.), pp. 37–64. Academic Press, New York.

Real, L. A. (1981a). Uncertainty and pollinator–plant interactions: The foraging behavior of bees and wasps on artificial flowers. *Ecology* **62,** 20–26.

Real, L. A. (1981b). Nectar availability and bee-foraging on *Ipomoea* (Convolvulaceae). *Biotropica* **13** (Suppl.), 64–69.

Real, L., Ott, J., and Silverfine, E. (1982). On the trade off between the mean and the variance in foraging: Effects of spatial distribution and color preference. *Ecology* **63,** 1617–1623.

Richards, K. W. (1975). Population ecology of bumblebees in southern Alberta. Ph.D. Thesis, Univ. of Kansas, Lawrence.

Rothschild, M., and Stiglitz, J. (1970). Increasing risks I: A definition. *J. Econ. Theory* **2,** 225–243.

Schaffer, W. M., Jensen, D. B., Hobbs, D. E., Gurevich, J., Todd, J. R., and Schaffer, M. J. (1979). Competition, foraging energetics, and the cost of sociality in three species of bees. *Ecology* **60,** 976–987.

Schelling, T. (1978). "Micromotives and Macrobehavin." Norton, New York.

Stevens, D. W. (1981). The logic of risk-sensitive foraging preferences. *Anim. Behav.* **29,** 628–629.

Stiles, F. G. (1977). Coadapted competitors: The flowering seasons of hummingbird pollinated plants in a tropical forest. *Science* **198,** 1177–1178.

Stuart, C. (1979). Search and the spatial organization of trading. *In* "Studies in the Economics of Search" (S. A. Lippman and J. J. McCall, eds.), pp. 17–33. North-Holland Publ., Amsterdam.

Waddington, K. D., Allen, T., and Heinrich, B. (1981). Floral preferences of bumblebees (*Bombus edwardsii*) in relation to intermittent versus continuous rewards. *Anim. Behav.* **29,** 779–784.

Waser, N. (1978). Competition for pollination and sequential flowering in two Colorado wild flowers. *Ecology* **59,** 934–44.

Waser, N. (1983). Competition for pollination and floral character differences among sympatric plant species: A review of evidence. *In* "Handbook of Experimental Pollination Biology" (C. E. Jones and R. J. Little, eds.), pp. 277–293. Van Nostrand-Reinhold, New York.

Waser, N., and Real, L. A. (1979). Effective mutualism between sequentially flowering plant species. *Nature (London)* **281,** 670–672.

Wolf, L. (1978). Aggressive social organization in nectarivorous birds. *Am. Zool.* **18,** 765–778.

CHAPTER 12

Competition and Facilitation among Plants for Pollination

BEVERLY RATHCKE

Division of Biological Sciences
University of Michigan
Ann Arbor, Michigan

I. Introduction

Plant–pollinator systems offer a rich diversity of possible interactions and adaptations for our study, but interactions among plants for pollination seem especially intriguing because together plants can support, attract, and share pol-

POLLINATION BIOLOGY

ISBN 0-12-583980-4 (hardcover)
ISBN 0-12-583982-0 (paperback)

linators, and they can interact negatively or positively. I use the term ''facilitation'' to connote positive interactions due to resource sharing within a guild (Lynch, 1978), such as plants interacting for pollination. I distinguish this term from ''mutualism'' which I argue should be reserved for mutually positive interactions among different types of organisms, such as plants and their pollinators. Although ''mutualism'' has been used to encompass both types of interactions (Boucher *et al.*, 1982), this broad usage can be confusing because it often carries the connotation of symbiosis and mutual adaptations. For pollination systems, the term ''mutalism'' has been applied extensively to plant–pollinator interactions and their coevolution, which I am not addressing here. In plant–pollinator systems both levels are possible: mutualism among plants and pollinators and facilitation among plants for pollinators. Facilitative interactions within a guild are much less well understood, and they may be less likely to promote coevolution because the species associations will often be loose and variable (Connell, 1980). However, Thompson (1982) argues that in evolutionary time shared mutualists will tend to promote coevolution among small sets of competing species. Here I present some models and empirical approaches for examining these interactions, and I review the evidence for their occurrence in natural populations and their effects on pollination and seed set. I also review the evidence for both divergent and convergent character displacement which could arise from selection for the avoidance or promotion of these interactions.

Plant–pollinator systems have a number of idiosyncracies which may limit generalizations to other interactive systems. On the other hand, the often specialized terminology developed for pollination may mask the general nature of some features. Some seemingly unique interactions may prove to be more prevalent than is currently realized, and I hope to encourage their consideration by briefly describing the general nature of plant–pollinator systems. Both plants and pollinators act as resource users as well as resources; here I will focus on plants as users. Plants acting together both support and attract pollinators in local areas, and yet individuals must vie with each other for the services of the pollinators. Facilitative and competitive interactions can exist concurrently and net effects may vary. Because pollinators are mobile, boundaries are not well defined and local visitation may reflect resource support from other areas. Pollinator populations will also be highly influenced by factors other than their floral resources. Because local populations are open systems, pollinator limitation may alternate with overabundance in different sites and years, lending unpredictability to interactions and their outcomes.

Because resource (pollinator) sharing within a plant species can be essential to reproduction, the evolution of competitive ability for pollinator attraction will be limited by the constraints of satisfying the foragers but not satiating them. Intraspecific competition for divergence in resource use will be opposed by selection for pollinator sharing and convergence. The rich diversity of floral types could result from plant–pollinator coevolution and selection for effective pollen transfer, as well as from competition or facilitation among plant species for pollinators. The contribution of these different selective regimes may be difficult to separate.

These interactions can directly affect seed set, a good index of fitness, but the relationship of seed set to recruitment of the next generation is complex. However, the greater seed set may increase probability of success, however low (e.g., Louda, 1982). By measuring the seed set of individuals, we do not have to make inferences about the effects of different processes based upon average population responses; instead we can detect differential selection upon individuals within the population (Thompson, 1982). In these systems, we may be able to translate ecological interactions quite directly into natural selection and the ecological and evolutionary outcomes.

II. Competitive and Facilitative Interactions: Definitions

At any one point in time, a plant can interact for pollination via pollinator visitation (PV) or improper pollen transfer (IPT), and the interaction of these two mechanisms will determine fertilization success. These processes must be distinguished because the ecological or evolutionary outcomes can be unique. For example, selection for differential pollen placement would occur through IPT but not through competition for pollinators. Also, the existence of IPT or effects on PV will not necessarily influence fertilization success. Even if one plant species attracts more pollinators, visitation to a competing species may remain sufficient to produce maximal seed set. Seed set could even increase with lower visitation if IPT is reduced and visitation quality increased. Final seed set will also be determined by factors such as nutrient limitation, and this may or may not reflect initial fertilization success. Clearly, interactions for PV and IPT must be separated from seed set reduction due to pollination limitation, and I will discuss these separately.

A. Improper Pollen Transfer

IPT will cause pollen loss by the donor and may result in the accumulation of improper pollen on stigmas. Experiments show that foreign pollen can lower seed set through stigma clogging, exploitation, chemical or physical interference by pollen, or production of inviable or sterile hybrids (Levin, 1972; Mulcahy, 1975; Thomson *et al.*, 1981; Waser, 1978). This mechanism has usually been termed "interspecific pollen transfer," but intraspecific pollen transfer could produce similar results if the pollen is genetically incompatible. Also, since it becomes difficult to draw a line between purported species, I propose the more general term "improper pollen transfer." IPT has also been called "interference competition," but I agree with Waser (1982) that this mechanism should not be forced into this zoologically derived term and that we should label its distinctive nature.

IPT is assumed to result always in a negative or neutral effect on seed set or quality, although within a species pollen grains can apparently show facilitative effects by increasing pollen tube growth and fertilization success (Brewbaker and Majumder, 1961; Jennings and Topham, 1971; Mulcahy, 1975). Whether this could happen for interspecific pollen is unknown. If deleterious hybridization occurs, the costs become more severe because of the loss of maternal effort in supporting the hybrid and possibly because of competition by the hybrid if it subsequently survives and grows. Selection should be stronger when deleterious hybridization is involved than in other cases of IPT.

Although both participating species would suffer some reproductive loss through pollen loss by the donor and possible IPT effects on the recipient, selection is likely to be stronger on the recipient because seed loss is probably a more severe cost than pollen loss. Pollen transfer can be different for the two participants, and crossability can also differ, placing the burden of hybridization disproportionately on one species (Levin and Schaal, 1970). The effects are not necessarily symmetrical, as Waser (1982) suggests. In evolutionary time, one species may tend to diverge from another species.

B. Interactions for Pollinator Visitation

Plants could interact for PV in three ways: competition (−,−; −,0), facilitation (+,+; +,0), or resource parasitism (+,−). Or we could find no measurable effects (0,0; neutralism). Competition for PV would occur when one species (or phenotype) attracts pollinators away from another (−,0) or when both suffer reduced visitation because they share pollinators (−,−). Waser (1982) refers to this as "competition for pollinator preference," but I would like to avoid invoking a specific behavior. Facilitation would occur when the presence of one species (or phenotype) increases visitation to another species at no cost to the second species. This has been termed "mutualism" (Brown and Kodric-Brown, 1979; Waser and Real, 1979) or "cooperation" (Thomson, 1978, 1981, 1982). I later discuss various ways in which facilitative effects could arise. For resource parasitism, one species would draw pollinators away from another and show enhanced visitation in the presence of this second species. Mimicry is one specific example of this type of interaction whereby the mimic gains by "parasitizing" the pollinators from the model species which loses visits. No term currently exists for such a general +,− interaction, so I have adopted Hazlett's term, "resource parasitism," which he has proposed for describing the +,− effects of shell trading by some hermit crabs (B. A. Hazlett, personal communication).

In these definitions I specifically include asymmetrical effects (e.g., −,0 or

+,0) because they are often disregarded. Their occurrence in numerous studies suggests that asymmetrical interactions are common (Lawton and Hassell, 1981), and they should be recognized because the outcomes of asymmetrical and symmetrical interactions will be quite different. I also distinguish these within-guild interactions from commensalism (+,0) and amensalism (−,0), which are more commonly applied to symbiotic relationships between different types of organisms.

In addition to attracting shared pollinators, plants can interact through supporting shared pollinators. By providing nutrients, they support the survival and reproduction of pollinators through the season and from year to year. At any one point in time, the local available nutrients (or rewards) will attract the local pollinators to various sites and plants. Any measurement of visitation rate per flower will reflect all these factors; however, pollinator support will set the absolute density of pollinators, and current attractiveness will determine the relative distribution of visits at any point in time measured as an instantaneous visitation rate.

C. Interactions for Pollination

Both IPT and PV will interact to determine the pollination success of flowers, and the final outcome on fertilization may be negative, positive, or neutral for one or both participants. Waser (1982) subsumes these two mechanisms into competition for pollination, which he defines as any interaction in which co-occurring plant species (or phenotypes) suffer reduced reproductive success because they share pollinators. If the interaction results in increased reproductive success, this could be termed "facilitation for pollination." But the effects could be asymmetrical or even mixed (+,−) and could result from numerous causes. The contributions of these different mechanisms to seed set are difficult to distinguish and often remain confounded. To establish pollinator limitation, the number of visits by any one pollinator type necessary for maximal seed set must be compared with the observed visitation rates. For IPT limitation, the observed levels of improper pollen must be shown to influence seed set; often the improper pollen cannot be distinguished from proper pollen.

III. Interactions for Pollination: Models

A. Density Effects on Pollination

Numerous studies show that various pollinators, especially bumblebees and hummingbirds, can assess the costs and rewards of various floral choices, and their visitation can be highly sensitive to resource density and dispersion (Waddington and Heinrich, 1981; Real, 1981; Real *et al.*, 1982). In dense floral patches, bumblebees tend to turn more and to visit more plants than in sparse patches (Heinrich, 1979; Schaal, 1980; Waddington and Heinrich, 1981; Roubik, 1982). Although quantitative data are lacking, visitation rates per flower would seem likely to change with density given the behavior of pollinators. Some evidence suggests that rare flowers have a minority disadvantage (Johnson and Hubbell, 1975; Silander, 1978; Levin and Anderson, 1970). Levin (1972) found that white flowers of *Phlox* were visited proportionately less by butterflies than pink flowers when they were rare. At the other extreme, as floral resources become very abundant, they may recruit pollinators from greater distances, but eventually they may saturate the available pollinator pool and visitation per flower will decrease.

Here I propose a density–visitation model which describes how pollinator visitation per flower may change as floral density increases within a plant species or in a mixture of plant species which share pollinators (Fig. 1). The combined densities of several species in a mixture may show the same type of functional density–visitation curve if they share pollinators (Fig. 1). In this graphic model, visitation rates are disproportionately low for rare flowers, increasing as floral density increases up to some maximum level, at which visitation declines as pollinators are saturated. If this functional relationship between visitation and floral density occurs, this model illustrates how interactions for pollinators can change from facilitation to competition as floral densities increase. For rare flowers, any floral increase will significantly increase their visitation rate, and interactions among flowers will be facilitative if pollinators are shared. As density increases, plants will continue to show positive, facilitative interactions as

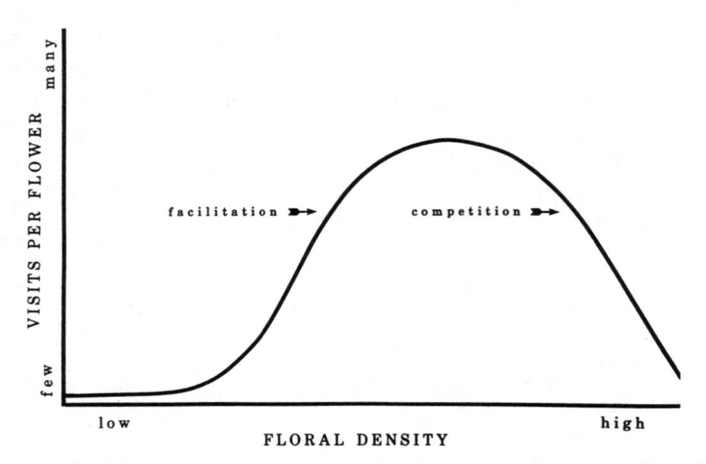

Fig. 1. Density–visitation curve showing floral visitation by pollinators as floral density increases. Interactions will be facilitative to the left of the maximum point on the visitation axis and competitive to the right as floral density increases.

visitations increase. After the maximum point on the density–visitation curve is reached, any further increase in floral density results in fewer visits and competition for pollinators occurs, becoming more severe as density continues to increase. No density–visitation curves have been reported, but some data suggest that this type of curve is realistic (B. Rathcke, unpublished data). Certainly, the shapes of these curves will vary with different species, and they may include threshold effects. Measurements need to be made to determine density–visitation curves for different species, and to determine whether plants in natural populations experience mainly facilitation or competition or whether their interactions change over space and time.

If this type of visitation curve is realistic, visitation rates and hence plant attractiveness will depend not only upon its own reward rates but upon those of its neighbors of the same or of different species. By acting together, the individual plants may be able to attract more pollinators per individual than each could independently (a facilitative effect), but within these mutually gained resources, individuals compete. The net effect will depend upon the relative effects of attraction and competition, or the "elasticity of attraction" (Real, 1983).

If visitation is sensitive to density, we must specify attractiveness and visitation under the specific set of populational conditions that exist in the local neighborhoods. If different pollinators are attracted from different distances because of their mobility, the scales of the interactive neighborhoods will vary with the pollinator type. For instance, honeybees may travel several kilometers, whereas most small bees may tend to remain within several hundred meters of their nests; the interactive neighborhoods for these two pollinators will be quite different (Thomson, 1981).

These density effects will also influence selection for "competitive" ability for attracting pollinators, which will also vary with the populational attributes of local neighborhoods. It is possible that a highly attractive plant will always be most attractive under any natural conditions of density, whereas other plants will always be at some disadvantage in most natural species mixtures. However, most plants will probably fall somewhere in between and show variable interactions. If attraction is strongly density dependent, a low-nectar species could be attractively equivalent to a high-nectar species if it were at a high density, and selection for attractive (competitive) ability of a plant may vary with the local neighborhood, i.e., the density and dispersion of conspecifics and other interacting species. A low-nectar species may suffer relatively more from pollination limitation at the limits of its distribution, where it is relatively rare, than would a high-nectar species.

If visitation is sensitive to floral density, it will change throughout the flowering of the individual plants as flowering begins, peaks, and ends. Such changes in visitation have been described for *Phlox* (Zimmerman, 1980) and *Diervilla* (J. D. Thomson). Pollination limitation was found to occur during the second half of flowering in *Phlox* but not during the first half, when flowers apparently were more sparse (Zimmerman, 1980). Visitation quality may also change and compensate for the density effect because outcrossing may be higher early or late in the flowering period as pollinator flights become longer between sparse flowers (Ellstrand *et al.*, 1978; J. Schmitt, 1983; Stephenson, 1982).

In this density–visitation model, I have only considered PV and potential effects on competition and facilitation, but IPT will also vary with density and dispersion of species or incompatible genotypes (Thomson, 1982). Both will act in concert to influence the final pollination success and seed set. At low floral densities, bees tend to show lower fidelity (Zimmerman, 1980; Waddington and Heinrich, 1981), and other pollinators such as hummingbirds are often indiscriminate. Sparse species will suffer doubly because most of the relatively few visits they do receive will be interspecific in mixtures (Levin and Anderson, 1970). At the other extreme, as flowers become dense, pollinators may become more sedentary and even territorial (Carpenter, 1976; Linhart, 1973; Feinsinger, 1976; Frankie *et al.*, 1976; Stephenson, 1982), and most visits may be local. This may increase IPT if visits occur mainly within one individual plant or among closely related plants (Carpenter, 1976; Linhart, 1973; Price and Waser, 1979; Augspurger, 1980; Stephenson, 1982), augmenting any pollinator limitation effects on seed set. But patchy dispersion within a species may be highly influential in reducing interspecific pollen exchange with other species (Levin, 1972).

The density effects on PV and IPT places rare species in double jeopardy and makes their persistence even more precarious if they are dependent upon outcrossing and seeds for persistence. This effect would be especially severe where other processes, such as nutrient competition, disturbance, or herbivory could limit the persistence of individual plants, making them highly dependent upon reestablishment by seeds. The obvious selective force may seem to be the one acting upon the vegetative phase, but poor seed set could be the critical factor preventing its reestablishment. As a result, pollination could be a nonobvious but crucial link in the persistence of rare species.

These density effects for IPT should be examined in the purported floral mimics, for if these tend to be rare and if they persist because pollinators cannot distinguish them, then IPT should be high. Here we would expect strong selection for differential pollen placement (Brown and Kodric-Brown, 1979). Interestingly, a number of purported mimics are either orchids or *Asclepias* which have pollinia and are not so likely to suffer from IPT from unrelated species (Bierzychudek, 1981b; Heinrich, 1975; but see Dafni and Ivri, 1979).

Floral abundance of species, even in perennial plant communities, can vary greatly and independently from year to year (see Tepedino and Stanton, 1980, for a review), and the degree of overlap among species can vary among years (Schemske *et al.*, 1978; Reader, 1975; Hocking, 1968) and in space (B.

Rathcke, unpublished data). Nectar production within a species can also vary with rainfall or insolation (Hocking, 1968; Feinsinger *et al.*, 1979; Brown and Kodric-Brown, 1979; B. Rathcke, unpublished data). How much these populational factors influence floral visitation and interactions for pollination in natural populations is unknown. Thompson (1982, p. 136) notes that "no studies have yet quantified how the interaction structure of a community changes over time." My studies on the pollination of shrubs in The Great Swamp in Rhode Island indicate that interactions for pollinators and pollination can vary significantly over space and time (B. Rathcke, unpublished data). We need to understand how prevalent such variation is before we can predict the ecological or evolutionary effects of plant interactions for pollination.

B. Removal Experiments and Diversity Effects

To determine the effects of a second species (or phenotype) on PV, floral visits can be compared for a species in a mixture and alone. However, if floral density is a major determinant of PV, the presence of a second species would also influence the overall density in the mixture, and its influence could result from a density effect or from its attractive ability. Assigning the interactive effects will depend upon whether we compare the effects in experiments with or without replacement. That is, we can remove one species and examine the resulting effect on visitation (without replacement), or we can remove one species and replace it with the remaining species so as to maintain the same original overall density (with replacement). For example, if we remove species B from an AB mixture, the floral density will decrease and pollinators may abandon the area. Visits to species A will drop, and species B would be assigned as having had a positive effect on A in the mixture. However, if we replaced species B with species A to maintain the overall density, we may find that visitation per flower to A increases and B would be labeled a competitor to A in the mixture. Either comparison could be biologically valid. A no-replacement comparison may be more reasonable if the abundance of A is not likely to be able to increase with B's removal, either because other factors will limit its abundance or because B will tend to be replaced by a species other than A. These complications arising from different comparisons are a problem not only to the experimenter but also to the plant. The ecological and evolutionary consequences may be difficult to predict in natural populations.

These replacement experiments are analogous to the de Wit replacement series (Harper, 1977), except that in this case the plants are not interacting for some set amount of resource but instead are attracting pollinators (resources) into the area. These replacement experiments allow for a comparison of interspecific and intraspecific effects and for measurement of relative attractiveness of two species at the same densities. These interactive effects are also likely to vary with the different proportions of the two species in mixtures as well as in absolute density (Inouye and Schaffer, 1981).

A mixture of species may also have special properties beyond those predicted by the densities and relative attractive abilities of the participating species. Salt (1979) terms these "emergent properties" if they cannot be predicted from a knowledge of each species' characteristics measured in isolation. These effects can be determined only by comparisons between replacement and no-replacement experiments. This approach has been used to distinguish the effects of plant density and diversity on insect herbivores in monocultures and polycultures (Bach, 1980), but this has not been done for pollination studies.

Two emergent effects due to diversity that could arise in plant–pollinator systems are pollinator enhancement or pollinator depression. These would not be predicted from the floral densities or relative attractive abilities of the species. One possible mechanism that could result in pollinator enhancement would be resource complementation whereby one species provides rich nectar rewards and another provides pollen. This could result in higher visitation rates due to attraction and to higher survival of pollinators. Conceivably, a mixture could produce pollinator depression through some mechanism such as a repellent or confusion effect similar to that described for plants and herbivores (Atsatt and O'Dowd, 1976; Tahvanainen and Root, 1972). This could be termed "interference" whereby one species is limiting access to resources (Birch, 1957), although this must be distinguished from IPT, which has also been termed "interference competition." Only replacement experiments would allow us to determine if these diversity effects occur among plants for pollinators, and such experiments have only recently been initiated.

C. Facilitation through Pollinator Support

The type of facilitation that may be most apparent and ubiquitous among plants is the maintenance of pollinators over time. Plant species flowering early in the season may support the initial pollinator populations which survive and reproduce and are then available to pollinate the later-flowering species. Waser and Real (1979) call this "sequential mutualism," and they present a dramatic example occurring between two species in Colorado. When the early-flowering *Delphinium* was sparse, migrating hummingbirds did not stay in the area and were unavailable to pollinate the later-flowering *Ipomopsis*. Such sequential support is necessary for the maintenance of long-lived pollinators such as hummingbirds or bats and for social bees which form long-lived colonies. In temperate areas bumblebee queens forage early, and their later colony production is dependent upon the initial support they gain from earlier-flowering species.

Support of pollinators from year to year will also be important. In areas where floral abundance varies greatly among years, a sparse species in one year that cannot provide much support may benefit the next year from the past support of other flowers. A number of studies show that floral densities can vary greatly even in stable perennial plant communities, and they may vary independently so that a good year for one species is a poor year for others (see Tepedino and Stanton, 1980, for a review). If they share pollinators, some facilitation through pollinator support is inevitable over the years.

D. Facilitation through Pollinator Sharing

Pollinators can vary in their effectiveness in pollination of different plant species, i.e., the number of visits necessary for maximum seed set of a flower (Primack and Silander, 1975; Motten *et al.*, 1981; Schemske, 1983). Plants can vary in their support of pollinators through pollen, nectar, and other rewards. If plants share pollinators, the interactions can be facilitative but their costs (via support) and benefits (via effectiveness gained) may be different and the trades between costs and benefits may be unequal. For example, one species may provide major support for a pollinator which is more beneficial (more effective) for another plant species. Here I present a graphic model for pollinator sharing which illustrates the possible costs and benefits of pollinator sharing. This general model was developed for shell exchanges among hermit crabs (D. Rubenstein and B. A. Hazlett, personal communication), and it seems applicable to plant interactions for pollinators because in both systems resource utilization does not destroy (consume) the resource and resources are both supported and shared. I present a brief, modified version of the model to stimulate further consideration of these interactions through resource sharing.

Most plants have a diversity of pollinators which may vary in their effectiveness. Ideally, we could draw a curve of relative expected effectiveness values (utility) of each pollinator type for each of two plant species in relation to some vector of pollinator types (Fig. 2). This could be considered a utility curve. The point where the utility curves for two species intersect represents a pollinator type which is of equal utility for both species. At some points (x_1 in Fig. 2), the second species is supporting a pollinator which is of greater benefit to the first species. In Fig. 2, this is balanced by the reciprocal support of another pollinator by the second species (at x_2). In reality, the utility curves are likely to be asymmetrical, and the costs and benefits to the species may be unequal. Individual hermit crabs that trade shells usually gain in most exchanges, and interactive effects will often be positive (Hazlett, 1981). Plants cannot guarantee this mutual reciprocal benefit and sharing may often be unequal, although overall the sharing may be facilitative.

This type of pollinator sharing may be more apparent over time. Even highly specialized, oligolectic Andrenid bees will collect pollen and nectar from other plant species if their speciality is sparse (Cruden, 1972; Eickwort and Ginsberg,

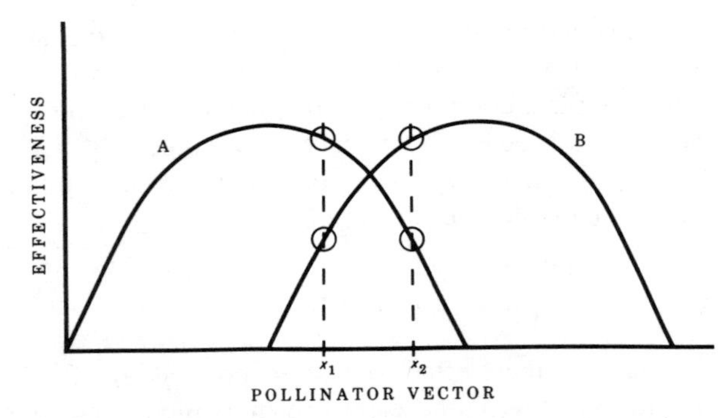

Fig. 2. A pollinator sharing model showing the effectiveness of pollinators for two plant species, A and B, and their overlap in sharing of pollinator types. At point x_1 species B may be supporting a pollinator that is more effective for species A; the reciprocal relationship exists at point x_2.

1980). If their host plants tend to vary in floral abundance, the support and maintenance of these specialists may be highly dependent upon other plant species which may or may not be effectively pollinated by these visitors. The effectiveness of pollinators is not generally known. Strickler (1979) found that specialized bees were more effective than generalists; other specialized bees are less effective as pollinators (Schemske, 1983).Motten *et al.* (1981) found that a highly generalized bee fly was as effective a pollinator of *Claytonia* as was the specialized Andrenid bee. Reciprocal pollinator sharing and the effects upon the participants have not been studied.

IV. Ecological and Evolutionary Evidence for Interactions

First, I will present evidence that many plants do experience pollination limitation of seed set in natural populations. This limitation could result from competition among plants for limited pollinators, from IPT, or from a paucity of pollinators that exists for other reasons. I will then review the evidence that plant interactions for pollination result in reduced seed set and affect the distribution or abundance of plants. I will also discuss the evidence for the different types of interactions inferred from various studies. Finally, I will review the evidence that these interactions have given rise to character displacement among plant species.

A. Pollination Limitation of Seed Set

Bawa and Beach (1981) tend to dismiss pollination as a significant factor limiting seed set and propose instead that maternal success in plants is usually limited by the amount of nutritional resources available for developing progeny, whereas paternal success is limited by pollen dispersal (see also Stephenson, 1981). However, even if significant abortion of pollinated flowers occurs, initial pollination must set the stage and its level may still affect the final outcome. Facilitative interactions among plants for pollinators could be an important reason why pollination limitation is not more common. The importance of maternal resources or choice in governing seed set does not necessarily denigrate the importance of interactions among plants for pollination.

To determine the pollination limitation of flowers, fruit or seed set per flower (fruit or seed effectiveness) must be compared with that of simultaneous hand-pollinated controls because of potential temporal variation in abortion rates (Stephenson, 1981), self-fertility (Pinthus, 1959), or pollen viability (Woodell *et al.*, 1977). Fruit and seed production per flower should be distinguished because they represent different levels of pollination. I define "fruit effectiveness" as the proportion of flowers producing fruit (using "fruit" as a general term for a propagule) and "seed effectiveness" as the proportion of ovules within a flower that develop into seeds. Seed effectiveness measured alone will omit any pollination limitation on fruit set, but it will more closely reflect the effectiveness of the pollen loads that are deposited on stigmas. Ideally, both should be examined.

To understand how selection might act, seed set for the entire plant (genet) should be measured. It is possible that the entire plant would not be able to maintain the effectiveness seen in a subsample of hand-pollinated flowers (Bierzychudek, 1981a; Stephenson, 1981). However, if hand pollination can increase effectiveness in some subset, natural pollination could have done the same, and I regard this as sufficient evidence of pollination limitation for these flowers. However, a plant may be able to compensate for this limitation by producing a superabundance of flowers and by adjusting abortion rates to achieve the physiologically maximal seed set (Bawa and Beach, 1981; Stephenson, 1981). As a result of this compensation, final seed set for the plant may not be pollination limited (K. S. Bawa, personal communication). But by determining pollination limitation of flowers throughout the flowering period, we can measure the timing and risk of pollination and this can give us evolutionary insights. For example, if early flowers were consistently pollination limited, selection could favor later flowering whether or not compensation is possible. More information is needed on the prevalence of mechanisms that would allow plants to compensate for inadequate pollination; their widespread occurrence would suggest that pollination limitation of flowers is common.

Methodological difficulties in determining pollination limitation of flowers include distinguishing nonfertilization of flowers from abortion of fertilized flowers and ensuring similar treatments of experimental and control flowers. Hand pollination as the standard is conservative and could underestimate pollination limitation because this may not result in the maximum possible seed set if the techniques are poor or harmful. Bierzychudek (1981a) cites three studies in which hand pollination resulted in lower seed set than natural pollination: *Catalpa speciosa* (Stephenson, 1979), *Geranium* (Willson *et al.*, 1979), and *Ipomopsis aggregata* and *Delphinium nelsoni* (Waser, 1978). The effect of pollination on seed and seedling quality should also be measured, but this has been done in only a few cases (Price and Waser, 1979; van Leevwen, 1981).

In a dramatic example of pollination limitation, Bierzychudek (1981a) increased seed production in jack-in-the-pulpit, *Arisaema triphyllum,* 40 times with hand pollination. In her review, she notes that hand pollination increased seed set in *Erythronium albidum* (Schemske *et al.*, (1978), *Phlox divaricata* (Willson *et al.*, 1979), *Combretum fruticosum* (Schemske, 1980b), *Brassavola nodosa* (Schemske, 1980a), *Lithospermum carolinense* (Weller, 1980), and *Encyclia cordigera* (Janzen *et al.*, 1980). For tree species, Bawa (1974) found that hand pollination increased seed set 2- to 90-fold over natural pollination for 26 of 34 tree species (81%) he examined in Costa Rica. Hand pollination increased seeds per flower in *Andira inermis* from 0.008 to 0.35 (Frankie *et al.*, 1976) and fruit set by 540% in *Chilopsis linearis* (Peterson *et al.*, 1982). Pollination limitation similarly occurs in four out of eight shrub species in the Great Swamp (B. Rathcke, unpublished data).

Other studies on herbaceous plants showing pollination limitation include Schemske *et al.* (1978) for six of seven spring flowering woodland herbs, Barrett (1980) for *Eichornia crassipes,* Melampy and Hayworth (1980) for *Isopyrum biternatum* in 2 of 3 years, and Gross and Werner (1983) for two of three *Solidago* species, with *S. graminifolia* showing limitation only in early flowering clones. Rust and Roth (1981) suggest that absence of pollination or improper pollination was responsible for the largest loss of potential seeds in mayapple, *Podophyllum peltatum.* Seed set in *Claytonia* was pollination limited only in rainy weather and early in the flowering season (Motten *et al.*, 1981). *Phlox foliossimum* was pollination limited only during the last half of the flowering period (Zimmerman, 1980).

Other results suggest pollination limitation, but hand pollination controls were not done. Augspurger (1980) found that seed set for individual *Hybanthus* shrubs varied significantly with differential visitation by *Melipona* bees. For *Cassia biflora* the percentage of ovules setting seed was significantly lower in isolated plants (Silander, 1978). Floral effectiveness in the tree *Metrosideros* was lower during peak flowering, possibly because honeycreepers became more sedentary (Carpenter, 1976).

Experimental studies have clearly demonstrated that the application of interspecific pollen can reduce seed set (Levin, 1972; Waser, 1978; Thomson *et al.*, 1981), but evidence that this reduces seed set in natural populations is sparse. Weller (1979) found that interspecific pollen on stigmas did not appear to reduce fecundity in *Oxalis alpina.* Interspecific pollen on stigmas has been reported for other species, but the seed set effects are unknown, i.e., *Ipomopsis*

(Brown and Kodric-Brown, 1979), *Delphinium* (Waser, 1978), and *Phlox* (Levin, 1972).

Interspecific pollen transfer will depend upon pollinator fidelity, which varies greatly among pollinators. Although individual bumblebees often show high fidelity to one plant species, Heinrich (1976) calculated from three extensive studies on pollen loads that an average of 57% were pure and the rest were mixtures of two (32%), 3 (13%) or four (5%) species. Hummingbirds are usually reported to be indiscriminate in their visitation (Brown and Kodric-Brown, 1979; Feinsinger, 1978; Waser, 1978). In experimental gardens of intermingled *Delphinium* and *Ipomopsis,* Waser (1978) found that 44% of the hummingbird flights were interspecific, 25% of the stigmas had foreign pollen, and seed set was typically reduced by about 25% over that in pure plots. In mixed experimental populations of two *Phlox* species, butterflies transferred low levels of interspecific pollen (0.27 grains/stigma), but effects on seed set were not determined (Levin, 1972).

Although negative results may tend to remain unreported, a few studies have shown no significant differences between hand and natural pollination: *O. alpina* (Weller, 1979), two *Phlox* species (Levin, 1972; Levin and Berube, 1972), and a New Zealand shrub, *Leptospermum scoparium* (Primack and Lloyd, 1980). Other studies demonstrate competition among pollinators, suggesting that flowers rather than pollinators are limiting, although the effectiveness of pollen transfer and seed set could still be poor (Inouye, 1978; Ford, 1979; Feinsinger, 1976, 1978; Frankie *et al.*, 1976; Mosquin, 1971; Roubik, 1982).

Evidence exists to suggest strongly that pollination limitation does occur in natural populations and does not appear to be a rare phenomenon. It is also apparent that this limitation does vary over space and time, even during the flowering period of a single individual. This cautions against using single sample surveys for making generalizations (Wiens, 1981) and suggests that interactive effects may vary as predicted by the density–visitation model.

B. Interactions among Plants for Pollination

Only a few studies indicate that exploitative competition for pollinators may significantly reduce seed set in some natural plant populations. Gross and Werner (1983) found that three *Solidago* species were visited less by honeybees when weedy species were flowering; *S. graminifolia* and *S. canadensis* experienced pollination limitation, whereas *S. Juncea* appeared to remain at its potential maximum seed set. Schemske *et al.* (1978) noted that although *Dentaria laciniata* attracted certain flower visitors out of proportion to its abundance in an Illinois woodland, no depression of seed set among simultaneously flowering species was evident. Melampy and Hayworth (1980), however, point out that the maximum seed set for *I. biternatum* occurred after the flowering peak of *D.*

laciniata and near the end of the peak for *Claytonia* in these woods and suggest that this resulted from increased visitation. Zimmerman (1980) suggests that intraspecific competition caused the pollination limitation observed in *Polemonium foliossimum* during the last half of the flowering period because bee visitation was lower and correlated with seed set, but he recognizes that other causes including interspecific competition cannot be discounted.

Other studies on pollinator visitation suggest that exploitative competition occurs, but effects on seed set were not measured. Mosquin (1971) stimulated the resurgence of pollination studies when he reported that willow (*Salix*) and dandelion (*Taraxacum*) appeared to attract pollinators away from other plants, but he presented no quantitative data. Free (1968) also noted that dandelions in orchards can attract pollinators away from apple trees (a note of irony since dandelions are largely apomictic). Frankie *et al.* (1976) saw bees move away from *A. inermis* when one large *Dalbergia retusa* tree came into flower unseasonally. Butterflies visited white morphs of *Phlox* less frequently than pink morphs (Levin, 1972). Lack (1976) observed that bumblebees tended to visit *Centaurea scabiosa* more than *C. nigra* even when *C. scabiosa* was rare. However, honeybees tended to visit *C. nigra* and probably eliminated any seed set effects of possible competition for bumblebees.

In contrast to these studies suggesting exploitative competition, Feinsinger (1978) could find no evidence that the flowering of one plant influences hummingbirds to abandon another; instead birds moved indiscriminately among different species. However, this may have resulted in lower visitation per flower or in IPT, but this is unknown.

In an excellent study, Waser (1978) suggests that interspecific pollen transfer rather than competition for pollinators may have been a potent selective force in determining the flowering times of *Ipomopsis* and *Delphinium*. He measured decreased seed set during the periods of flowering overlap of these two species and within experimental gardens where these two species were intermingled. Hand application of interspecific pollen showed similar decreases in seed set.

Documentation that these negative plant interactions for pollination result in ecological sorting whereby one species is precluded from coexisting with another in any local area is lacking and would demand some careful reciprocal transplants of likely species. Results from experiments with *Clarkia* and *Melandrium* have been interpreted as evidence that hybridization effects could influence population size, but alternative explanations are possible (Lewis, 1961; Levin, 1970). In synthetic mixed populations of two *Clarkia* species, *C. biloba* was generally more successful, although both species declined. Lewis (1961) suggested that *C. lingulata* was hybridized out of the populations because it was initially a minority species (2:5) and suffered more from loss to sterile hybrids. However, no nearby monospecific control populations were available for comparison and environmental conditions changed drastically, although *C. lingulata* may have been favored. Vegetative competition was not considered. Lewis (1961) also noted that natural populations of these two species tend to form discrete, monospecific patches and that hybridization is rare. A similar result was proposed for *Melandrium dioicum* and *M. album* in England (Baker, 1948). Although hybridization and introgression did occur, it appears that the disappearance of *M. dioicum* was due more to deforestation than hybridization. Also, *M. dioicum* is a forest diurnal flowering species and *M. album* is a weedy nocturnal species, and their contact is due to recent disturbances. In general, hybrids often persist only in ''hybrid'' habitats and both parental species remain separated to some degree, precluding the possibility of local extinction due to hybridization. Hybridization is a special case of plant interactions, and no studies purport to show population reduction resulting from other plant interactions for pollination.

Other studies suggest the possibility of facilitative effects, but convincing quantitative data are sparse. Thomson (1978) first proposed the existence of a cooperative ''magnet'' species effect when he found that the yellow *Hieracium florentinum* received more visits per stand when mixed with the orange *H. aurantiacum* than it did in pure stands. However, visitation rates at the two sites may have been different for other reasons, and Thomson admits that he cannot assume *ceteris paribus*. Both species are apomictic, so that visitation is irrelevant to seed set. Thomson has since found that visitation rates tended to be positively correlated for two *Potentilla* species (Thomson 1981) and for two of three subalpine plant species examined in artificial and natural competition experiments (Thomson, 1982).

To date no studies have been published in which controlled experiments with or without replacements of two species have been done and effects on visitation and seed set measured. No evidence exists for separating diversity effects and density effects. Data are lacking for making any inferences about the prevalence or importance of different types of interactions among plants for pollination and subsequent seed set.

C. Character Displacement for Pollination

In the search for evidence of character displacement, both Levin (1978) and Waser (1982) have reviewed potential examples of character divergence for shifts in flowering time, morphology, or color. I reiterate their findings here mostly for completeness and review the evidence for convergence. I agree with their conclusions that evidence is sparse and found mostly among closely related species (Grant, 1972; Levin, 1978). This may reflect the difficulties of finding both allopatric and sympatric comparisons and of knowing where to look among unrelated species. Examinations of related species are more common, and so perhaps it is not surprising that more examples are found here.

Establishment of any character displacement pattern should include rigorous analysis by comparing allopatric and sympatric populations (Grant, 1972) and by doing experiments showing that the trait indeed will act to alleviate the negative effects of competition or IPT or to promote facilitative effects (Connell, 1980). The trait should be shown to be genetically heritable. The assumption that allopatric populations show the precontact traits rather than character release should be examined, and alternative explanations such as adaptations to different available pollinators should be tested. The difficulty of establishing selection for reproductive isolation among plants is well illustrated by the studies on *Aquilegia* (Chase and Raven, 1975; Grant, 1976).

Separating reproductive and competitive mechanisms of character displacement may be especially difficult for plants because both mechanisms could occur in any instance. For example, if two species hybridize, they could also compete for pollinator visitations. Which effect was stronger can never be determined. However, some types of shifts are assignable to certain mechanisms. Color differences can enforce pollinator constancy in some pollinators, such as bees and butterflies, even with equal floral rewards (Levin and Kerster, 1967; Marden and Waddington, 1981), and therefore reduce IPT, but a color shift will not reduce competition. Morphological shifts may result in differences in pollen placement which could reduce IPT but not competition. Alternatively, these could enforce pollinator specialization, which would reduce both IPT and competition. Shifts in flowering times would also reduce both IPT and competition.

Only one example of character divergence has been reported between unrelated species, and this appears assignable to IPT avoidance. Waser (1982) documents the occurrence of flip-flops in the flowering times of *Ipomopsis* and *Delphinium* in different geographic areas, although they always show low flowering overlap. His earlier studies show that these two species share hummingbirds, which are indiscriminate visitors, and that interspecific pollen transfer can significantly reduce seed set in intermingled populations and during flowering overlap (Waser, 1978). Allopatric comparisons are not available, but the different flowering sequences strongly support local divergences for IPT avoidance.

All other possible examples of character divergence involve closely related species, and only two species examples have the supplementary experiments necessary to establish the effects of the displaced traits: *Phlox* (Levin and Kerster, 1967; Levin and Schaal, 1970) and *Talinum* (Carter and Murdy, 1982). *Phlox pilosa* shows sporadic shifts from pink- to white-flowered phases in the presence of *P. glaberimma,* and transplant experiments show that the white forms receive less interspecific pollen than do the pink forms (Levin and Kerster, 1967). However, the white forms are always visited less by butterflies, and they always show reduced seed set. They also show evidence of a better genetic barrier to interspecific crossing than the pink forms, so clearly this color shift is not an ideal solution and factors are complex (Levin and Berube, 1972; Levin and Kerster, 1967; Levin and Schaal, 1970). Carter and Murdy (1982) document that *Talinum mengesii* populations show different style lengths and times of diurnal flowering in areas of contact with *T. teretifolium,* and their experiments show that both traits decrease the incidence of sterile, triploid hybrids. Both of these examples probably reflect selection by IPT and hybridization leading to asymmetrical shifts by one species.

Other possible examples of character divergence lack the experimental evidence establishing the efficacy of the "shifted" traits for avoiding IPT or competition. Flowering time differences between heavy-metal adapted populations of *Agrostis tenuis* and *Anthoxanthum odoratum* and conspecific populations on uncontaminated soils may be maintained by selection to reduce wind-mediated pollen transfer and deleterious intraspecific hybrids (McNeilly and Antonovics, 1968). Shifts in floral size and morphology in areas of sympatry have been reported for *Solanum* species (Whalen, 1978) and *Fuchsia* (Breedlove, 1969), and these shifts result in visits by different pollinators in the areas of sympatry versus allopatry. Either IPT or competition could be implicated. Hummingbirds pollinate *Fuchsia*, and IPT may tend to be a potent selective force for these indiscriminate pollinators. Other examples of species shifts which lack good evidence on the pollination system include color shifts in *Clarkia* (Levin, 1970; Lewis and Lewis, 1955), ultraviolet reflectance patterns in *Rudbeckia* (McCrea, 1981), and morphological shifts in *Polansia* (Iltis, 1958) as reviewed in Waser (1982).

Character convergence for pollination could arise if facilitative interactions are effective. Again, allopatric and sympatric comparisons and experimental evidence should be obtained, but here we must separate convergence due to plant interactions from incidental convergence due to similar adaptations to similar pollinators, an independent coevolutionary phenomenon that does not require the existence of facilitative plant interactions among plants for pollination. Convergence among flowers has usually been examined as cases of mimicry rather than character displacement, but both require similar types of evidence to determine if evolutionary adaptations have occurred because of the interactions among the plants.

Two types of floral mimicry have been proposed: (1) Müllerian mimicry, in which the mimics have similar rewards and together promote visitation rates by pollinators, and (2) Batesian mimicry, in which a no-reward mimic resembles a rewarding model and "parasitizes" its pollinators (a specific case of resource parasitism, +,−). Batesian mimicry among flowers of different species has been suggested for a number of plants, but neither allopatric-sympatric comparisons nor experimental studies have been done (Brown and Kodric-Brown, 1979, 1981; Heinrich, 1975; Macior, 1974; Williamson and Black, 1981). Only Bierzychudek (1981b) has examined a purported case of mimicry by doing the necessary critical tests of geographic comparisons and experiments on visitation rates and seed set. She could find no evidence to support the hypothesis that *Asclepias curassavica, Lantana camara,* and an orchid, *Epidendrom radicans,* form a Batesian mimicry complex in Costa Rica. More such critical studies need to be done.

Müllerian mimicry has been proposed for eight hummingbird-pollinated plants species in Arizona (Brown and Kodric-Brown, 1979) and two *Costus* species in Panama (Schemske, 1981), but neither have measurements which could show increased visitation due to mimicry. However, both studies document the existence of shared pollinators and IPT. The fact that they have not diverged to reduce pollinator sharing and IPT suggests that traits are difficult to change under selection, that competition is not a strong selective force (it may be weak or nondirectional), or that facilitation is important in maintaining the similarities in floral designs.

The few possible examples of character displacement among plants for pollination suggest that IPT or hybridization was the likely mechanism and that both of these could be considered reproductive character displacements. Competitive character displacement appears to be as elusive to establish for plants as it has been for animals (Grant, 1975). It may be that IPT is a more ubiquitous and directional selective force, whereas interactions for pollinators may tend to be more sporadic and variable, perhaps even alternating between competition and facilitation.

Conclusions

The density–visitation model and facilitative models proposed here suggest that many factors may intervene to influence the strength and direction of different interactions among plants for pollination. Competitive and facilitative effects may even alternate as the populational attributes of local neighborhoods change. The possibility of concurrent facilitative and competitive effects through aggregate attractiveness and individualistic competition suggests that the final outcomes may be difficult to predict for any individual plant. If these interactions are highly variable, we might expect the appearance of individual characteristics to buffer plants against these vagaries. Because pollinator populations may often be more limiting in harsh and unpredictable environments and because flowering abundance may also vary more in these areas, both competition and facilitation may be more common and influential than they are in more benign environments, where pollinator populations may be more assured and where pollinators may compete for limited floral resources.

Evidence from experiments and from character displacement patterns suggests that IPT is a more ubiquitous and stronger directional selective force than is competition for pollinators (Waser, 1982). Different interactive neighborhoods would probably be involved. IPT may be highly localized and sensitive to the density and dispersion of two species. On the other hand, pollinator attractiveness and competition or facilitation could act over large areas. Bees can fly several kilometers, and hummingbirds migrate to seek out areas of high nectar rewards. Plants in different habitats could interact for mobile pollinators. It becomes more difficult to draw the boundaries of these neighborhoods, and hence more difficult to predict where we might find evolutionary adjustments due to interactions for pollinators than for IPT.

Plant–pollinator systems may be especially dynamic and sensitive to local neighborhoods, but this may not be so unusual; it may only be more obvious than in other systems. The influence and importance of a process in any ecological system is likely to be specific to the situation in time and space. We must measure processes and their effects under different conditions to understand how species interact and evolve in natural systems.

Acknowledgments

I wish to thank Brian Hazlett, Tom Getty, and Catherine Bach for their stimulating ideas and discussions about mutualisms and for their comments on this manuscript. I am also grateful to many other people who entered into lively arguments about mutualism, facilitation, and ecological semantics. This chapter was written under the support of National Science Foundation Grant G-DEB78-24678.

References

Atsatt, P. R., and O'Dowd, D. J. (1976). Plant defense guilds. *Science* **193,** 24–29.

Augspurger, C. K. (1980). Mass-flowering of a tropical shrub (*Hybanthus prunifolius*): Influence on pollinator attraction and movement. *Evolution (Lawrence, Kans.)* **34,** 475–488.

Bach, C. E. (1980). Effects of plant density and diversity on the population dynamics of a specialist herbivore, the striped cucumber beetle, *Acalymma vittata* (Fab.). *Ecology* **61,** 1515–1530.

Baker, H. G. (1948). Stages in the invasion and replacement demonstrated by species of *Melandrium. J. Ecol.* **36,** 96–119.

Bawa, K. S. (1974). Breeding systems of tree species of a lowland tropical community. *Evolution (Lawrence, Kans.)* **28,** 85–92.

Bawa, K. S., and Beach, J. H. (1981). Evolution of sexual systems in flowering plants. *Ann. M. Bot. Gard.* **68,** 254–274.

Barrett, S. C. H. (1980). Sexual reproduction in *Eichhornia crassipes* (water hyacinth). II. Seed production in natural populations. *J. Appl. Ecol.* **17,** 113–124.

Bierzychudek, P. (1981a). Pollinator limitation of plant reproductive effort. *Am. Nat.* **117,** 838–840.

Bierzychudek, P. (1981b). *Asclepias, Lantana,* and *Epidendrum*: A floral mimicry complex? *Biotropica, Suppl.* **13,** 54–58.

Birch, L. C. (1957). The meanings of competition. *Am. Nat.* **91,** 5–18.

Boucher, D. H., James, S., and Keeler, K. H. (1982). The ecology of mutualism. *Ann. Rev. Ecol. Syst.* **13,** 315–347.

Breedlove, D. E. (1969). The systematics of *Fuchsia* section *Encliandra* (Onagraceae). *Univ. Calif. Publ. Bot.* **53,** 1–69.

Brewbaker, J. L., and Majumder, S. K. (1961). Cultural studies on the pollen population effect and the self-incompatibility inhibition. *Am. J. Bot.* **48,** 457–464.

Brown, J. H., and Kodric-Brown, A. (1979). Convergence, competition, and mimicry in a temperate community of hummingbird-pollinated flowers. *Ecology* **60,** 1022–1035.

Brown, J. H., and Kodric-Brown, A. (1981). Reply to Williamson and Black's comment. *Ecology* **62,** 497–498.

Carpenter, F. L. (1976). Plant-pollinator interactions in Hawaii: pollination energetics of *Metrosideros collina* (Myrtaceae). *Ecology* **57,** 1125–1144.

Carter, M. E. B., and Murdy, W. H. (1982). Divergence of diurnal flowering time and floral characters between populations of *Talinum mengesii* with and without contact with *T. teretifolium*. *Bull. Ecol. Soc. Am.* **63,** 132.

Chase, V. C., and Raven, P. H. (1975). Evolutionary and ecological relationships between *Aquilegia formosa* and *A. pubescens* (Ranunculaceae), two perennial plants. *Evolution (Lawrence, Kans.)* **29,** 474–486.

Connell, J. H. (1980). Diversity and the coevolution of competitors, or the ghost of competition past. *Oikos* **35,** 131–138.

Cruden, R. W. (1972). Pollination biology of *Nemophila menziesii* (Hydrophyllaceae) with comments on the evolution of oligolectic bees. *Evolution (Lawrence, Kans.)* **26,** 373–389.

Dafni, A., and Ivri, Y. (1979). Pollination ecology of, and hybridization between, *Orchis coriophora* L. and *O. collina* Sol. ex Russ. (Orchidaceae) in Israel. *New Phytol.* **83,** 181–187.

Eickwort, G. C., and Ginsberg, H. S. (1980). Foraging and mating behavior in Apoidea. *Ann. Rev. Entomol.* **25,** 421–426.

Ellstrand, N. C., Torres, A. M., and Levin, D. A. (1978). Density and the rate of apparent outcrossing in *Helianthus annuus* (Asteraceae). *Syst. Bot.* **3,** 403–407.

Feinsinger, P. (1976). Organization of a tropical guild of nectarivorous birds. *Ecol. Monogr.* **46,** 257–291.

Feinsinger, P. (1978). Ecological interactions between plants and hummingbirds in a successional tropical community. *Ecol. Monogr.* **48,** 269–287.

Feinsinger, P., Linhart, Y. B., Swarm, L. A., and Wolfe, J. A. (1979). Aspects of the pollination biology of three *Erythrina* species on Trinidad and Tobago. *Ann. M. Bot. Gard.* **66,** 451–471.

Ford, H. A. (1979). Interspecific competition in Australian honeyeaters—depletion of common resources. *Aust. J. Ecol.* **4,** 145–164.

Frankie, G. W., Opler, P. A., and Bawa, K. S. (1976). Foraging behaviour of solitary bees: Implications for outcrossing of a neotropical forest tree species. *J. Ecol.* **64,** 1049–1057.

Free, J. B. (1968). Dandelion as a competitor to fruit trees for bee visits. *Appl. Ecol.* **5,** 169–178.

Grant, P. R. (1972). Convergent and divergent character displacement. *Biol. J. Linn. Soc.* **4,** 39–68.

Grant, P. R. (1975). The classical case of character displacement. *Evol. Biol.* **8,** 237–337.

Grant, V. (1976). Isolation between *Aquilegia formosa* and *A. pubescens*: A reply and reconsideration. *Evolution (Lawrence, Kans.)* **30,** 625–628.

Gross, R. S., and Werner, P. A. (1983). Relationships among flowering phenology, insect visitors, and seed set: Experimental studies on four co-occurring species of goldenrod (*Solidago*: Compositae). *Ecology* (in press).

Harper, J. L. (1977). "Population Biology of Plants." Academic Press, New York.

Hazlett, B. A. (1981). The behavioral ecology of hermit crabs. *Ann. Rev. Ecol. Syst.* **12,** 1–22.

Heinrich, B. (1975). Bee flowers: A hypothesis on flower variety and blooming times. *Evolution (Lawrence, Kans.)* **29,** 325–334.

Heinrich, B. (1976). The foraging specializations of individual bumblebees. *Ecol. Monogr.* **46,** 105–128.

Heinrich, B. (1979). Resource heterogeneity and patterns of movement in foraging bumblebees. *Oecologia* **140,** 235–245.
Hocking, B. (1968). Insect-flower associations in the high Arctic with special reference to nectar. *Oikos* **19,** 359–387.
Iltis, H. H. (1958). Studies in the Capparidaceae IV. *Polanisia* Raf. *Brittonia* **10,** 33–58.
Inouye, D. W. (1978). Resource partitioning in bumblebees: Experimental studies of foraging behavior. *Ecology* **59,** 672–678.
Inouye, R. S., and Schaffer, W. M. (1981). On the ecological meaning of ratio (de Wit) diagrams in plant ecology. *Ecology* **62,** 1679–1681.
Janzen, D. H., DeVries, P., Gladstone, D. E., Higgins, M. L., and Lewisohn, T. M. (1980). Self- and cross-pollination of *Encyclia cordigera* (Orchidaceae) in Santa Rosa National Park, Costa Rica. *Biotropica* **12,** 72–74.
Jennings, D. L., and Topham, P. B. (1971). Some consequences of raspberry pollen dilution for its germination and for fruit development. *New Phytol.* **70,** 371–380.
Johnson, L. K., and Hubbell, S. P. (1975). Contrasting foraging strategies and coexistence of two bee species on a single resource. *Ecology* **56,** 1398–1406.
Lack, A. (1976). Competition for pollinators and evolution in *Centaurea. New Phytol.* **77,** 787–792.
Lawton, J. H., and Hassell, M. P. (1981). Asymmetrical competition in insects. *Nature (London)* **289,** 793–795.
Levin, D. A. (1970). Reinforcement of reproductive isolation: Plants versus animals. *Am. Nat.* **104,** 571–581.
Levin, D. A. (1972). Pollen exchange as a function of species proximity in Phlox. *Evolution (Lawrence, Kans.)* **26,** 251–258.
Levin, D. A. (1978). The origin of isolating mechanisms in flowering plants. *Evol. Biol.* **11,** 185–317.
Levin, D. A., and Anderson, W. W. (1970). Competition for pollinators between simultaneously flowering species. *Am. Nat.* **104,** 455–467.
Levin, D. A., and Berube, D. E. (1972). Phlox and Colias: The efficiency of a pollination system. *Evolution (Lawrence, Kans.)* **26,** 242–250.
Levin, D. A., and Kerster, H. W. (1967). Natural selection for reproductive isolation in Phlox. *Evolution (Lawrence, Kans.)* **21,** 679–687.
Levin, D. A., and Schaal, B. A. (1970). Corolla color as an inhibitor of interspecific hybridization in Phlox. *Am. Nat.* **104,** 273–283.
Lewis, H. (1961). Experimental sympatric populations of *Clarkia. Am. Nat.* **95,** 155–168.
Lewis, H., and Lewis, M. E. (1955). The genus *Clarkia. Univ. Calif. Publ. Bot.* **20,** 241–392.
Linhart, Y. B. (1973). Ecological and behavioral determinants of pollen dispersal in hummingbird-pollinated *Heliconia. Am. Nat.* **107,** 511–523.
Louda, S. M. (1982). Distribution ecology: Variation in plant recruitment over a gradient in relation to insect seed predation. *Ecol. Monogr.* **52,** 25–41.
Lynch, M. (1978). Complex interactions between natural coexploiters — *Daphnia* and *Ceriodaphnia. Ecology* **59,** 552–564.
Macior, L. W. (1974). Behavioral aspects of co-adaptations between flowers and insect pollinators. *Ann. M. Bot. Gard.* **61,** 760–769.
McCrea, K. D. (1981). Ultraviolet floral patterning, reproductive isolation and character displacement in the genus *Rudbeckia* (Compositae). Doctoral Dissertation, Purdue University, West Lafayette, Indiana.
McNeilly, T., and Antonovics, J. (1968). Evolution in closely adjacent plant populations. IV. Barriers to gene flow. *Heredity* **23,** 205–218.
Marden, J. H., and Waddington, D. D. (1981). Floral choices by honeybees in relation to the relative distances to flowers. *Physiol. Entomol.* **6,** 431–435.

Melampy, M. N., and Hayworth, A. M. (1980). Seed production and pollen vectors in several nectarless plants. *Evolution (Lawrence, Kans.)* **34,** 1144–1154.

Mosquin, T. (1971). Competition for pollinators as a stimulus for the evolution of flowering time. *Oikos* **22,** 398–402.

Motten, A. F., Campbell, D. R., Alexander, D. E., and Miller, H. L. (1981). Pollination effectiveness of specialist and generalist visitors to a North Carolina population of *Claytonia virginica*. *Ecology* **62,** 1278–1287.

Mulcahy, D. L., ed. (1975). "Gamete Competition in Plants and Animals." North-Holland Publ., Amsterdam.

Peterson, D., Brown, J. H., and Kodric-Brown, A. (1982). An experimental study of floral display and fruit set in *Chilopsis linearis* (Bignoniaceae). *Oecologia* **55,** 7–11.

Pinthus, M. J. (1959). Seed set of self-fertilized sunflower heads. *Agron. J.* **51,** 626.

Price, M. V., and Waser, N. M. (1979). Pollen dispersal and optimal outcrossing in *Delphinium nelsoni. Nature (London)* **277,** 294–297.

Primack, R. B., and Lloyd, D. B. (1980). Andromonoecy in the New Zealand shrub manuka, *Leptospermum scoparium* (Myrtaceae). *Am. J. Bot.* **67,** 361–368.

Primack, R. B., and Silander, J. A. (1975). Measuring the relative importance of different pollinators to plants. *Nature (London)* **255;** 143–144.

Reader, R. J. (1975). Competitive relationships of some bog ericads for major insect pollinators. *Can. J. Bot.* **53,** 1300–1305.

Real, L. A. (1981). Uncertainty and pollinator-plant interactions: The foraging behavior of bees and wasps on artificial flowers. *Ecology* **62,** 20–26.

Real, L. A. (1983). Microbehavior and macrostructure in pollinator-plant interactions. *In* "Pollination Biology" (L. A. Real, ed.). Academic Press, New York.

Real, L., Ott, J., and Silverfine, E. (1982). On the tradeoff between the mean and the variance in foraging: Effect of spatial distribution and color preference. *Ecology* **63,** 1617–1623.

Roubik, D. W. (1982). The ecological impact of nectar-robbing bees and pollinating hummingbirds on a tropical shrub. *Ecology* **63,** 354–360.

Rust, R. W., and Roth, R. R. (1981). Seed production and seedling establishment in the mayapple *Podophyllum peltatum* L. *Am. Midl. Nat.* **105,** 51–60.

Salt, G. W. (1979). A comment on the use of the term emergent properties. *Am. Nat.* **113,** 145–148.

Schaal, B. A. (1980). Measurement of gene flow in *Lupinus texensis. Nature (London)* **284,** 450–451.

Schemske, D. E. (1980a). Evolution of floral display in the orchid *Brassavola nodosa. Evolution (Lawrence, Kans.)* **34,** 489–493.

Schemske, D. E. (1980b). Floral ecology and hummingbird pollination of *Combretum fruticosum* in Costa Rica. *Biotropica* **12,** 169–181.

Schemske, D. M., Willson, M. F., Melampy, M. M., Miller, L. J., Verner, L., Schemske, K. M., and Best, L. B. (1978). Flowering ecology of some spring woodland herbs. *Ecology* **59,** 351–366.

Schemske, D. W. (1981). Floral convergence and pollinator sharing in two bee-pollinated tropical herbs. *Ecology* **62,** 946–954.

Schemske, D. W. (1983). Limits to the specialization and coevolution in plant-animal mutualism. *In* "Coevolution" (M. H. Nitecki, ed.). Univer. of Chicago Press, Chicago, Illinois (in press).

Schmitt, J. (1983). Individual flowering phenology, plant size, and reproductive success in *Linanthus androsaceus,* a California annual. *Oecologia* (in press).

Silander, J. A. (1978). Density-dependent control of reproductive success in *Cassia biflora. Biotropica* **10,** 292–296.

Stephenson, A. G. (1979). An evolutionary examination of the floral display of *Catalpa speciosa* (Bignoniaceae). *Evolution (Lawrence, Kans.)* **33,** 1200–1209.

Stephenson, A. G. (1981). Flower and fruit abortion: Proximate causes and ultimate functions. *Annu. Rev. Ecol. Syst.* **12,** 253–279.

Stephenson, A. G. (1982). When does outcrossing occur in a mass-flowering plant? *Evolution (Lawrence, Kans.)* **36,** 762–767.

Strickler, K. (1979). Specialization and foraging efficiency of solitary bees. *Ecology* **60,** 998–1009.

Tahvanainen, J. O., and Root, R. B. (1972). The influence of vegetational diversity on the population ecology of a specialized herbivore, *Phyllotreta cruciferae* (Coleoptera: Chrysomelidae). *Oecologia* **10;** 321–346.

Tepedino, V. J., and Stanton, N. L. (1980). Spatiotemporal variation in phenology and abundance of floral resources on shortgrass prairie. The *Great Basin Nat.* **40,** 197–215.

Thompson, J. N. (1982). "Interaction and Coevolution." Wiley, New York.

Thomson, J. D. (1978). Effect of stand composition on insect visitation in two-species mixtures of *Hieracium. Am. Midl. Nat.* **100,** 431–440.

Thomson, J. D. (1981). Spatial and temporal components of resource assessment of flower-feeding insects. *J. Anim. Ecol.* **50,** 49–59.

Thomson, J. D. (1982). Patterns of visitation by animal pollinators. *Oikos* **39,** 241–250.

Thomson, J. D., Andrews, B. J., and Plowright, R. C. (1981). The effect of a foreign pollen on ovule development in *Diervilla lonicera* (Caprifoliaceae). *New Phytol.* **90,** 777–783.

van Leevwen, B. H. (1981). The role of pollination in the population biology of the monocarpic species *Cirsium palustre* and *Circium vulgare. Oecologia* **51,** 28–32.

Waddington, K. D., and Heinrich, B. (1981). Patterns of movement and floral choice by foraging bees. *In* "Foraging Behavior: Ecological, Ethological, and Psychological Approaches" (A. Kamil and T. Sargent, eds.), pp. 230–245. Garland STPM Press, New York.

Waser, N. M. (1978). Competition for hummingbird pollination and sequential flowering in two Colorado wildflowers. *Ecology* **59,** 934–944.

Waser, N. M. (1982). Competition for pollination and floral character differences among sympatric plant species: A review of evidence. *In* "Handbook of Experimental Pollination Ecology" (C. E. Jones and R. J. Little, eds.). Van Nostrand–Reinhold, New York.

Waser, N. M., and Real, L. A. (1979). Effective mutualism between sequentially flowering plant species. *Nature (London)* **281,** 670–672.

Weller, S. G. (1979). Variation in heterostylous reproductive systems among populations of *Oxalis alpina* in S. E. Arizona. *Syst. Bot.* **4,** 57–71.

Weller, S. G. (1980). Pollen flow and fecundity in populations of *Lithospermum carolinense. Am. J. Bot.* **67,** 1334–1341.

Whalen, M. D. (1978). Reproductive character displacement and floral diversity in *Solanum* section *Androceras. Syst. Bot.* **3,** 77–86.

Wiens, J. A. (1981). Single-sample surveys of communities: Are the revealed patterns real? *Am. Nat.* **117,** 90–98.

Williamson, G. B., and Black, E. M. (1981). Mimicry in hummingbird-pollinated plants? *Ecology* **62,** 494–496.

Willson, M. F., Miller, L. J., and Rathcke, B. J. (1979). Floral display in *Phlox* and *Geranium*: Adaptive aspects. *Evolution (Lawrence, Kans.)* **33,** 52–63.

Woodell, S. R. J., Mattsson, O., and Philipp, M. (1977). A study in the seasonal reproductive and morphological variation in five Danish populations of *Armeria martimia. Bot. Tidsk.* **72,** 15–30.

Zimmerman, M. (1980). Reproduction in *Polemonium*: Competition for pollinators. *Ecology* **61,** 497–501.

Index

F

G

H

N

O

P

T

U

V

W

X

Y

Z